Lecture Notes in Artificial Intelligence   3301

Edited by J. G. Carbonell and J. Siekmann

Subseries of Lecture Notes in Co

T0230277

Gabriele Kern-Isberner   Wilhelm Rödder
Friedhelm Kulmann (Eds.)

Conditionals, Information, and Inference

International Workshop, WCII 2002
Hagen, Germany, May 13-15, 2002
Revised Selected Papers

 Springer

Series Editors

Jaime G. Carbonell, Carnegie Mellon University, Pittsburgh, PA, USA
Jörg Siekmann, University of Saarland, Saarbrücken, Germany

Volume Editors

Gabriele Kern-Isberner
Universität Dortmund
Fachbereich Informatik
Lehrstuhl VI - Information Engineering
44221 Dortmund, Germany
E-mail: gabriele.kern-isberner@cs.uni-dortmund.de

Wilhelm Rödder
FernUniversität in Hagen
Fachbereich Wirtschaftswissenschaft
Lehrstuhl für BWL, insb. Operations Research
58084 Hagen, Germany
E-mail: wilhelm.roedder@fernuni-hagen.de

Friedhelm Kulmann
FernUniversität in Hagen
Fachbereich Wirtschaftswissenschaft
Lehrstuhl BWL, insb. Operations Research
58084 Hagen, Germany
E-mail: friedhelm.kulmann@fernuni-hagen.de

Library of Congress Control Number: 2005925863

CR Subject Classification (1998): I.2, F.4

ISSN   0302-9743
ISBN-10   3-540-25332-7 Springer Berlin Heidelberg New York
ISBN-13   978-3-540-25332-7 Springer Berlin Heidelberg New York

Springer is a part of Springer Science+Business Media

springeronline.com

© Springer-Verlag Berlin Heidelberg 2005
Printed in Germany

Typesetting: Camera-ready by author, data conversion by Scientific Publishing Services, Chennai, India
Printed on acid-free paper   SPIN: 11408017   06/3142   5 4 3 2 1 0

Preface

Conditionals are fascinating and versatile objects of knowledge representation. On the one hand, they may express rules in a very general sense, representing, for example, plausible relationships, physical laws, and social norms. On the other hand, as default rules or general implications, they constitute a basic tool for reasoning, even in the presence of uncertainty. In this sense, conditionals are intimately connected both to information and inference.

Due to their non-Boolean nature, however, conditionals are not easily dealt with. They are not simply true or false — rather, a conditional "if A then B" provides a context, A, for B to be plausible (or true) and must not be confused with "A entails B" or with the material implication "not A or B." This illustrates how conditionals represent information, understood in its strict sense as reduction of uncertainty. To learn that, in the context A, the proposition B is plausible, may reduce uncertainty about B and hence is information. The ability to predict such conditioned propositions is knowledge and as such (earlier) acquired information.

The first work on conditional objects dates back to Boole in the 19th century, and the interest in conditionals was revived in the second half of the 20th century, when the emerging Artificial Intelligence made claims for appropriate formal tools to handle "generalized rules." Since then, conditionals have been the topic of countless publications, each emphasizing their relevance for knowledge representation, plausible reasoning, nonmonotonic inference, and belief revision. Indeed, conditionals have raised a considerable amount of scientific work, not only in Artificial Intelligence and computer science, but also in philosophy and the cognitive sciences. To date, diverse default and conditional theories have been brought forth, in quantitative as well as in qualitative frameworks, but clear benchmarks are still in discussion. Therefore, the proper handling of conditional information is still a challenge both for theoretical issues and practical applications.

The papers presented in this volume are extended and revised versions of contributions to the *Workshop on Conditionals, Information, and Inference – WCII 2002* which took place on the campus of the FernUniversität in Hagen in May 2002. The intention of this workshop was to bring together researchers interested in and working with conditionals and information processing, in order to present new results, discuss open problems and intensify cooperation. Logicians, philosophers, computer scientists, and scientists concerned with the cognitive meaning of conditionals for processing information all contributed to realizing this aim, ensuring that conditionals were discussed in an adequately broad scope. The topic of conditionals, in particular their relevance for information and inference, proved to bear an amazingly vigorous power to provide a unifying forum for studying and comparing different approaches to the crucial questions of knowledge and reasoning, for discussing different viewpoints, and for bridging different areas researchers are working in. To give an impression of this creative and fruitful atmosphere, and to show its results are the principal aims of this volume.

Ernest W. Adams points out clearly the insufficiencies of classical logic in handling conditionals adequately, emphasizing the dynamic nature of conditionals and their crucial meaning for practical reasoning. *Philip G. Calabrese* investigates deductive relations on conditionals which can be built from four basic implication relations. Moreover, he considers the question of how much confidence can be attached to probability values computed by using the principle of maximum entropy. *Didier Dubois* sketches a unifying framework for plausible reasoning under incomplete information, based on confidence relations. Inference is achieved by conditioning, thereby imitating standard probabilistic inference in an ordinal environment.

In the section with regular papers, *Jean-François Bonnefon* and *Dennis Hilton* begin with presenting an argumentative approach to interpreting conditional statements, offering a broader view on conditional human reasoning by taking a speaker's intention into account. In *Emil Weydert*'s paper, conditionals both serve as building blocks for epistemic states and provide new evidence which has to be incorporated by revision. He proposes a method to perform such a revision of epistemic states by new conditional beliefs which can also be used for iterated revision. Belief revision is also the concern of *Richard Booth*'s paper. Its main purpose is to make a start on a theory of iterated non prioritized revision, differentiating between regular beliefs and core beliefs.

Rainer Osswald makes use of conditional relationships to express observation categoricals, and to represent default assumptions in observational logic. He interprets defaults in two ways: first, as intuitionistic conditionals, and, second, in a Reiter-style way. *Piotr Chrzastowski-Wachtel* and *Jerzy Tyszkiewicz* present a Maple package for experimenting with conditional event algebras. Their implementation uses the correspondence of conditional events to Moore machines and Markov chains, and offers the possibility of deriving symbolic results. *František Matúš* uses discrete structures built from matrices to discover conditional independences in Gaussian vectors. *Christoph Beierle* and *Gabriele Kern-Isberner* address the question of how probabilistic logic and probabilistic conditional logic can be formalized as abstract logical systems, using the framework of institutions. They further investigate the formal logical relationships between these two logics and propositional logic.

Jeff B. Paris and *A. Vencovská* apply the maximum entropy paradigm to problems in inductive reasoning. They show not only that this yields results in accordance with common sense, but also that "reasons" can be found to explain these results. *Manfred Schramm* and *Bertram Fronhöfer* use modified maximum entropy techniques to make incomplete Bayesian networks complete. They deal with two concepts of incompleteness, and consider two different MaxEnt-modifications.

Our special thanks go to our invited speakers Ernest Adams, Phil Calabrese, and Didier Dubois, for enhancing the workshop and the present volume with their ideas, but we are also indebted to all participants for joining us in Hagen for an inspiring workshop. We thank all authors for revising and extending their papers for this volume, and all referees for their critical, but helpful work. We are especially grateful to Jörg Siekmann and Springer for making this publication possible. Finally, this volume would not have come into existence without the FernUniversität in Hagen's generous supporting of the *WCII 2002*-workshop.

October 2004

Gabriele Kern-Isberner
Wilhelm Rödder
Friedhelm Kulmann

FernUniversität in Hagen

Organization

The Workshop on Conditionals, Information, and Inference 2002 (WCII 2002) was initiated and organized by the Department of Computer Science and the Department of Operations Research, University in Hagen.

Program Committee

Conference Co-chairs	Gabriele Kern-Isberner
	Wilhelm Rödder
Organizing Chair	Friedhelm Kulmann

Referees

Joachim Baumeister	Universität Würzburg, Germany
Salem Benferhat	Université Paul Sabatier, France
Alexander Bochman	Holon Academic Institute of Technology, Israel
Richard Booth	Universität Leipzig, Germany
Gerhard Brewka	Universität Leipzig, Germany
Philip G. Calabrese	SPAWAR/NAVY San Diego, USA
Jim Delgrande	Simon Fraser University, Canada
Georg Dorn	University of Salzburg, Austria
Didier Dubois	Université Paul Sabatier, France
Angelo Gilio	Università La Sapienza, Italy
I.R. Goodman	SPAWAR/NAVY San Diego, USA
Simon Handley	University of Plymouth, UK
Andreas Herzig	Université Paul Sabatier, France
Achim Hoffmann	University of New South Wales, Australia
Hannes Leitgeb	Universität Salzburg, Austria
Thomas Lukasiewicz	TU Wien, Austria
David Makinson	King's College London, UK
Frantisek Matús	Academy of Sciences, Czech Republic
Rainer Osswald	FernUniversität in Hagen, Germany
Jeff Paris	University of Manchester, UK
Simon Parsons	University of Liverpool, UK
Hans Rott	Universität Regensburg, Germany
Karl Schlechta	Université de Provence Marseille, France
Gerhard Schurz	Heinrich-Heine-Universität Düsseldorf, Germany

Marco Schorlemmer	University of Edinburgh, UK
Guillermo R. Simari	Universidad Nacional del Sur, Argentina
Andrea Weidenfeld	University of Potsdam, Germany
Emil Weydert	Max Planck Institute for Computer Science, Germany
Guo-Qiang Zhang	Case Western Reserve University, USA

Table of Contents

What Is at Stake in the Controversy over Conditionals

Ernest W. Adams

University of California, Berkeley
eadams@surfnetusa.com

1   Introduction

The following remarks are going to suggest that there is more at stake in the current controversy over conditionals than merely the logic of conditionals. Among the more general issues involved are the limits of truth-conditionality, the implications of this for certain programs in the philosophy of language, the values of truth and probability, and the rationale of logic [and, incidentally, the foundations of probability and decision theory as well, but I won't get into that].

2   An Example

A lot of my points can be made in terms of the following hopefully transparent example. A person, Jones, is about to be dealt a five card poker hand from a shuffled deck of 52 cards, and someone asserts:

$F \rightarrow S$: If Jones's first card is an ace, then his second card will be too.

Here 'F' stands for "Jones's first card will be an ace" and 'S' stands for "his second card will be too", and the arrow, $\rightarrow$, stands for the ordinary language "if..then..".

According to the orthodox 'material' or 'Philonean' analysis, the conditional is logically equivalent to a disjunction. That is, it should be logically equivalent to:

$\sim F \vee S$: Either Jones's first card won't be an ace or his second one will be.

I will call the claim that the conditional and the disjunction are logically equivalent the MATERIAL CONDITIONAL THESIS, abbreviated 'MCT'. Superficially, the controversy over conditionals concerns the validity of this thesis. Of course this has been debated ever since the time of the Stoics, but one thing that has given special impetus to reopening it recently has been the introduction of probabilistic considerations. Callimachus is supposed to have said that even the crows on the rooftops are cawing about which conditionals are true (cf. [14–p. 203]), but now they are beginning to crow about which ones are probable. Of course that isn't the only new wrinkle, since Professor Stalnaker (cf. [17]) and others insist that modal dimensions are even more important, but it is the probabilistic dimension that I will be talking about. So we will turn to that.

G. Kern-Isberner, W. Rödder, and F. Kulmann (Eds.): WCII 2002, LNAI 3301, pp. 1–11, 2005.
© Springer-Verlag Berlin Heidelberg 2005

3   Probabilities

What is most controversial about the probabilistic approach to the logic of conditionals is what I call the PROBABILITY CONDITIONAL THESIS (PCT), which asserts that the probability of a conditional is a conditional probability. I won't state the thesis formally, but I will illustrate it in application to the conditional in the example. According to this thesis, the probability that if Jones's first card is an ace then his second one will be too is equal to the proportion of aces in the 51 cards that are left after one ace has already been dealt. This is approximately 0.0588, which is a fairly low probability.

The reason that PCT is controversial is because it is inconsistent with certain principles that have almost the status of dogma in present day logic − − i.e., in first courses in Logic that are taught in most departments of philosophy. In fact, the most superficial of these is MCT, and you can see how it conflicts with PCT by calculating the probability of the disjunction "Either Jones's first card won't be an ace or his second one will be" which is equivalent to the conditional according to MCT. This is approximately 0.9276, which is obviously a lot higher than the conditional probability. Since logically equivalent propositions should be equiprobable it follows immediately that MCT is inconsistent with PCT. Given the long-standing controversies over the material conditional, the orthodox philosopher's initial impulse could be to give that up, since the low probability of the conditional seems more intuitively right than the high probability that it would have if MCT were right. But things aren't that simple. David Lewis's famous Triviality Results imply that PCT is inconsistent with more fundamental logical principles.

4   Implications of the Triviality Results

Let me start by stating one last thesis. That is the PROBABILITY OF TRUTH THESIS (PTT) that *the probability of a proposition is equal to the probability that it is true.* The new thesis seems almost self-evident and it is closely related to traditional logical principles, such as that what is known must be true, and to believe it is to believe that it is true. But self-evident as these theses seem to be, one important implication of the Triviality Results is that each of them is inconsistent with PCT (David Lewis points this out, and it is also pointed out in my book *The Logic of Conditionals*, cf. [13], [1]). Since PTT seems solider than MCT, these results put the question of which thesis has to be given up in a different light. There is something still solider that it is inconsistent with, but before coming to that I want to point out another implication of the Triviality Results.

That is that assuming PCT, not only cannot the conditional be equivalent to a disjunction; it cannot be equivalent to any other bivalent truth-conditional proposition. This applies even to various possible-worlds theories, relevance theories, necessary and sufficient condition theories, and so forth. Therefore if it is right then conditionals are *sui generis*, not truth-conditional at all, and not reducible to non-conditionals.

The final point is this. I said that self-evident as PTT may be, it rests on still solider foundations. In fact it is an immediate consequence of Convention T, and it follows that PCT is inconsistent with Convention T itself (cf. [2]). Therefore either Convention T or PCT must be given up, and when the question arises which one, just about anyone would say it has to be the latter. I disagree with this for reasons that will be outlined shortly, but first I will remind the reader of some consequences of giving up Convention T and truth-conditionality in application to conditionals.

5 Consequences of Giving Up Convention T in Application to Conditionals

Start with *knowledge*, and the traditional principle that for persons to know propositions they must be true, the persons must believe them, and they must have good reasons for this. Giving up Convention T calls in question all three of these when the propositions are conditional. That is because they cannot be true in the Tarskian, disquotational sense, they cannot be believed if believing them is believing that they are true, and there cannot be good reasons for believing them if they cannot be believed. Of course I don't believe that. I think that we can believe that if Jones's first card is an ace then his second one won't be, we can have good reason for this belief, and we can even be rightly held to know that if Jones's first four cards are aces his fifth one won't be. It is even proper to say that certain conditionals are *logically true*, without being committed to truth in a Tarskian sense.

Actually I think that what I have just said applies to a lot more than conditionals, and I will say something about broader implications later, but coming to philosophy of language, it is obvious that giving up Convention T has implications for theories in that field. The most obvious ones have to do with Donald Davidson's programme for basing an analysis of meaning in natural language on the requirement that the analysis must yield 'equivalences of form T' for all meaningful sentences (cf. [9]).[1] If I am right and equivalences of form T don't hold for natural language conditionals, then Davidson's programme cannot satisfy a criterion of adequacy for theories of meaning, which is that they should apply to all assertive sentence forms. Similar things can be said about other semantic theories of meaning, and to a lesser extent about speech act theories that formulate preparatory or adequacy conditions for statements in terms of reasons for the truth of what is stated. There isn't space to go into this here, but I will just mention the somewhat paradoxical fact that while Paul Grice has been one of the most forceful advocates of MCT in recent years (cf. [10–Chapter 4]), his own theories of utterance meaning and conversational implicature actually fit in with my views on conditionals a good deal better than the semantic and speech act theories I spoke of. These ideas are developed in my paper "On the meaning of the conditional," ([10], [3]).

---

[1] E.g., statements like "if Jones's first card is an ace then his second one won't be" is true if and only if, if Jones's first card is an ace then his second one won't be.

Giving up truth-conditionality has the most obvious implications for logic, but they ought to be noted anyway. MCT has to be given up, but that's only a start. If truth itself is called in question in application to conditionals, then all logical concepts that depend on it have to be reexamined. There are two groups of such concepts. The first are the meanings of 'truth-conditional compounds' of conditionals like negations and conjunctions, and the second are the meanings of logical consequence, consistency, and logical truth, which are normally defined truth-conditionally. I will make a couple of comments on each, more to show what the problems are than to discuss how I and others are attacking them.

Negations of conditionals, such as "It is not the case that if Jones's first card is an ace his second card will be too", are common, but if what they negate isn't true or false it's unclear what the 'logical values' of their negations ought to be. I am going to argue that it is a methodological error to stipulate either truth or probability values arbitrarily, just to round out theories, and it is really a new problem of analysis to describe the meanings of grammatical negations of conditionals as they are used in everyday life. That the meanings of these constructions aren't automatically determined by the meanings of the things negated may seem to be a weakness in the probability approach, but I think it is an advantage that this approach doesn't hide real problems under a formalist rug.

It is obvious that if the concepts of logical truth, consistency, and logical consequence have any meaning in application to conditionals they cannot be defined in the usual ways, in terms of possible combinations of truth values. Not only that, we cannot any longer assume that they are interdefinable in standard ways. For instance, according to standard theory, a conditional is a logical consequence of a disjunction if its negation is inconsistent with the disjunction. But if the very meaning of negation is in doubt this principle cannot any longer be accepted without question. I am not saying that the logical concepts cannot be analyzed; in fact certain 'first-approximation' analyses have been proposed and I have been guilty of proposing some of them. But the point is that these concepts are now seen to be relatively autonomous, and each requires its own analysis.

The above is a quick run-down of problems that have to be faced if Convention T and truth-conditionality are given up in application to conditionals, and the reader is not to be blamed if she reacts in the way David Lewis did when he wrote in his paper "Probabilities of Conditionals and Conditional Probabilities":

> I have no objection to the hypothesis that indicative conditionals are non-truth-valued sentences, governed by a special rule of assertability that does not involve their nonexistent probabilities of truth. I have an inconclusive objection, however: the hypothesis requires too much of a fresh start. It burdens us with too much work to be done, and wastes too much that has been done already. ([13], see also [11–p.136].

Of course I disagree with Lewis. There is a lot of work to be done, but I think it is there to *be* done, and it is not so much wasting as misinterpreting what has been done already to hold that it applies to the issues at hand. Anyway, I will now sketch my reasons for giving up Convention T and truth-conditionality in application to conditionals. The main reason is that I think that the conditional probability approach explains a lot of phenomena relating to conditionals more easily and naturally than the truth-conditional approach does. I will illustrate with some comments on MCT. In spite of objections to it, MCT has considerable intuitive plausibility that has to be explained.

6   Conditionals Deficits

The formula below, the CONDITIONAL DEFICIT FORMULA (CDF), states an exact relation between the probabilities of 'if F then S' according to MCT and according to PCT, where the former is written as $Prob(F \supset S)$ and the latter as $Prob(F \Rightarrow S)$:

$$(\text{CDF}) \; \text{Prob}(F \supset S) - Prob(F \Rightarrow S)$$
$$= [1 - Prob(F \supset S)] \times [Prob(\sim F)/Prob(F)].$$

The material conditional's probability is always at least as high as that of the probability conditional, and their difference is given by the conditional deficit that appears as the product of two 'deficit factors' on the right side of CDF. For instance, in the card example, where 'F' and 'S' symbolize "The first card will be an ace" and "The second card will be ace",

$$Prob(F \supset S) - Prob(F \Rightarrow S) = [1 - Prob(F \supset S)] \times [Prob(\sim F)/Prob(F)]$$
$$= .0724 \times (12/13 \div 1/13) = .9276 - .0588$$
$$= .8768,$$

which conforms to CDF.

This throws light on MCT from two directions. If PCT is right, then the material conditional's probability is always at least as high as that of the probability conditional, which means that in a certain sense it is always rational to infer the material conditional from the conditional, hence half of MCT is valid. Of course, the card example shows that the probability conditional's probability can be much less than the material conditional's, but CDF also shows that in certain situations the deficit has to be small, and persons may infer probability conditionals from material conditionals. Here are two of them. One is that in which the material conditional is certain and not just highly probable. In that case the first factor in the deficit, $1 - Prob(F \supset S)$, equals zero so the probability conditional must also be certain. Thus, there is one valid interpretation of MCT: while the conditional and the material conditional aren't necessarily equiprobable, one is certain if and only if the other is.

The other situation is one in which the deficit has to be small even though the material conditional isn't certain. That is where the disjunction to which the material conditional is equivalent doesn't derive its probability from its first disjunct. In that case the second component of the deficit, $[Prob(\sim F)/Prob(F)]$, isn't high, while the first component is low, so their product has to be low. Conversational implicature considerations suggest that that is the normal case when disjunctions are asserted,[2] since it might violate a 'maxim of conversational quality' to assert a disjunction in a situation in which a person is in a position to assert one of its disjuncts. Incidentally, one can easily see that the probabilities in the card example are ones in which the disjunction that is equivalent to $F \rightarrow S$ according to MCT does derive most of its probability from its first disjunct, and so asserting it would violate the maxim.

This hardly scratches the surface so far as concerns implications of the CDF, but there isn't space to go farther into this here. It can be said, though, that this formula and others like it play a role in probability logic that is similar to the role of equivalences like the distributive laws in truth-conditional logic, but with an important difference. There are far more combinations of probability values than of there are of truth values, and therefore the analysis of probabilities involved in inference patterns is much more fine-grained than the analysis of possible truth combinations is. We are not just confined to saying that a pattern of inference is unsound because it is possible for its premises to be true while its conclusion is false.[3] It is also possible to examine in detail special circumstances in which it is rational to draw conclusions even when inferences aren't strictly valid probabilistically. That is important in everyday life and even in science, where it is rare to find 'deductive reasoning' that conforms to the rigorous standards of formal logic, and even rarer to find reasoning in which the premises are certainties. It would be an error to dismiss all other reasoning as if it were just as fallacious as, say, inferring a disjunct from a disjunction, and the probabilistic approach seems to make a promising start on a more fine-grained and realistic analysis of this real-life reasoning.[4]

But I cannot go farther into this, since I want to talk about a quite different reason for adopting a probabilist as against a truth-conditionalist approach to conditionals. This has to do with the utilitarian values of truth and probability. I should say that this is where I part company with most other probabilists, since my philosophy of logic is essentially a Pragmatist one, which runs counter to the Fregian, anti-pragmatist philosophy that dominates current logical thinking even among probabilists. My general views are very close to ones set forth by Ramsey in the last two sections of "Truth and Probability", though I differ from him about conditionals.

Let me start with a couple of comments on the value of truth, and on the 'rationale' of logical theory.

---

[2] Of course, material conditionals are rarely if ever <u>asserted</u>, hence conversational implicatures don't apply directly to them.

[3] (Cf. [4]), which investigates four 'criteria of validity' that inferences involving conditionals might be required to satisfy.

[4] Bamber's work on 'entailment with near certainty' ([8]) seems to me to be a particularly promising approach to the study of inferences that are not strictly probabilistically valid, but which are valid in most circumstances.

7 Pragmatist Considerations

Modern logic scarcely concerns itself with the question of why persons should conform to its precepts or even what conformity with them would be, but it seems to me that, along with natural science, it presupposes that truth is something worth seeking and falsehood is something to avoid. If it didn't presuppose these things it wouldn't make sense to hold that it is rational to reason according to logical principles that lead to true conclusions. This presupposition constitutes the rationale of logic, conceived as a theory of good reasoning.

Now I suggest that proposers of truth-conditional theories of conditionals have forgotten this rationale, because the 'truth-values' they propose do not have the values of truth and falsehood. Their truth and falsehood are properties of statements, but it is no longer the case that truth is to be sought and falsity avoided.[5] Here is an example, which is close to one that appeared in my book *The Logic of Conditionals* (cf. [1]).

Consider the following 'logical transaction', involving a person, A, who points to some non-poisonous mushrooms, and asks another person, B, "If I eat these mushrooms will I be poisoned?" If B answers "yes, if you eat the mushrooms you will be poisoned," A is very unlikely to eat the mushrooms, and therefore according to MCT, B's statement is true because it is a conditional with a false antecedent (cf. Figure 1, below).

But most people feel that there is something wrong with this, and the man in the street usually says that what is wrong is that B's statement is plainly false because the mushrooms were actually non-poisonous. I think that things are not as simple as that, but what I would like to point out here is that there is something more practical involved. That is that the material truth of "if you eat the mushrooms you will be poisoned" does not have the value it ought to have. A was 'practically mistaken' to have accepted the statement, because doing so made him decide not to eat the mushrooms in circumstances in which he would have preferred to eat them. This is not the sort of truth that A has reason to seek, and there is no reason to follow logical rules that lead to it.

Now, I think it can plausibly be argued that any bivalent truth-conditional characterization of conditionals must suffer from the same sort of defect that the material truth-conditions do, but I will confine myself to suggesting that there is no reason to expect any of the current truth-conditional theories to do any better. That is because the proponents of these theories seem oblivious to the fact that the truths and falsities that they characterize should have values that make them worth seeking or avoiding. This is obvious in Quine's justification of the material conditional in Methods of Logic, where he wrote:

> Where the antecedent is false, on the other hand, the adoption of a truth value for the conditional seems rather more arbitrary; but the decision that seems most convenient is to regard all conditionals with false antecedents as true ([15–p.13]).

---

[5] (cf. [7]), develops this point in detail.

PRACTICAL REASON

Fig. 1. Practical Reasoning

How can one expect 'truth-values' that are adopted for convenience to have any value other than theoretical convenience? The rationales for non-standard truth-conditional theories of conditionals are scarcely more reassuring. For instance, Stalnaker's theory (cf. [17]) has it that B's statement is true if A is poisoned in the nearest possible world in which he eats the mushrooms. Is this truth any more worth seeking than orthodox material truth? That is doubtful. There is no reason to expect a truth that is defined without regard to whether it is worth seeking to be worth seeking. Certainly no argument has been given that possible-worlds, relevance, strict-implication, or any other of the myriad non-standard 'logics' of conditionals have any utility whatever, and unless or until such arguments are given the presumption is against them.

But the same kinds of questions can be raised about the probabilistic theory. It focuses on probability values rather than truth values, but one has the right to ask whether these values are any more worth seeking than the *ersatz* truth values of the theories I have just been criticizing. This is a very difficult question, which I have tried to deal with in one chapter of my Logic of Conditionals book (cf. [1]; see also [5]) and I can only make a couple of vague comments on it. I will return to the mushroom example, and argue that what was wrong with B's statement was that it was improbable, and the practical disutility of this gives some indication of the utilitarian value of probability.

8   The Utility of Probability

Go back to A's original question "Will I be poisoned if I eat the mushrooms?" It is very plausible that in asking it A was seeking information that would help him decide whether or not to eat the mushrooms. So we have to ask how B's answer helps A to make this decision, and modern decision theory throws light on that (cf. especially [12]).

It pictures A's decision as based on the strengths of his preferences for the possible consequences of eating the mushrooms, and the probabilities of these happening if he eats them. Therefore A needs information about these probabilities, and he interprets B's answer as giving him this information: namely that if he eats the mushrooms he will probably be poisoned. I argue in [3] that this interpretation is legitimate because it is part of the conventional meaning of B's sentence, and it is the interpretation that leads A not to eat the mushrooms. Moreover, on this interpretation the probability of "You will be poisoned if you eat the mushrooms" is a conditional probability, in accord with PCT.

But this also hints at the utilitarian value of probabilities. Actually, A was led to the wrong decision because he thought it highly probable that he would be poisoned if he ate the mushrooms, and he would have been better off if he had thought that this was improbable. The wrong probabilities led to the wrong decision, and that's why we feel that B's answer was wrong. What he should have said was "No, you won't be poisoned if you eat them" because that was probable, and I will add not because it was true or even probably true.

Let us generalize. I suggest that attaching 'right' conditional probabilities to conditional statements has the same practical utility as attaching right nonconditional probabilities to nonconditional statements. Right probability values are like 'right' or 'real' truth values to this extent: the utilitarian reason for wanting to hold beliefs that have these values is that acting on them has desired consequences. The logic of conditionals ought to be concerned with right probabilities, and that should give it its rationale. But there is a problem connected with this that I should mention.

While truth and probability values are similar in that there are utilitarian reasons for seeking them, they differ in a very important way. The advantage of holding true beliefs is for the short run, while I argue that the advantage of holding probable ones is for the long run (cf. [1–Chapter III]). But the problem I haven't resolved to my own satisfaction is that of explaining why it is that in the non-conditional case truth and probability are so closely tied that probability is probability of truth, while in the conditional case the probability that has a utilitarian value isn't connected to any bivalent truth-characterization of conditionals, whatever its utilitarian value might be. I believe that the almost-too-close connection between truth and probability in the non-conditional case is part of the reason why up to now logicians haven't paid much attention to the importance of probability in logic. That is one of the things that I will comment on in some final *obiter dicta* concerning very general issues in the philosophy of logic. These opinions will be stated without argument or elaboration, and I hope that my rashness in uttering them will be forgiven.

9   Obiter Dicta

Considering the way in which logicians have tried to deal with conditionals suggests that truth-conditionality has become a kind of Procrustean Bed, into which logicians fit statement forms by imposing truth-values on them whether or not they 'fit naturally'. They forget that the values they impose should have the value of truth, which would make them worth seeking, and Quine was only unusually frank about this when he wrote the passage quoted above. That is part of the reason why logic often seems to be a game whose rules have to be learned just to pass a course.

I feel that logic ought not to play this game, and it should concern itself with what is worth seeking. And, it should also recognize that truth is not the only logical value. Probability is another, and it is only overlooked because in the non-conditional case it is probability of truth.

There is another dimension of probability which I haven't mentioned, which I think is also important to logic. Reasoning, including deductive reasoning, is a matter of justifying conclusions, and this involves changing opinions and beliefs. Now, probabilities have a special relation to belief change because, unlike 'static' truth values, they are 'dynamic'. The same proposition cannot be false on one occasion and true on another, but it can be improbable on one occasion and probable on another, and that is what happens when it gets justified.

There are logically important things that truth-conditional statics cannot account for, which I believe probabilistic dynamics can. One of them is the principle that anything follows from a contradiction. I think that this is accepted in current logical theory because it reduces the dynamics of logical consequence to the statics of logical consistency. But from the point of view of probabilistic dynamics the principle is absurd. It doesn't follow from the laws of probability that believing contradictory propositions, even on the same occasion if that is possible, should lead to attaching a high probability to anything whatever (this goes back to the earlier point about the independence of logical consequence and consistency when they are viewed probabilistically).

One final point. I said that Logic should be concerned with what is worth seeking. But to establish that certain truths or probabilities are worth seeking we have to connect them with human concerns, as was suggested earlier in arguing that B's statement shouldn't be held to be true just because A didn't eat the mushrooms. Going farther, this suggests that connecting truth and probability with human concerns should involve breaking down the barrier separating pure from practical reason, and seeing them as two sides of a broader subject called 'Logic' with a capital 'L'. What was suggested above suggests that there are dangers inherent in focusing exclusively on the static, pure-reason side of Logic and ignoring its dynamic, practical side. The rationale for theories of pure reason is to be found in practical reason, but abstracting and focusing exclusively on pure reason is apt to make one forget this, and that can lead to logical sterility.

This brings us back to the stakes involved in the controversy over conditionals. The most important thing is that conditionals bring to the fore what I regard as very fundamental defects in current logical methodology. I see parallels between current logical methodology and Medieval Scholasticism. Both build towering edifices of formal theory on the shakiest of foundations, though now they are logical rather than theological dogmas, and both lose themselves in sterile disputations between 'schools' over matters that are of no concern to the rest of the world. For my part, I think that a renaissance is in order, which would sweep away the extravagances of current theory. Then it might return its attention to what really does matter, namely how people ought to reason.

References

1. Adams, E. W.: The Logic of Conditionals: An Application of Probability to Deductive Logic, (Reidel, Boston, 1975).
2. Adams, E. W.: Truth, proof, and Conditionals, Pacific Philosophical Quarterly, 62 (1981) 323-339.
3. Adams, E.W.: On the Meaning of the Conditional, Philosophical Topics, Vol. XV No.1 (1987) 5-22.
4. Adams, E.W.: Four Probability-preserving Properties of Inferences, Journal of Philosophical Logic, 25 (1996) 1-24.
5. Adams, E.W.: The Utility of Truth and Probability, in: P. Weingartner, G. Schurz, and G. Dorn, (eds.): Die Rolle der Pragmatik in der Gegenwartsphilosophie, (Verlag Holder-Pichler-Tampsky, Wien, 1998) 176-194.
6. Adams, E.W.: A Primer of Probability Logic, CSLI Publications, Stanford, California (1998).
7. Adams, E.W.: Truth Values and the Value of Truth, Pacific Philosophical Quarterly, Volume 83, No.3 (2002) 207-222.
8. Bamber, D.: Entailment with Near Surety of Scaled Assertions of High Conditional Probability, Journal of Philosophical Logic, 29 (2000) 1-74.
9. Davidson, D.: Truth and Meaning, Synthese, 17 (1967) 304-323.
10. Grice, P.: Studies in the Ways of Words, (Oxford University Press, London, 1989).
11. Harper, W.L.; Stalnaker, R.; Pearce, G. (eds.): Ifs, (D. Reidel Publishing Co., Dordrecht, 1980).
12. Jeffrey, R.C.: The Logic of Decision, Second Edition, (University of Chicago Press, Chicago and London, 1983).
13. Lewis, D.: Probabilities of Conditionals and Conditional Probabilities, Philosophical Review, 85 (1976) 297-315.
14. Mates, B.: Elementary Logic, (Oxford University Press, New York, 1965).
15. Quine, W.V.: Methods of Logic, Revised Edition, (Henry Holt and Co., New York, 1959).
16. Ramsey, F.P.: Truth and Probability, in: Braithwaite, R. (ed.): The Foundations of Mathematics and other Logical Essays, (Routledge and Kegan-Paul, London, 1930) 156-198.
17. Stalnaker, R.: A theory of Conditionals, Studies in Logical Theory, American Philosophical Quarterly, Monograph Series No. 2, Blackwell, Oxford, 1968.

Reflections on Logic and Probability
in the Context of Conditionals

Philip G. Calabrese

Data Synthesis,
2919 Luna Ave., San Diego, CA 92117
pc@datasynthesis.org

Abstract. Various controversies surrounding the meaning of conditionals like "A given B" or "if B then A" are discussed including that they can non-trivially carry the standard conditional probability, are truth functional but have three rather than two truth values, are logically and probabilistically non-monotonic, and can be combined with operations that extend the standard Boolean operations. A new theory of deduction with uncertain conditionals extends the familiar equations that define deduction between Boolean propositions. Several different deductive relations arise leading to different sets of implications. New methods to determine these implications for one or more conditionals are described. An absent-minded coffee drinker example containing two subjunctive (counter-factual) conditionals is solved. The use of information entropy to cut through complexity, and the question of the confidence to be attached to a maximum entropy probability distribution, are discussed including the results of E. Jaynes concerning the concentration of distributions at maximum entropy.

1 Introduction

Thirty–five years ago the theories of logic and of probability were conspicuously separated, missing a division operation to represent conditional statements. Even today many people still reduce all conditional statements such as "if B then A" to the unconditioned (or universally conditioned) statement "A or not B", the so-called "material conditional", even though it has long been recognized that the material conditional is of no use in estimating the probability of A in the context of the truth of B. The latter probability is the well-known conditional probability of A given B, the ratio of the probability of both A and B to the probability of B. The conditional probability is never greater than, and is generally much less than, the probability that A is true or B is false. Only when B is certain or when A is certain given the truth of B do the two expressions yield essentially the same result. Even when B is false, they differ since the ratio is undefined while the material conditional has probability 1. This has all been quantified for instance in [1]. Yet for purposes of doing 2-valued logic, the material conditional works just fine. Mathematicians have long proved their theorems of the form "if B then A" by proving that in all cases either A is true or B is false.

G. Kern-Isberner, W. Rödder, and F. Kulmann (Eds.): WCII 2002, LNAI 3301, pp. 12–37, 2005.
© Springer-Verlag Berlin Heidelberg 2005

However when B is uncertain or when A is uncertain given the truth of B, the material conditional is not an appropriate Boolean proposition to represent the conditional statement "if B then A". Nor is there any other Boolean proposition that can serve the purpose of both logic and probability as early shown by D. Lewis [2]. This non-existence is reminiscent of results throughout the history of mathematics that preceded the invention of new numbers needed to satisfy some relationships that naturally arose. The irrational numbers were needed to represent the length of the hypotenuse of a square in terms of the length of a side of that square; complex numbers were invented to solve polynomial equations such as $x^2 + 1 = 0$ and integer fractions were invented to have numbers that could solve equations like $3x = 20$. In each case, mathematicians didn't stop with the declaration that there were no such numbers in the existing system; they instead invented new numbers that included the old ones but also solved the desired equations. The same thing has worked in the case of events and propositions, (see [1] & [3]), and the result is no less profound. The more surprising thing is that it has taken so long for the development to occur in the case of events and propositions. Apart from Boole himself, such a system of ordered pairs was envisioned by a few researchers including De Finetti [14], G. Schay [4] and Z. Domotor [5], but these developments didn't go far enough in the right direction before getting bogged down. It is now clear, however, that a system of *ordered pairs* of probabilistic events or of logical propositions can be defined to represent conditional statements, avoid the triviality results of Lewis [2], and be assigned the standard conditional probability.

The operations on the ordered pairs of events or propositions (A|B), "A given B", have been extensively analyzed and motivated in [1], [3], & [6]. Using ' to denote "not" and juxtaposition to denote "and" these operations on conditionals are:

$$(A|B)' = (A' | B). \qquad (1.1)$$

That is, "not (A given B)" is equivalent to "(not A) given B".

$$(A|B) \text{ or } (C|D) = ((AB \text{ or } CD) | (B \text{ or } D)) \qquad (1.2)$$

The right hand side is "given either conditional is applicable, at least one is true".

$$(A|B) \text{ and } (C|D) = [ABD' \text{ or } ABCD \text{ or } B'CD] | (B \text{ or } D) \qquad (1.3)$$

The right hand side is "given either conditional is applicable, at least one is true while the other is not false". It can be rewritten as [AB(CD or D') or CD(AB or B') | B or D)] and also as [(AB or B')(CD or D') | B or D)].

$$(A|B) | (C|D) = (A | (B)(C|D)) \qquad (1.4)$$

The right hand side is "given B and (C|D) are *not false*, A is true."

By writing B as a conditional (B | Ω) with the universe Ω as condition the conjunction (B)(C|D) in (1.4) reduces to B(C $\vee$ D') using operation (1.3).

This system of "Boolean fractions" ($\mathcal{B}|\mathcal{B}$) includes the original events or propositions $\mathcal{B}$ as a subsystem and also satisfies the essential needs of both logic and conditional probability. Two conditionals (A|B) and (C|D) are *equivalent* (=) if and only if B=D and AB = CD. As with the past extensions of existing number systems, some properties no longer hold in the new system. For instance, the new system is not wholly distributive as are Boolean propositions.

As with any new system of numbers there has been quite a lot of resistance to this new algebra of conditionals. Some researchers (see [7], [8], [31], [32]), recognizing the virtue of a system of ordered pairs of events to represent conditional events, have nevertheless disputed the choice of extended operations on those ordered pairs. However, the operations for "or" and "and" in [1] were independent rediscoveries of the two so-called "quasi" operations for "or" and "and" early employed by E. Adams [9], [30], [10], a pioneer researcher of conditionals writing in the philosophical literature. Adams calls these operations "quasi" merely because they are not "monotonic". That is, combining two conditionals with "and" does not always result in a new conditional that implies each of the component conditionals. Nor does combining two conditionals with "or" always result in a conditional that is implied by each of the component conditionals. This seems rather counter intuitive when considered in the abstract because we are all so imbued with equal-condition thinking. But when two conditionals with different conditions are combined as in operations (1.2) or (1.3), the result is a conditional whose condition is the disjunction ("or") of the two original conditions. By expanding the context in this way probabilities have more freedom to change up or down. Deduction is also much more complicated when dealing with conditionals with different conditions, but now a successful extension of Boolean deduction for uncertain conditionals has been developed [11], [12], [6] by the author. Dubois and Prade [29] have also developed deduction in detail along similar lines.

Another issue that arises with conditionals is their truth functionality. Are conditionals "true" or "false" like ordinary propositions or events? Even the ancient Greeks were troubled by this question. For some reason Adams seems to take the attitude ([13], p.65, footnote) that "inapplicable" is not really a 3rd truth-value that can be assigned to a conditional. On the other hand, B. De Finetti [14], [15] early asserted that a conditional has three, rather than two, truth-values: If the condition B is true, then "A given B" is true or false depending on the truth of A. But when B is false, De Finetti asserted that the conditional was neither true nor false, but instead required a third truth-value, which he unfortunately identified with "unknown" and therefore assigned a numerical value somewhere between 0 and 1. But a conditional with a false condition is not "unknown"; it is "inapplicable". For instance, if I am asked, "if you had military service, in which branch did you serve?" I don't answer "unknown". I answer "inapplicable" because I haven't had military service. The question and its answer are not assigned a truth-value between 0 and 1; they are essentially ignored. The answer "unknown" would be appropriate by someone who thought I had military service but did not know in which branch I served.

While it is not immediately obvious, the question of what operations are used to combine conditional propositions is essentially equivalent to the question of which of the three truth-values should be assigned to the nine combinations of the truth (T), falsity (F) or inapplicability (I) for two different conditionals. (See [16], p.7 for a proof.) This approach was taken by A. Walker [17] to determine those few operations on conditionals that satisfy natural requirements such as being commutative and idempotent. This approach was also employed in [6] to provide careful motivations and a complete characterization of the 4 operations on conditionals (1.1) – (1.4) listed above and originally grouped together in [1]. Three of these operations in the form of

3-valued truth tables were identified by B. Sobocinski [18], [19], but his 4th operation was very different from the operation (1.4) in [1]. Similarly, Adams easily identified the negation operation for conditionals, but passed over the 4th iterated conditioning operation employed here because he interprets a conditional as an implication instead of as a new object - an event or proposition in a given context.

Recently, Adams reconsidered the issue of "embedded" or iterated conditionals ([13], p.269) and the so-called "import-export" principle which asserts that ((A | B) | C) = (A | B and C) for any expressions A, B and C. Operation (1.4) is a restricted form of this principle, which can be used to reduce any iterated conditional to a simple conditional with Boolean components. For propositions A, B, C, D, E, and F, a more general form of the import-export law follows from operations (1.1) – (1.4):

$$[(A|B) \mid (C|D)] \mid (E|F) = (A|B) \mid [(C|D) \, (E|F)] \tag{1.5}$$

Using "import-export" Adams cites the following example as a counter example of the basic logical principle of modus ponens that A is always a logical consequence of B and (A|B). Noting that by import-export, ((HD | H) | D) = (HD | HD), and that the latter is a logical necessity, Adams gives the example

$$D \text{ and } ((HD \mid H) \mid D) \text{ implies } (HD \mid H), \tag{1.6}$$

which, according to Adams, should be valid by modus ponens. For instance, interpreting D as "it is a dog" and H as "it is heavy (500 pounds)" modus ponens seems to fail because the implication (HD | H), that "it is a heavy dog given that it is heavy" should not logically follow from D and "(HD | H) given D". Adams mentions three authors who each take a different direction here, one accepting "import-export", one accepting modus ponens, and one accepting both with reservations about modus ponens.

But the difficulties raised by this example disappear when it is remembered that with modus ponens, it is not just "A" that is a logical consequence of "B and (A|B)", but rather "A and B" that is the logical consequence. And since conditionals are not logically monotonic, "A and B" does not necessarily imply "A" alone, as Adams has elsewhere shown. For conditionals, "A and B" may no longer imply "A" and may also have larger probability than "A" alone.

Therefore, the logical implication of the left side of equation (1.6) is "D and (HD | H)", which by operation (1.3) reduces to just D, and D is certainly a valid implication of the left side of (1.6). So the "paradox" arises because the notion that "B and A" must logically imply B is false for conditionals.

For example, consider a single roll of a fair die with faces numbered 1 through 6. The conditional (2 | even) representing "2 comes up given an even number comes up" has conditional probability 1/3, and it surely logically implies itself by any intuitive concept of implication. Now conjoin the conditional (1 or 3 | < 5), representing "1 or 3 comes up given the roll is less than 5", with (2 | even) and the result by operation (1.3) is (1 or 3 | not 5), which obviously does not logically imply (2 | even) by any intuitive concept of logical implication. Note also that (1 or 3 | not 5) has conditional probability 2/5, which is larger than 1/3, the conditional probability of (2 | even). All of these situations have been analyzed in [6]. Adams gives a similar example ([13], p. 273) that can be handled in the same way.

Concerning embedded conditionals, Adams claims [13, p. 274) that, "So far no one has come up with a pragmatics that corresponds to the truth-conditional or probabilistic semantics of the theories that they propose ...". However Adams has too quickly passed over the 4-operation system of Boolean fractions (conditionals events) recounted here, and he has not yet examined the additional theory of deduction defined in terms of the operations on those conditionals.

To repeat, most if not all of these so-called paradoxes of embedded conditionals and logical deduction arise from the unwarranted identification of the conditional (A|B) with the logical implication of A by B. Others arise by forgetting that conditionals are logically non-monotonic. However, when (A|B) is taken as a new object and deduction is defined in terms of the operations (1.1) – (1.4), these paradoxes disappear. Just as it is in general impossible to force Boolean propositions to carry the conditional probability, so too is it impossible to force conditionals to serve as implication relations. The latter must be separately defined in terms of, or at least consistent with, the chosen operations on conditionals.

In Section 2.1 and 2.2 the essentials of the theory of deduction with uncertain conditionals are recounted including some refinements such as Definition 2.2.4 on the "conjunction property". Section 2.2 also has new results on the deductively closed systems generated by two exceptional deductive relations. Section 2.3 provides three new illustrative examples of deduction with uncertain conditionals. Section 2.3.1 addresses the familiar question of what can be deduced by transitivity with conditionals. That is, what can be deduced from "A given B" and "B given C"? Section 2.3.2 analyzes a set of three rather convoluted conditionals concerning an absent-minded coffee drinker. Two of the three conditionals are so-called non-indicative, also called subjunctive or counter-factual conditionals. Such conditionals seem to pose no additional difficulty for this theory of deduction. In Section 2.3.3 the absent-minded coffee drinker example is modified to make it a valid deduction in the two-valued Boolean logic. The implications with respect to various deductive relations are again determined. Section 3 addresses the issue of practical computation of combinations of conditionals and deductions with conditionals. Section 3.1 illustrates the difficulties and complexities of pure Bayesian analysis when applied to the "transitivity example" of Section 2.3.1. Section 3.2 discusses the use of entropy in information processing as a reasonable and principled way to cut through complexity and solve for unknown probabilities and conditional probabilities. This idea has already been successfully implemented in the computer program SPIRIT developed at FernUniversität in Hagen Germany by a team headed by W. Rödder. Section 3.3 addresses the question of the confidence that can be attached to probabilities determined by the maximum entropy solution. In this regard the separate work of E.T. Jaynes, S. Amari, and A. Caticha are described, especially that of Jaynes, who proves an entropy concentration theorem that provides a statistical measure of the fraction of eligible probability distributions whose entropy falls below a specified critical value. Section 3.4 discusses the possibility of somehow using conditional independence, the basis of maximum entropy simplifications of probability calculation, to simplify the underlying logical calculations with conditional events.

My thanks go to the referees for a careful reading and many helpful suggestions.

2 Deduction with Uncertain Conditionals

Deduction for uncertain conditionals must be defined in terms of the operations (1.1) – (1.4) on conditionals listed in the introduction. For instance, if (A|B) and (C|D) are two conditionals, we may wish to define deduction of (C|D) by (A|B) to mean that the conjunction (A|B)(C|D) of the two conditionals should be equivalent to (A|B) as is the case with Boolean propositions. Recall that for Boolean propositions "p implies q" can be defined with the conjunction operation by the equation "(p and q) = p". Alternately, we could use the disjunction operation and define this same implication as "(p or q) = q". Still other ways exist such as "(q or not p) = 1 (true)". Surprisingly, in the realm of conditionals none of these definitions of implication are equivalent to one another! This has all been extensively developed in [11], [12], [3] and especially [6]. This development has been summarized, supplemented and streamlined in sections 2.1 and 2.2.

2.1 Deductive Relations

The expression "B ≤ A" is used to signify "B implies A" because for Boolean propositions this implication is equivalent to saying that "the instances of B are a subset of the instances of A". This is also the appropriate interpretation in case that A and B are probabilistic events. Some readers may wish to mentally substitute the entailment arrow ⇒ for ≤ to connote deduction.

Definition 2.1.1. An implication or deductive relation, ≤, on conditionals is a reflexive and transitive relation on the set of conditionals.

For instance, one such deductive relation is $\leq_{bo}$:

$$(A|B) \leq_{bo} (C|D) \text{ if and only if } B = D \text{ and } AB \leq CD \tag{2.1.1}$$

That is, conditional (A|B) implies conditional (C|D) with respect to this deductive relation if and only if the conditions B and D are equivalent propositions or events, and within this common condition, proposition A implies proposition C. This is called *Boolean deduction* because it is just ordinary Boolean deduction when applied to conditionals with the same condition, and a conditional can only imply another conditional provided they have equivalent conditions.

Using conjunction (∧) to define implication yields *conjunctive implication* ($\leq_\wedge$):

$$(A|B) \leq_\wedge (C|D) \text{ if \& only if } (A|B) \wedge (C|D) = (A|B) \tag{2.1.2}$$

For $\leq_\wedge$ the conjunction of two or more conditionals always implies each of its components.

Using disjunction (∨) to define implication yields *disjunctive implication* ($\leq_\vee$):

$$(A|B) \leq_\vee (C|D) \text{ if \& only if } (A|B) \vee (C|D) = (C|D) \tag{2.1.3}$$

For $\leq_\lor$ the disjunction of two or more conditionals is always implied by each of the component conditionals.

Applying the material conditional equation "q or not p = 1" to conditionals yields what is called *probabilistically monotonic implication* ($\leq_{pm}$):

$$(A|B) \leq_{pm} (C|D) \text{ if \& only if } (C|D) \lor (A|B)' = (\Omega \mid D \lor B) \qquad (2.1.4)$$

For $\leq_{pm}$ any conditional (C|D) implied by (A|B) has conditional probability no less than P(A|B). Here, the universal proposition is denoted "1" and the universal event is Ω. Dubois and Prade [29] have studied this deductive implication in connection with their development of non-monotonic consequence relations.

In [12] & [6] the defining equations on the right side of the definitions (2.1.1 – 2.1.4) have been reduced to Boolean deductive relations between the component Boolean propositions. For instance, 2.1.2 reduces to the two Boolean implications, (A $\lor$ B' $\leq$ C $\lor$ D') and (B' $\leq$ D'); 2.1.3 reduces to (AB $\leq$ CD) and (B $\leq$ D); and 2.1.4 reduces to (A $\lor$ B' $\leq$ C $\lor$ D') and (AB $\leq$ CD). Thus between two conditionals (A|B) and (C|D) four elementary Boolean deductive relations arise: B $\leq$ D, AB $\leq$ CD, A $\lor$ B' $\leq$ C $\lor$ D' and B' $\leq$ D'. What is implied by these implication relations is applicability, truth, non-falsity and inapplicability respectively. They have been denoted $\leq_{ap}$, $\leq_{tr}$, $\leq_{nf}$, and $\leq_{ip}$ respectively where "ap" means "applicable", "tr" means "truth'", "nf" means "non-falsity" and "ip" means "inapplicable". This leads to a hierarchy of deductive relations on conditionals as one, two, three or all four of these different Boolean rela-

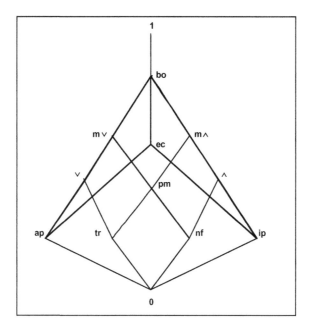

Fig. 1. Hierarchy of Implications (Deductive Relations) for Conditionals

tions are assumed necessary for a deductive relation $(A|B) \leq_x (C|D)$ to hold between two conditionals $(A|B)$ and $(C|D)$. See Figure 2.1. Actually, except for $\leq_{bo}$ all of these deductive relations can be defined in terms of just one or two of the four elementary ones because, for instance, the combined properties of $\leq_{ap}$ and $\leq_{nf}$ are equivalent to those of $\leq_{mv}$. Similarly, the combined properties of $\leq_{tr}$ and $\leq_{ip}$ are equivalent to those of $\leq_{m\wedge}$.

Although $(a|b) \leq_{pm} (c|d)$ implies $P(a|b) \leq P(c|d)$, the latter is a probabilistic, not a logical relationship. $(a|b) \leq_{pm} (c|d)$ logically means that the truth of $(a|b)$ implies the truth of $(c|d)$ and similarly for the non-falsity of $(a|b)$. But if $(a|b)$ is applicable and false then $(c|d)$ can be anything - inapplicable, true or false. By adding the requirement of $\leq_{ap}$ to $\leq_{pm}$, namely that $b \leq d$, the applicability of $(c|d)$ is assured by the applicability of $(a|b)$. This is the deductive relation $\leq_{mv}$, which also ensures that $(a|b) \vee (c|d) =$ $(c|d)$. Similarly, by itself $(a|b) \leq_{pm} (c|d)$ does not ensure that $(a|b) \wedge (c|d) = (a|b)$. But by adding to $\leq_{pm}$ the requirement of $\leq_{ip}$, to get $\leq_{m\wedge}$, this latter equation is assured. On the other hand, for some logical purposes just part of $\leq_{pm}$ may be appropriate; $\leq_{tr}$ preserves only truth; $\leq_{nf}$ preserves only non-falsity.

Trivial Implications

1 - Implication of Identity ($\leq_1$)
$(a|b) \leq_1 (c|d)$ iff $(a|b) = (c|d)$

0 - Universal Implication
$(a|b) \leq_0 (c|d)$ for all $(a|b)$ & $(c|d)$

Two Elementaries Combined

$\vee$ - Disjunctive Implication ($\leq_\vee$)
$(a|b) \leq_\vee (c|d)$ iff $b \leq d$ and $ab \leq cd$

pm - Probabilistically Monotonic Implication ($\leq_{pm}$)
$(a|b) \leq_{pm} (c|d)$ iff $ab \leq cd$ and
$(a \vee b') \leq_{nf} (c \vee d')$

$\wedge$ - Conjunctive Implication ($\leq_\wedge$)
$(a|b) \leq_\wedge (c|d)$ iff $d \leq b$ and
$(a \vee b') \leq_{nf} (c \vee d')$

ec - Implication of Equal Conditions ($\leq_{ec}$)
$(a|b) \leq_{ec} (c|d)$ iff $b = d$

Elementary Implications

tr - Implication of Truth ($\leq_{tr}$)
$(a|b) \leq_{tr} (c|d)$ iff $ab \leq cd$

nf - Implication of Non-Falsity ($\leq_{nf}$)
$(a|b) \leq_{nf} (c|d)$ iff $(a \vee b') \leq_{nf} (c \vee d')$

ap - Implication of Applicability ($\leq_{ap}$)
$(a|b) \leq_{ap} (c|d)$ iff $b \leq d$

ip - Implication of Inapplicability ($\leq_{ip}$)
$(a|b) \leq_{ip} (c|d)$ iff $d \leq b$

Three Elementaries Combined

mv - (Probabilistically) Monotonic and Applicability Implication ($\leq_{mv}$)

m$\wedge$ - (Probabilistically) Monotonic and Inapplicability Implication ($\leq_{m\wedge}$)

Four Elementaries Combined

bo - Boolean Deduction ($\leq_{bo}$)
$(a|b) \leq_{bo} (c|d)$ iff $b = d$ and $ab \leq cd$

2.2 Deductively Closed Sets of Conditionals

Having defined the idea of a deductive relation on conditionals it is now possible to define the set of implications of a set of conditionals with respect to such a deductive relation.

Definition 2.2.1. A subset $\mathcal{H}$ of conditionals is said to be a *deductively closed set* (DCS) with respect to a deductive relation $\leq_x$ if and only if $\mathcal{H}$ has both of the following properties:

If $(A|B) \in \mathcal{H}$ and $(C|D) \in \mathcal{H}$ then $(A|B) \wedge (C|D) \in \mathcal{H}$, and

If $(A|B) \in \mathcal{H}$ and $(A|B) \leq_x (C|D)$ then $(C|D) \in \mathcal{H}$.

A set of conditionals with the first property is said to have the *conjunction property* and a set of conditionals satisfying the second property is said to have the *deduction property*.

The following theorem states that the intersection of two DCS's with respect to two different deductive relations is a DCS with respect to that deductive relation formed by combining the requirements of the initial two deductive relations.

Theorem 2.2.2. Conjunction Theorem for Deductively Closed Sets with respect to two Deductive Relations. If $\mathcal{H}_x$ is a deductively closed set of conditionals with respect to a deductive relation $\leq_x$, and $\mathcal{H}_y$ is a deductively closed set of conditionals with respect to a deductive relation $\leq_y$, then the intersection $\mathcal{H}_x \cap \mathcal{H}_y$ is a DCS, $\mathcal{H}_{x \cap y}$, with respect to the *combined deductive relation* $\leq_{x \cap y}$ defined by:

$(A|B) \leq_{x \cap y} (C|D)$ if and only if $(A|B) \leq_x (C|D)$ and $(A|B) \leq_y (C|D)$.

The proof is very straightforward including showing that $\leq_{x \cap y}$ is a deductive relation. However, in general not all DCS's with respect to $\leq_{x \cap y}$ are intersections of DCS's with respect to the component deductive relations $\leq_x$ and $\leq_y$.

Definition 2.2.3. Deductive Implications of a Set J of Conditionals. If J is any subset of conditionals, $\mathcal{H}_x(J)$ will denote the smallest deductively closed subset with respect to $\leq_x$ that includes J. (Note that this is unique because it follows easily from Definition 2.2.1 that the intersection of any number of deductively closed sets of conditionals with respect to a deductive relation is deductively closed with respect to that relation.) We say that $\mathcal{H}_x(J)$ is the deductive **extension** of J with respect to $\leq_x$, or that J generates or implies $\mathcal{H}_x(J)$ with respect to $\leq_x$. A DCS is *principal* if it is generated by a single conditional.

Definition 2.2.4. Conjunction Property for Deductive Relations. A deductive relation $\leq_x$ has the *conjunction property* if and only if

$(A|B) \leq_x (C|D)$ and $(A|B) \leq_x (E|F)$ implies $(A|B) \leq_x (C|D) \wedge (E|F)$.

(Note: This is different from the conjunction property for a set of conditionals.)

Theorem 2.2.5. Principal Deductively Closed Sets. With respect to any deductive relation $\leq_x$ having the conjunction property the deductively closed set generated by a single conditional $(A|B)$ is the set of conditionals that subsume it with respect to the deductive relation. That is, $\mathcal{H}_x\{(A|B)\} = \{(Y|Z): (A|B) \leq_x (Y|Z)\}$. $\mathcal{H}_x\{(A|B)\}$ will be denoted by $\mathcal{H}_x(A|B)$.

Proof of Theorem 2.2.5. $\mathcal{H}_X(A|B)$ has the conjunction property. For suppose that $(C|D)$ and $(E|F)$ are in $\mathcal{H}_X(A|B)$. So $(A|B) \leq_X (C|D)$ and $(A|B) \leq_X (E|F)$. Therefore $(A|B) \leq_X (C|D)(E|F)$, by the conjunction property of $\leq_X$. So $(C|D)(E|F) \in \mathcal{H}_X(A|B)$. $\mathcal{H}_X(A|B)$ obviously also has the deduction property by the transitivity of any deductive relation $\leq_X$. Therefore $\mathcal{H}_X(A|B)$ is a DCS of conditionals. Clearly any DCS containing $(A|B)$ must also include $\mathcal{H}_X(A|B)$. So $\mathcal{H}_X(A|B)$ is the smallest DCS containing $(A|B)$.   □

Theorem 2.2.6. The four elementary deductive relations $\leq_{ap}$, $\leq_{tr}$, $\leq_{nf}$, and $\leq_{ip}$ on conditionals and their combinations, have the conjunction property of Definition 2.2.4.

Proof of Theorem 2.2.6. Suppose that $(A|B) \leq_{ap} (C|D)$ and $(A|B) \leq_{ap} (E|F)$. So $B \leq D$ and $B \leq F$. So $B \leq (D \wedge F) \leq (D \vee F)$. Therefore $(A|B) \leq_{ap} (C|D) \wedge (E|F) = (CDF'$ $\vee D'EF \vee CDEF \mid D \vee F)$ because $B \leq D \vee F$. Suppose next that $(A|B) \leq_{tr} (C|D)$ and $(A|B) \leq_{tr} (E|F)$. So $AB \leq CD$ and $AB \leq EF$. Therefore $(A|B) \leq_{tr} (C|D) \wedge (E|F)$ because $AB \leq (CD) \wedge (EF) \leq (CDF' \vee D'EF \vee CDEF) \wedge (D \vee F)$. Suppose next that $(A|B) \leq_{nf} (C|D)$ and $(A|B) \leq_{nf} (E|F)$. So $(A \vee B') \leq (C \vee D')$ and $(A \vee B') \leq (E \vee F')$. Therefore $(A|B) \leq_{nf} (C|D) \wedge (E|F)$ because $(A \vee B') \leq (C \vee D') \wedge (E \vee F') = (CD \vee D') \wedge (EF \vee F') = (CDEF \vee D'EF \vee CDF') \vee D'F'$, which is just $(CDF' \vee D'EF \vee CDEF) \vee (D \vee F)'$. Fourthly, suppose that $(A|B) \leq_{ip} (C|D)$ and $(A|B) \leq_{ip} (E|F)$. So $B' \leq D'$ and $B' \leq F'$. Therefore $(A|B) \leq_{ip} (C|D) \wedge (E|F)$ because $B' \leq D' \wedge F' = (D \vee F)'$. Finally, Suppose that $(A|B) \leq_{x \cap y} (C|D)$ and $(A|B) \leq_{x \cap y} (E|F)$ where x and y are in $\{ap, tr, nf, ip\}$. So $(A|B) \leq_x (C|D)$ and $(A|B) \leq_y (C|D)$ and $(A|B) \leq_x (E|F)$ and $(A|B) \leq_y (E|F)$. Therefore $(A|B) \leq_x (C|D)$ and $(A|B) \leq_x (E|F)$ and so $(A|B) \leq_x (C|D) \wedge (E|F)$. Similarly $(A|B) \leq_y (C|D) \wedge (E|F)$. Therefore $(A|B) \leq_{x \cap y} (C|D) \wedge (E|F)$.   □

Corollary 2.2.7. If $\leq_X$ is one of the elementary deductive relations $\leq_{ap}$, $\leq_{tr}$, $\leq_{nf}$, and $\leq_{ip}$ or a deductive relation combining two or more of these, then the DCS generated by $(A|B)$ with respect to $\leq_X$ is $\mathcal{H}_X(A|B) = \{(Y|Z): (A|B) \leq_X (Y|Z)\}$.

Proof of Corollary 2.2.7. The proof follows immediately from Theorems 2.2.5 and 2.2.6.   □

These results allow the principal DCS's with respect the four elementary deductive relations and their combinations to be explicitly expressed in terms of Boolean relations. See [6] for details. For instance, $\mathcal{H}_{ap}(A|B) = \{(Y|Z)$: Y any event or proposition and Z any event or proposition with $B \leq Z\} = \{(Y \mid B \vee Z)$: Y and Z any events or propositions$\}$. For the elementary deductive relations these solutions are

$$\mathcal{H}_{ap}(A|B) = \{(Y \mid B \vee Z): \text{any events or propositions Y and Z in } \mathcal{B}\} \qquad (2.2.1)$$
$$\mathcal{H}_{tr}(A|B) = \{(AB \vee Y \mid AB \vee Z): \text{any events or propositions Y and Z in } \mathcal{B}\} \quad (2.2.2)$$
$$\mathcal{H}_{nf}(A|B) = \{(AB \vee B' \vee Y \mid Z): \text{any Y, Z in } \mathcal{B}\} \qquad (2.2.3)$$
$$\mathcal{H}_{ip}(a|b) = \{(Y \mid BZ): \text{any Y, Z in } \mathcal{B}\} \qquad (2.2.4)$$

The following result allows the principal DCS's of the deductive relations formed by combining two or more of the elementary deductive relations to be expressed as an intersection of principal DCS's of the elementary deductive relations. This result does not extend to DCS's generated by a set of conditionals.

Theorem 2.2.8. The principal DCS $\mathcal{H}_{x \cap y}(A|B)$ of a single conditional $(A|B)$ with respect to a combination deductive relation $\leq_{x \cap y}$ is the intersection of the DCS's with respect to the component deductive relations $\leq_x$ and $\leq_y$. That is, $\mathcal{H}_{x \cap y}(A|B) = \mathcal{H}_x(A|B) \cap \mathcal{H}_y(A|B)$.

Proof of Theorem 2.2.8. $\mathcal{H}_{x \cap y}(A|B) = \{(C|D): (A|B) \leq_{x \cap y} (C|D)\} = \{(C|D): (A|B) \leq_x (C|D) \text{ and } (A|B) \leq_y (C|D)\} = \mathcal{H}_x(A|B) \cap \mathcal{H}_y(A|B).$ □

Using the formulas for the principal DCS's with respect to the elementary deductive relations, the principal DCS's with respect to the combined deductive relations have been calculated in [6]. For the deductive relations mentioned above, the principal DCS's are:

$$\mathcal{H}_v(A|B) = \{ (AB \vee Y \mid B \vee Z): \text{any } Y, Z \text{ in } \mathcal{B}\} \tag{2.2.5}$$

$$\mathcal{H}_{pm}(A|B) = \{ (AB \vee B' \vee Y \mid AB \vee Z): \text{any } Y, Z \text{ in } \mathcal{B}\} \tag{2.2.6}$$

$$\mathcal{H}_\wedge(A|B) = \{ (AB \vee Y \mid BZ): \text{any } Y, Z \text{ in } \mathcal{B}\} \tag{2.2.7}$$

Having described the principal DCS's of the elementary deductive relations and their combination deductive relations, these results can be used to describe the DCS's of a set of conditionals with respect to these deductive relations.

For Boolean deduction, the implications of a finite set of propositions or events is simply the implications of the single proposition or event formed by conjoining the members of that initial finite set of conditionals. One of the counter-intuitive features of deduction with a set conditionals is the necessity of considering the deductive implications of all possible conjunctions of the members of that initial set of conditionals. Both Adams [10] and Dubois & Prade [29] found a need for the following definition.

Definition 2.2.9. Conjunctive Closure of a Set of Conditionals. If J is a set of conditionals then the *conjunctive closure* $C(J)$ of J is the set of all conjunctions of any finite subset of J.

Theorem 2.2.10. Deduction Theorem. For all the elementary deductive relations $\leq_x$ and their combinations, except for $\leq_{tr}$ and $\leq_\vee$, the DCS $\mathcal{H}_x(J)$ with respect to $\leq_x$ of a set J of conditionals is the set of all conditionals implied with respect to $\leq_x$ by some member of the conjunctive closure $C(J)$ of J. That is,

$$\mathcal{H}_x(J) = \{ (Y|Z): (A|B) \leq_x (Y|Z), (A|B) \in C(J)\}$$

For a proof see subsection 3.4.3 of [6].

Corollary 2.2.11. Under the hypotheses of the Deduction Theorem, it follows from Theorem 2.2.5. (Principal Deductively Closed Sets) that

$$\mathcal{H}_X(J) \; = \; \cup \; \mathcal{H}_X(A|B)$$
$$(A|B) \; \in \; C(J)$$

That is, the deductively closed set with respect to $\leq_X$ generated by a subset J of conditionals is the set of all conditionals implied with respect to $\leq_X$ by some member of the conjunctive closure $C(J)$ of J.

For most deductive relations $\leq_X$ it is necessary in general to first determine the conjunctive closure $C(J)$ of a finite set of conditionals J in order to determine the DCS $\mathcal{H}_X(J)$ of J. However for the non-falsity, inapplicability and conjunctive deductive relations, that is for $x \in \{nf, ip, \wedge\}$, the DCS of J is $\mathcal{H}_X(J) = \mathcal{H}_X(A|B)$, where $(A|B)$ is the single conditional formed by conjoining all the conditionals in J.

Corollary 2.2.12. With respect to the three deductive relations $\leq_{nf}$, $\leq_{ip}$, and $\leq_\wedge$ the DCS of a finite set of conditionals J is principal and is generated by the single conditional formed by conjoining all the conditionals in J.

Proof of Corollary 2.2.12. Let $x \in \{nf, ip, \wedge\}$, and suppose $(A|B)$ is the conjunction of all the conditionals in the set J of conditionals. Then $\mathcal{H}_X(J) = \mathcal{H}_X(A|B)$ because with respect to $\leq_X$, $(A|B) \leq_X (Y|Z)$ for all $(Y|Z)$ in $C(J)$. This follows from the fact, which is easily checked, that for these deductive relations the conjunction of two conditionals always implies each of the component conditionals. $\qquad \square$

Having described the DCS's of all but two of the deductive relations in the hierarchy, it remains to solve for the DCS's of the exceptional deductive relations $\leq_{tr}$ and $\leq_\vee$.

Theorem 2.2.13. For the deductive relation $\leq_{tr}$, the DCS generated by two conditionals $(A|B)$ and $(C|D)$ is principal and generated by the conjunction ABCD of the components of the conditionals. That is,

$$\mathcal{H}_{tr}\{(A|B), (C|D)\} \; = \; \mathcal{H}_{tr}(ABCD).$$

Proof of Theorem 2.2.13. Let $\mathcal{H} = \mathcal{H}_{tr}\{(A|B), (C|D)\}$. Note that $(A|B) =_{tr} AB$ and $(C|D) =_{tr} CD$. Since $(A|B) \leq_{tr} AB$, therefore $AB \in \mathcal{H}$. Since $(C|D) \leq_{tr} CD$, therefore $CD \in \mathcal{H}$. Since both AB and CD are in $\mathcal{H}$ therefore $(AB)(CD) \in \mathcal{H}$. So $\mathcal{H}_{tr}(ABCD) \subseteq \mathcal{H}$. Conversely, since $ABCD \leq_{tr} AB$ and $AB \leq_{tr} (A|B)$ therefore $(A|B) \in \mathcal{H}_{tr}(ABCD)$. Similarly, $(C|D) \in \mathcal{H}_{tr}(ABCD)$. Since both $(A|B)$ and $(C|D)$ are in $\mathcal{H}_{tr}(ABCD)$ therefore $\mathcal{H} = \mathcal{H}_{tr}\{(A|B), (C|D)\} \subseteq \mathcal{H}_{tr}(ABCD)$. So $\mathcal{H} = \mathcal{H}_{tr}(ABCD)$. $\qquad \square$

Corollary 2.2.14. $\mathcal{H}_{tr}\{AB, CD\} = \mathcal{H}_{tr}(ABCD)$. In general, $\mathcal{H}_{tr}\{A,C\} = \mathcal{H}_{tr}(AC)$.

Proof of Corollary 2.2.14. Since $AC \leq A$ and $AC \leq C$, both A and C are in $\mathcal{H}_{tr}(AC)$. So $\mathcal{H}_{tr}\{A,C\} \subseteq \mathcal{H}_{tr}(AC)$. Conversely, $AC \in \mathcal{H}_{tr}\{A,C\}$. So $\mathcal{H}_{tr}\{A,C\} = \mathcal{H}_{tr}(AC)$. Replacing A with AB and C with CD yields that $\mathcal{H}_{tr}\{AB, CD\} = \mathcal{H}_{tr}(ABCD)$. $\qquad \square$

Corollary 2.2.15. If $J = \{(A_i \mid B_i): i = 1, 2, ..., n\}$ then

$$\mathcal{H}_{tr}(J) = \mathcal{H}_{tr}(\prod_{i=1}^{n} A_i B_i) = \mathcal{H}_{tr}\{A_1 B_1, A_2 B_2, ..., A_n B_n\}.$$

Proof of Corollary 2.2.15. The same argument works as in Theorem 2.2.13 and Corollary 2.2.14. $\square$

Note that $\mathcal{H}_{tr}(AB) \wedge \mathcal{H}_{tr}(CD) \subseteq \mathcal{H}_{tr}(ABCD)$ but in general $\mathcal{H}_{tr}(AB) \wedge \mathcal{H}_{tr}(CD) \neq \mathcal{H}_{tr}(ABCD)$ since conjunctions don't necessarily imply their conjuncts.

Note also that while $\{AB, CD\}$ and $(ABCD)$ both generate $\mathcal{H}_{tr}\{(A\mid B), (C\mid D)\}$, the conjunction $(A\mid B)(C\mid D)$ may not. In general, $\mathcal{H}_{tr}((A\mid B)(C\mid D)) \subseteq \mathcal{H}_{tr}\{(A\mid B), (C\mid D)\}$.

Corollary 2.2.16. $\mathcal{H}_{tr}((A\mid B)(C\mid D)) = \mathcal{H}_{tr}\{(A\mid B), (C\mid D)\}$ if and only if $AB \leq D$ and $CD \leq B$.

Proof of Corollary 2.2.16. $\mathcal{H}_{tr}((A\mid B)(C\mid D)\} = \mathcal{H}_{tr}(ABD' \vee B'CD \vee ABCD \mid B \vee D)$ $= \mathcal{H}_{tr}(ABD' \vee B'CD \vee ABCD)$. Since $ABCD \leq ABD' \vee B'CD \vee ABCD$, $\mathcal{H}_{tr}(ABD' \vee B'CD \vee ABCD \mid B \vee D) \subseteq \mathcal{H}_{tr}(ABCD)$ with equality if $ABD' = 0 = B'CD$, that is, if $AB \leq D$ and $CD \leq B$. But if either $ABD' \neq 0$ or $B'CD \neq 0$ then $\mathcal{H}_{tr}(ABD' \vee B'CD \vee ABCD \mid B \vee D)$ does not contain $ABCD$ since all members of $\mathcal{H}_{tr}(ABD' \vee B'CD \vee ABCD \mid B \vee D)$ have both components above $(ABD' \vee B'CD \vee ABCD)$ and $(ABD' \vee B'CD \vee ABCD) > ABCD$. $\square$

For completeness there is also the following theorem concerning the DCS's with respect to the deductive relation $\leq_{ap}$.

Theorem 2.2.17. For the deductive relation $\leq_{ap}$, the DCS generated by two conditionals $(A\mid B)$ and $(C\mid D)$ is $\mathcal{H}_{ap}\{(A\mid B), (C\mid D)\} = \mathcal{H}_{ap}(A\mid B) \cup \mathcal{H}_{ap}(C\mid D)$. In general, if $J = \{(A_i\mid B_i): i = 1, 2, ..., n\}$ then $\mathcal{H}_{ap}(J) = \mathcal{H}_{ap}(A_1\mid B_1) \cup \mathcal{H}_{ap}(A_2\mid B_2) \cup ... \cup \mathcal{H}_{ap}(A_n\mid B_n)$.

Proof of Theorem 2.2.17. By the Deduction Theorem, or directly, $\mathcal{H}_{ap}\{(A\mid B), (C\mid D)\} = \mathcal{H}_{ap}(A\mid B) \cup \mathcal{H}_{ap}(C\mid D) \cup \mathcal{H}_{ap}((A\mid B)(C\mid D))$. But $(A\mid B) \leq_{ap} (A\mid B)(C\mid D)$ because $B \leq B \vee D$. Therefore $\mathcal{H}_{ap}((A\mid B)(C\mid D)) \subseteq \mathcal{H}_{ap}(A\mid B)$. So $\mathcal{H}_{ap}\{(A\mid B), (C\mid D)\} = \mathcal{H}_{ap}(A\mid B) \cup \mathcal{H}_{ap}(C\mid D)$. In general, all conjunctions of members of J have antecedents that are implied by the antecedent of any one of the conjuncts. So all conjunctions of members of J are in the union $\mathcal{H}_{ap}(A_1\mid B_1) \cup \mathcal{H}_{ap}(A_2\mid B_2) \cup ... \cup \mathcal{H}_{ap}(A_n\mid B_n)$, and this union is easily a DCS. $\square$

Theorem 2.2.18. For the deductive relation $\leq_\vee$ the DCS generated by two conditionals $(A\mid B)$ and $(C\mid D)$ is

$$H_\vee\{(A\mid B), (C\mid D)\} = H_\vee(A\mid B) \cup H_\vee(C\mid D) \cup H_\vee(ABCD \mid B \vee D).$$

Proof of Theorem 2.2.18. Let $\mathcal{H}$ denote the right hand side of the above equation. It will be shown that $\mathcal{H}$ is the smallest DCS containing both (A|B) and (C|D). Clearly $\mathcal{H}$ contains both (A|B) and (C|D) since it includes $\mathcal{H}_\vee$(A|B) and $\mathcal{H}_\vee$(C|D). Concerning the deduction property, suppose that (W|Z) $\in$ $\mathcal{H}$ and (W|Z) $\leq_\vee$ (S|T). Then (W|Z) is in $\mathcal{H}_\vee$(A|B) or $\mathcal{H}_\vee$(C|D) or $\mathcal{H}_\vee$(ABCD | B $\vee$ D), and by the deduction property for these DCS's, so too is (S|T) in that DCS. So (S|T) $\in$ $\mathcal{H}$. Concerning the conjunction property, suppose (E|F) and (G|H) are both in $\mathcal{H}$. To show that their conjunction must also be in $\mathcal{H}$ consider cases. If both are in any one of $\mathcal{H}_\vee$(A|B), $\mathcal{H}_\vee$(C|D) or $\mathcal{H}_\vee$(ABCD | B $\vee$ D), then their conjunction (E|F)(G|H) will also be in that one since they are each DCS's. So in those cases (E|F)(G|H) $\in$ $\mathcal{H}$. If instead, (E|F) $\in$ $\mathcal{H}_\vee$(A|B) and (G|H) $\in$ $\mathcal{H}_\vee$(C|D), then (E|F) = (AB $\vee$ Y1 | B $\vee$ Z1) and (G|H) = (CD $\vee$ Y2 | D $\vee$ Z2) for some events Y1, Z1, Y2, Z2. (E|F) reduces to (AB $\vee$ Y1(B $\vee$ Z1) | B $\vee$ Z1) and (G|H) reduces to (CD $\vee$ Y2(D $\vee$ Z2) | D $\vee$ Z2). So

$$(E|F)(G|H) = [(AB \vee Y1(B \vee Z1))(D \vee Z2)' \vee (CD \vee Y2(D \vee Z2))(B \vee Z1)'$$
$$\vee \quad (AB \vee Y1(B \vee Z1))(CD \vee Y2(D \vee Z2)) \ | \ B \vee D \vee Z1 \vee Z2 \ ]$$

$$= [(AB)(CD) \vee \ldots \ | \ B \vee D \vee Z1 \vee Z2 \ ] \in \mathcal{H}_\vee(ABCD \ | \ B \vee D) \subseteq \mathcal{H}.$$

So again (E|F)(G|H) is in $\mathcal{H}$. If instead (E|F) $\in$ $\mathcal{H}_\vee$(A|B) and (G|H) $\in$ $\mathcal{H}_\vee$(ABCD | B $\vee$ D) then a similar computation yields that (E|F)(G|H) $\in$ $\mathcal{H}_\vee$(ABCD | B $\vee$ D) $\subseteq$ $\mathcal{H}$. By symmetry, the same result holds in case (E|F) $\in$ $\mathcal{H}_\vee$(C|D) and (G|H) $\in$ $\mathcal{H}_\vee$(ABCD | B $\vee$ D). So in all cases, (E|F)(G|H) $\in$ $\mathcal{H}$. Therefore $\mathcal{H}$ is a DCS. Finally, $\mathcal{H}$ is the smallest DCS containing (A|B) and (C|D) because any DCS containing them must contain (AB | B $\vee$ D) since (A|B) $\leq_\vee$ (AB | B $\vee$ D). Similarly (CD | B $\vee$ D) is in any DCS containing (A|B) and (C|D). Therefore (AB | B $\vee$ D)(CD | B $\vee$ D) = (ABCD | B $\vee$ D) must be in any DCS containing (A|B) and (C|D). Therefore $\mathcal{H}$ = $\mathcal{H}_\vee$(A|B) $\cup$ $\mathcal{H}_\vee$(C|D) $\cup$ $\mathcal{H}_\vee$(ABCD | B $\vee$ D) must be the smallest DCS with respect to $\leq_\vee$ containing both (A|B) and (C|D). That completes the proof of the theorem. $\square$

Corollary 2.2.19. For three conditionals (A|B), (C|D) and (E|F) the DCS generated with respect to the deductive relation $\leq_\vee$ is

$$\mathcal{H}_\vee\{(A|B), (C|D), (E|F)\}$$
$$= \mathcal{H}_\vee(A|B) \cup \mathcal{H}_\vee(C|D) \cup \mathcal{H}_\vee(E|F)$$
$$\cup \mathcal{H}_\vee(ABCD \ | \ B \vee D) \cup \mathcal{H}_\vee(ABEF \ | \ B \vee F) \cup \mathcal{H}_\vee(CDEF \ | \ D \vee F)$$
$$\cup \mathcal{H}_\vee(ABCDEF \ | \ B \vee D \vee F).$$

Proof of Corollary 2.2.19. Let $\mathcal{H}$ denote the right hand side of the above equation. Cleary $\mathcal{H}$ contains the three conditionals (A|B), (C|D) and (E|F). $\mathcal{H}$ is also a DCS since it has the deduction and conjunction properties. It has the deduction property since if (G|H) $\in$ $\mathcal{H}$ and (G|H) $\leq_\vee$ (R|S) then (G|H) is at least one of the seven DCS's whose union is $\mathcal{H}$. So (R|S) is in at least one of the seven DCS's by the deduction property applied to that DCS. So $\mathcal{H}$ has the deduction property. Concerning the conjunction property, suppose (G|H) and (R|S) are in $\mathcal{H}$. Then by considering cases similar to those in the proof of the theorem it follows that (G|H)(E|F) is also in $\mathcal{H}$. It is also clear from the theorem that any DCS containing the three conditionals must also contain the conjunctions (A|B)(C|D), (C|D)(E|F), and (C|D)(E|F) and so must contain (ABCD | B $\vee$ D), (ABEF | B $\vee$ F) and (CDEF | D $\vee$ F) as shown in the proof of the theorem. Furthermore, if a DCS contains these latter three conditionals, then it must contain the conjunction of any two of them, which is (ABCDEF | B $\vee$ D $\vee$ F). So $\mathcal{H}$ is the smallest DCS that contains the original three conditionals.   □

2.3 Examples of Deduction with Uncertain Conditionals

In [6] the implications of the three well known "penguin postulates" have been completely described with respect to the elementary deductive relations and their combinations. In this section two more examples will be given. First the implications of the set J of the two conditionals {(A|B), (B|C)} will be determined. Of interest is the conditional (A|C), which is easily true when the initial two conditionals are certainties, but may be false when one or the other is uncertain. We are often interested in chaining deductions and inferences in this way. What are the implications and inferences to be made from knowing "A given B" and "B given C", allowing for the lack of certainty of these conditionals?

Example 2.3.1. Transitivity. Consider the set J consisting of two uncertain conditionals (A|B) and (B|C). Then the conjunctive closure $C(J)$ = {(A|B), (B|C), (A|B)(B|C)} = {(A|B), (B|C), (AB | B $\vee$ C)}. For $x \in$ {nf, ip, $\wedge$}, by Theorem 2.2.8 on principal DCS's, the DCS generated by J is $\mathcal{H}_x(J) = \mathcal{H}_x(AB | B \vee C)$. So using equations (2.2.3-4), and (2.2.7),

$$\mathcal{H}_{ip}(J) = \{(Y \mid (B \vee C)Z)\text{: any } Y, Z \text{ in } \mathcal{B}\}$$
$$\mathcal{H}_{nf}(J) = \{(AB \vee B'C' \vee Y \mid Z)\text{: any } Y, Z \text{ in } \mathcal{B}\}$$
$$\mathcal{H}_{\wedge}(J) = \{(AB \vee Y \mid (B \vee C)Z)\text{: any } Y, Z \text{ in } \mathcal{B}\}$$

Notice that (A|C) $\in \mathcal{H}_{nf}(J)$ by setting Y = AB' and Z = C. In that case (AB $\vee$ B'C' $\vee$ Y | Z) = (AB $\vee$ AB' $\vee$ B'C' | C) = (A $\vee$ B'C' | C) = (A|C). Thus with respect to the non-falsity deductive relation $\leq_{nf}$, the conditional (A|C), as expected, is implied by (A|B) and (B|C). When (A|B) and (B|C) are non-false then so is (A|C). $\mathcal{H}_{nf}(J)$ is the set of all conditionals whose conclusion includes the truth of (A|B) and also the inapplicability of both (A|B) and (B|C). By similar arguments (A|C) is in $\mathcal{H}_{ip}(J)$ and also in $\mathcal{H}_{\wedge}(J)$.

For the elementary deductive relations $\leq_x$ or some combination of them except for $\leq_{tr}$ and $\leq_{\vee}$, by Corollary 2.2.11 the DCS generated by J is $\mathcal{H}_x(J) = \mathcal{H}_x(A|B) \cup \mathcal{H}_x(B|C) \cup \mathcal{H}_x(AB | B \vee C)$.

Now let x = pm. That is, consider the deductions of J with respect to the probabilistically monotonic deductive relation $\leq_{pm}$. Since (AB | B $\vee$ C) $\leq_{pm}$ (A|B), therefore

$\mathcal{H}_{pm}(AB \mid B \vee C) \supseteq \mathcal{H}_{pm}(A \mid B)$. Thus, $\mathcal{H}_{pm}(J) = \mathcal{H}_{pm}(B \mid C) \cup \mathcal{H}_{pm}(AB \mid B \vee C)$. So by equation (2.2.6) $\mathcal{H}_{pm}(J) = \{BC \vee C' \vee Y \mid BC \vee Z)$: any Y, Z in $\mathcal{B}\} \cup \{AB \vee B'C' \vee Z \mid AB \vee Z)$: any Y, Z in $\mathcal{B}\}$. Note that $(A \mid C)$ is not necessarily a member of $\mathcal{H}_{pm}(J)$.

Furthermore, since $\leq_{pm}$ is probabilistically monotonic, all the conditionals in $\mathcal{H}_{pm}(B \mid C) = \{BC \vee C' \vee Y \mid BC \vee Z)$: any Y, Z in $\mathcal{B}\}$ have conditional probability no less than $P(B \mid C)$, and all the conditionals in $\mathcal{H}_{pm}(AB \mid B \vee C) = \{AB \vee B'C' \vee Z \mid AB \vee Z)$: any Y, Z in $\mathcal{B}\}$ have conditional probability no less than $P(AB \mid B \vee C)$.

Dubois & Prade [29] have studied a number of similar transitivity problems. They showed that with respect to $\leq_{pm}$ the set of conditionals $J = \{(A \mid BC), (B \mid C)\}$ does imply the conditional $(A \mid C)$. This follows easily by conjoining the two conditionals of J to get $(AB \mid C)$, which easily implies $(A \mid C)$ with respect to $\leq_{pm}$.

Example 2.3.2. Absent-minded Coffee Drinker. The second example by H. Pospesel [20] is a typical inference problem called the "absent-minded coffee drinker": "Since my spoon is dry I must not have sugared my coffee, because the spoon would be wet if I had stirred the coffee, and I wouldn't have stirred it unless I had put sugar in it."

This is not a valid argument in the 2-valued logic, but there are still deductions and inferences to be drawn from these conditional premises. Letting D denote "my spoon is dry", G denote "I sugared my coffee", and R denote "I stirred my coffee", all deductions with respect to various deductive relations will now be determined. Translating into this terminology the set of premises is $J = \{D, (D' \mid R), (R' \mid G')\}$. Therefore the conjunctive closure of J is

$$C(J) = \{D, (D' \mid R), (R' \mid G'), D(D' \mid R), D(R' \mid G'), (D' \mid R)(R' \mid G'), D(D' \mid R)(R' \mid G')\}.$$

Using the operations on conditionals (1.1)–(1.4) this reduces to

$$C(J) = \{D, (D' \mid R), (R' \mid G'), DR', DG \vee DR'G', (D'RG \vee R'G' \mid R \vee G'), DR'\}.$$

So according to the Corollary 2.2.11, for any of the elementary deductive relations $\leq_x$ or their combinations, except for $\leq_{tr}$ and $\leq_\vee$,

$$\mathcal{H}_x(J) = \mathcal{H}_x(D) \cup \mathcal{H}_x(D' \mid R) \cup \mathcal{H}_x(R' \mid G') \cup \mathcal{H}_x(DR') \cup \mathcal{H}_x(DG \vee DR'G')$$
$$\cup \, \mathcal{H}_x(D'RG \vee R'G' \mid R \vee G').$$

Now this union can be simplified because some of these DCS's are included in the others. For instance, since all of these deductive relations satisfy $DR' \leq D$, therefore $\mathcal{H}_x(DR') \supseteq \mathcal{H}_x(D)$. Similarly, $DR' \leq DG \vee DR' = D(G \vee R') = D(G \vee R'G') = DG \vee DR'G'$. So $\mathcal{H}_x(DR') \supseteq \mathcal{H}_x(DG \vee DR'G')$. Thus,

$$\mathcal{H}_x(J) = \mathcal{H}_x(D' \mid R) \cup \mathcal{H}_x(R' \mid G') \cup \mathcal{H}_x(DR') \cup \mathcal{H}_x(D'RG \vee R'G' \mid R \vee G').$$

For x = ip, nf or $\wedge$, by Corollary 2.2.12, $\mathcal{H}_x(J) = \mathcal{H}_x(D(D'|R)(R'|G')) = \mathcal{H}_x(DR')$. Therefore

$$\mathcal{H}_{nf}(J) = \mathcal{H}_{nf}(DR') = \{(DR' \vee Y \mid Z): \text{any } Y, Z \text{ in } \mathcal{B}\}.$$

That is, the implications of J when its conditionals are regarded as merely non-false, are all those conditionals with any antecedent and whose consequent includes the event DR', that "my spoon is dry" and "I did not stir my coffee". Notice that G', "I did not sugar my coffee", is not an implication of J with respect to the non-falsity deductive relation, and neither is it a valid consequence of J in the 2-valued Boolean logic. In the 2-valued logic the implications of J are the universally conditioned events that include DR', that the spoon is dry and my coffee is not stirred. But the implications with respect to the "non-falsity" deductive relation $\leq_{nf}$ include all those with any other condition attached.

Similarly, by Corollary 2.2.12,

$$\mathcal{H}_\wedge(J) = \mathcal{H}_\wedge(DR') = \{(DR' \vee Y \mid Z): \text{any } Y, Z \text{ in } \mathcal{B}\} = \mathcal{H}_{nf}(J).$$

So in this case the implications with respect to $\leq_\wedge$ are equal to the implications with respect to $\leq_{nf}$.

Turning to $\leq_{pm}$, there are no further simplifications to the union of the four sets. So

$$\mathcal{H}_{pm}(J) = \mathcal{H}_{pm}(D'|R) \cup \mathcal{H}_{pm}(R'|G') \cup \mathcal{H}_{pm}(DR') \cup \mathcal{H}_{pm}(D'RG \vee R'G' \mid R \vee G').$$

Therefore, a minimum generating set, J_{min}, for $\mathcal{H}_{pm}(J)$ is

$$J_{min} = \{(D'|R), (R'|G'), DR', (D'RG \vee R'G' \mid R \vee G')\}.$$

This is "minimum" in the sense that the union of the implications of individual conditionals in J_{min} provides the set of all implications of J, but no smaller subset of J_{min} will suffice. By equation (2.2.6) for deductions from a single conditional,

$$\mathcal{H}_{pm}(D'|R) = \{(D'R \vee R' \vee Y \mid D'R \vee Z): \text{any } Y, Z \text{ in } \mathcal{B}\},$$
$$\mathcal{H}_{pm}(R'|G') = \{(R'G' \vee G \vee Y \mid R'G' \vee Z): \text{any } Y, Z \text{ in } \mathcal{B}\},$$
$$\mathcal{H}_{pm}(DR') = \{(DR' \vee Y \mid DR' \vee Z): \text{any } Y, Z \text{ in } \mathcal{B}\}, \text{ and}$$
$$\mathcal{H}_{pm}(D'RG \vee R'G' \mid R \vee G')$$
$$= \{(D'RG \vee R' \vee Y \mid D'RG \vee R'G' \vee Z): \text{any } Y, Z \text{ in } \mathcal{B}\}.$$

The first two sets listed above are the deductive implications with respect to $\leq_{pm}$ of (D'|R) and (R'|G'), two of the three original conditionals in J. These implications were apparent from J without further analysis. Together with these the other two conditionals of J_{min} provide a minimum basis for expressing all deductions of J with respect to $\leq_{pm}$ in terms of the implications of individual conditionals.

$\mathcal{H}_{pm}(DR')$ is the set of all conditionals whose antecedents and consequents include my not stirring my coffee and my spoon being dry. $\mathcal{H}_{pm}(D'RG \vee R'G' \mid R \vee G')$ is all conditionals whose premise includes my stirring my coffee or not sugaring my coffee and whose conclusion includes my stirring and sugaring my coffee and wetting my spoon or neither stirring nor sugaring my coffee.

All the conditionals in $\mathcal{H}_{pm}(DR')$ have conditional probability no less than $P(DR')$, and those conditionals in $\mathcal{H}_{pm}(D'RG \vee R'G' \mid R \vee G')$ have conditional probability no less than $P(D'RG \vee R'G' \mid R \vee G')$.

If the spoon is observed to be dry (D=1) then

$$\mathcal{H}_{pm}(J) = \mathcal{H}_{pm}(0 \mid R) \cup \mathcal{H}_{pm}(R' \mid G') \cup \mathcal{H}_{pm}(R') \cup \mathcal{H}_{pm}(R'G' \mid R \vee G').$$

Since $(0 \mid R) \leq_{pm} (R' \mid G')$ is true, therefore $\mathcal{H}_{pm}(0 \mid R) \supseteq \mathcal{H}_{pm}(R' \mid G')$, and since $(0 \mid R) \leq_{pm} R'$ is true, therefore $\mathcal{H}_{pm}(0 \mid R) \supseteq \mathcal{H}_{pm}(R')$, and since $(0 \mid R) \leq_{pm} (R'G' \mid R \vee G')$ is true, therefore $\mathcal{H}_{pm}(0 \mid R) \supseteq \mathcal{H}_{pm}(R'G' \mid R \vee G')$. Thus $\mathcal{H}_{pm}(J)$ reduces to a single set of conditionals. $\mathcal{H}_{pm}(J) =$

$$\mathcal{H}_{pm}(0 \mid R) = \{ (0 \vee R' \vee Y \mid 0 \vee Z) \colon \text{any } Y, Z \} = \{ (R' \vee Y \mid Z) \colon \text{any events } Y, Z \}.$$

So if I observe my spoon to be dry (D=1), the implications of J with respect to $\leq_{pm}$ are just all conditionals whose conclusion includes R', that I didn't stir my coffee. There are no other implications. Thus the probability that I sugared my coffee is not further constrained.

Example 2.3.3. Absent-minded Coffee Drinker Revisited. It interesting to see what happens with this example when the conditional $(R' \mid G')$ in J is replaced by $(R \mid G)$. Instead of saying "I wouldn't have stirred my coffee unless I had put sugar in it" suppose it was "if I sugared my coffee then I stirred it." Thus $J = \{D, (D' \mid R), (R \mid G)\}$.

In the Boolean 2-valued logic, the implications of J are those of the conjunction $D(D'|R)(R|G)$ where the conditionals are equated to their material conditionals and have a conjunction $D(D' \vee R')(R \vee G') = DR'G'$. So G', that I didn't sugar my coffee is an implication in standard 2-valued logic.

Concerning the non-falsity deductive relation $\leq_{nf}$, by Corollary 2.2.12,

$$\mathcal{H}_{nf}(J) = \mathcal{H}_{nf}(DR'G') = \{ (DR'G' \vee Y \mid Z) \colon \text{any } Y, Z \text{ in } \mathcal{B} \},$$

namely all conditionals whose conclusion includes DR'G'.

More generally the conjunctive closure of J is

$$C(J) = \{D, (D'|R), (R|G), D(D'|R), D(R|G), (D'|R)(R|G), D(D'|R)(R|G)\}$$
$$= \{D, (D'|R), (R|G), DR', DG' \vee DRG, (D'RG' \vee D'RG \mid R \vee G), DR'G'\}.$$

Obviously, the propositions D and DR' are implications with respect to all deductive relations of DR'G', and so for all deductive relations $\leq_x$ their implications are included in

$$\mathcal{H}_x(J) = \mathcal{H}_x(D'|R) \cup \mathcal{H}_x(R|G) \cup \mathcal{H}_x(DG' \vee DRG) \cup \mathcal{H}_x(D'R \mid R \vee G) \cup \mathcal{H}_x(DR'G').$$

Furthermore, $(DG' \vee DRG) = D(G' \vee RG) = D(G' \vee R) = D(R'G' \vee R) = DR'G' \vee R)$ is also an implication of DR'G'. So dropping $\mathcal{H}_x(DG' \vee DRG)$ from the union,

$$\mathcal{H}_x(J) = \mathcal{H}_x(D'|R) \cup \mathcal{H}_x(R|G) \cup \mathcal{H}_x(D'R \mid R \vee G) \cup \mathcal{H}_x(DR'G').$$

Note that the proposition DR'G' (having a dry spoon, unstirred coffee, and unsugared coffee) which is the conjunction of the three original conditionals of $J = \{D, (D' \mid R), (R \mid G)\}$, is an implication with respect to all these deductive relations. It is a logical consequence of J. By rearranging the conditioning, its probability is given by

$$P(DR'G') = P(D)P(R'G' \mid D) = P(D)P((G' \mid R') \mid D)P(R' \mid D) = P(D)P(G' \mid DR')P(R' \mid D).$$

This latter product has easily estimated conditionals probabilities. $P(D) = 1$ by observation, and both $P(G' \mid DR')$ and $P(R' \mid D)$ are also close to or equal to 1. This is one way the reasoning can proceed even though the initial phrasing was in terms of conditionals whose probabilities are not so easily estimated.

In addition, $(D'R \mid R \vee G) \leq_{pm} (D' \mid R)$ because $(D'R)(R \vee G) \leq (D'R)$ and $D'R \vee R'G' \leq D'R \vee R'$. Thus all the implications of J with respect to $\leq_{pm}$ are:

$$\mathcal{H}_{pm}(J) = \mathcal{H}_{pm}(R \mid G) \cup \mathcal{H}_{pm}(D'R \mid R \vee G) \cup \mathcal{H}_{pm}(DR'G').$$
$$= \{(RG \vee G' \vee Y \mid RG \vee Z): \text{any } Y, Z \text{ in } \mathcal{B}\}$$
$$\cup \{(D'R \vee R'G' \vee Y \mid D'R \vee Z): \text{any } Y, Z \text{ in } \mathcal{B}\}$$
$$\cup \{(DR'G' \vee Y \mid DR'G' \vee Z): \text{any } Y, Z \text{ in } \mathcal{B}\}.$$

Furthermore, the conditionals in $\mathcal{H}_{pm}(R \mid G)$ all have conditional probability no less than $P(R \mid G)$, and similarly for the conditionals in $\mathcal{H}_{pm}(D'R \mid R \vee G)$ and in $\mathcal{H}_{pm}(DR'G')$.

3   Computations with Conditionals

In formulating his "high conditional probability" logic Adams [30] found it useful to define operations on the underlying pairs of events. Now while the preceding sections provide an adequate theoretical basis for calculating and reasoning with conditional propositions or events, the problem of the complexity of information is no less daunting. Indeed, even without the added computational burden of operating with explicit conditionals, just operating with Boolean expressions in practical situations with, say, a dozen variables, is already too complex for practical pure Bayesian analysis. The reason for this is that in most situations the available information is insufficient to determine a single probability distribution satisfying the known constraints of the situation. Various possibilities concerning unknown dependences between subsets of variables result in complicated solutions to relatively simple problems.

3.1   Pure Bayesian Analysis

For example, consider again the transitivity problem of Section 2.3.1. If "A given B" and "B given C" are both certain, then it follows that "A given C" is also a certainty. But if they are not certain, then by pure Bayesian analysis, $P(A \mid C)$ can be zero no matter how high are the conditional probabilities of $(A \mid B)$ and $(B \mid C)$. This happens because $P(B \mid C)$ and $P(A \mid B)$ can be almost 1 while $P(A \mid BC)$ is zero, and it is the latter probability that appears in the Bayesian solution: $P(A \mid C) = P(AB \text{ or } AB' \mid C) = P(AB \mid C) + P(AB' \mid C) = P(ABC)/P(C) + P(AB'C)/P(C) = P(ABC \mid BC) P(BC \mid C) +$

P(AB'C | B'C) P(B'C|C) = P(A|BC)P(B|C) + P(A|B'C)P(B'|C). Without knowing anything about P(A|BC) or P(A|B'C), nothing more can be said about P(A|C).

3.2 Choosing a Bayesian Solution Using Maximum Information Entropy

Continuing the example of Section 3.1, knowing that C is true might dramatically change P(A|B) up or down. But if nothing is known one way or the other, the choice of the maximum information entropy distribution assumes that P(A|BC) = P(A|B). This latter equation is called the *conditional independence* of A and C given B. It can also be expressed as P(AC|B) = P(A|B)P(C|B) or as P(C|AB) = P(C|B). Using this principle P(A|C) = P(A|B)P(B|C) + P(A|B'C)P(B'|C). So if P(A|B) and P(B|C) are 0.9 and 0.8 respectively, then P(A|C) is at least 0.72. Additionally, since nothing is known one way or the other about the occurrence of A when B is false and C is true, this principle of "maximum indifference" implies that P(A|B'C) should be taken to be ½. So the term P(A|B'C)P(B'|C) contributes (1/2)P(B'|C) = (1/2)(1 − 0.8) = 0.1 to P(A|C) bringing the total to 0.82.

In effect the principle of maximum information entropy chooses that probability distribution P that assumes conditional independence of any two variables that are not explicitly known to have some dependence under the condition. This greatly simplifies computations and often allows situations of several dozen variables to be rapidly analyzed as long as the clusters of dependent variables are not too large and not too numerous. The maximum entropy solution is always one of the possible Bayesian solutions of the situation. If there is just one Bayesian solution, then the two solutions will always agree.

It is a remarkable fact that such a function as the entropy function exists, and it is now clear that it has wide application to information processing under uncertainty. If the n outcomes of some experiment are to be assigned probabilities p_i for i=1 to n subject to some set of constraints, then the distribution of probabilities that assumes conditional independence unless dependence is explicitly known is the one that maximizes the entropy function

$$H(p_1, p_2, p_3, \ldots, p_n) = - \sum_{i=1}^{n} p_i \log p_i \tag{3.1}$$

and also satisfies the known constraints. If there is an a priori distribution $q_1, q_2, q_3, \ldots, q_n$ then H is given by

$$H(p_1, p_2, p_3, \ldots, p_n, q_1, q_2, q_3, \ldots, q_n) = - \sum_{i=1}^{n} p_i \log (p_i/q_i) \tag{3.2}$$

This allows maximum entropy updates when additional information is available. See J. E. Shore [21] for a derivation.

W. Rödder [22], [23] and his colleagues at FernUniversität in Hagen, Germany are continuing to develop a very impressive interactive computer program SPIRIT that implements this practical approach to the computation of propositions and conditional propositions and their probabilities. Starting with an initially defined set of variables and their values, the user can input statements and conditionals statements about these variables taking various values, and can also assign conditional probabilities to them. The *utility* of having a variable take one of its values can also be incorporated.

An even more general approach is that of Bamber and Goodman [33], [34], and together with Nguyen [35]. They take a purely second order Bayesian approach by assuming a distribution on the set of solution distributions. Needless to say, computational difficulties are even greater with this approach.

3.3 Confidence in Maximum Entropy Solutions

While in general the maximum entropy solution may provide the most plausible or "most likely" probability distribution for a situation among all of the Bayesian solutions, it does not immediately provide a means for estimating how much confidence to attach to that solution. This issue has been taken up by E. Jaynes [24], S. Amari [25], C. Robert [36], A. Caticha [26] and recently by W. Rödder [37].

Jaynes puts the matter as follows: "Granted that the distribution of maximum entropy has a favored status, in exactly what sense, and how strongly, are alternative distributions of lower entropy ruled out?" He proves an entropy "concentration theorem" in the context of an generalized experiment of N independent trials each having n possible results and satisfying a set of m ($<$ n) linearly independent, linear constraints on the observed frequencies of the experiment. Jaynes shows that in the limit as the number N of trials approaches infinity, the fraction F of probability distributions satisfying the m constraints and whose entropy H differs from the maximum by no more than ΔH is given by the Chi-square distribution χ_k^2 with k = n − m −1 degrees of freedom as

$$2N(\Delta H) = \chi_k^2(F). \tag{3.3}$$

That is, the critical, threshold entropy value H_α for which only the fraction α of the probability distributions satisfying the m constraints have smaller entropy is given by

$$H_\alpha = H_{max} - \chi_k^2(1-\alpha) / 2N. \tag{3.4}$$

For N = 1000 independent trials of tossing a 6-sided die and with a significance level $\alpha = 0.05$ and degrees of freedom k = 6 - 1 - 1 = 4, 95% of the eligible probability distributions have entropy no less than $H_\alpha = H_{max} - 9.49 / 2N = H_{max} - 0.0047$. H_{max} is on the order of 1.7918 for a fair die and 1.6136 for a die with average die value of 4.5 instead of 3.5. Letting $\alpha = 0.005$ it follows that 99.5% of the eligible distributions will have entropy no less than $H_{max} - 14.9 / 2000 = H_{max} - 0.00745$.

Clearly eligible distributions that significantly deviate in entropy from the maximum value are very rare. However this result does not directly answer the question of how much confidence to have in the individual probabilities associated with distributions having maximum or almost maximum entropy. That is, can a probability distribution with close to maximum entropy assign probabilities that are significantly different from the probabilities of the maximum entropy distribution?

For instance, a 6-sided die having two faces with probabilities 1/12 and 1/4 respectively and four faces each having 1/6 probability has entropy 0.0436 less than the maximum of 1.7918 for a fair die. So for N=1000 independent trials and a significance level of $\alpha = 0.05$ such a distribution would differ from the maximum entropy value for a fair die by considerably more than 0.0047. However for N=100, ΔH = 9.49/200 = 0.047, which is large enough to include such a distribution.

It should be remembered that maximum entropy works best where it is needed most - in situations involving a relatively large number of variables or values. In small sample situations it can produce obviously erroneous results.

Furthermore, how does the confidence in the probabilities determined by a maximum entropy solution depend upon the amount of under-specification of the situation that produced that solution? Surely a maximum entropy distribution that relies upon a great deal of ignorance about a situation offers less confidence about the probabilities determined than does a maximum entropy solution that is based upon a minimum of ignorance about the situation. Put another way, confidence about the maximum entropy distribution should be higher when conditional independencies are positively known than when they are merely provisionally assumed.

Amari [25], [27] takes up these issues in the context of differential geometry. Under-specification of information gives rise to a manifold of possible probability distributions. A Riemannian metric on these distributions early introduced by C. R. Rao [28] allows a very general approach to quantifying the distance between distributions. This development provides a very general approach to these problems of multiple possible distributions, but so far the results don't seem to directly apply to the issue of the confidence to be attached to the individual probabilities dictated by a maximum entropy distribution. Unfortunately Amari offers no numerical example to illustrate how these results might be applied to allow a confidence measure to be put upon the probabilities associated with distributions having maximum or close to maximum entropy.

Caticha [26] frames the question along the same lines as Jaynes: "Once one accepts that the maximum entropy distribution is to be preferred over all others, the question is to what extent are distributions with lower entropy supposed to be ruled out?" Using a parameterized family of distributions Caticha shows how this question can be rephrased as another maximum entropy problem, but he too offers no simple illustrative example of how his results can be applied to the question of how much confidence to have in any one probability value associated with the maximum entropy distribution.

What seems to be needed is a way to solve for the probabilities of specified outcomes in terms of entropies equal to or close to the maximum entropy. If 95% of the eligible probability distributions have entropy H no less than H_{max} - ΔH, then what confidence limits are implied for the individual probabilities of those distributions?

W. Rödder [37] has recently addressed these questions using his maximum entropy methods but now applied to second order probabilities of "relevance".

3.4 Conditional Event Algebra and Conditional Independence

Maximum entropy techniques utilize the concept of conditional independence to simplify computations involving the probabilities and conditional probabilities of events and conditional events. In the simplest case P(A|B) can be computed as P(A) when A and B are independent. However, such methods are not directly applicable to the underlying conditional events, (A|B) and A. Since they have different conditions one cannot simply identify (A|B) with A just because they have equal probabilities. On the other hand, such simplifications do seem reasonable in many situations.

For instance, the sample space for a roll of one die might be S = {1,2,3,4,5,6} with equal probabilities. If a coin is independently flipped with sample space T={H,T}, a combined sample space of SxT = {1H, 1T, 2H, 2T, 3H, 3T, 4H, 4T, 5H, 5T, 6H, 6T} might represent the 12 possible outcomes with equal probabilities. Obviously, in this case any conditional event like "given an odd roll, a number less than 3 comes up" can be adequately represented either in (S|S) or in (SxT| SxT), but (S|S) is simpler and so preferred. Along these lines perhaps conditional independence can be used to simplify computations involving events and conditional events.

Such an approach to simplifying logical computations needs a theory of homomorphic images of probability spaces to support the identification of events and conditional events with different conditions. Sample space S is the homomorphic image of SxT under the mapping $\sigma(nX) = n$, where n is an outcome of the die roll, and X is either H or T. Event {1H, 1T} in SxT must be paired with event {1} in S, and {2H, 2T} paired with {2}, and so forth, and their probabilities must also be equal. These homomorphisms would need to be extended to the algebra of conditionals. Such a theory could be very useful in simplifying operations on conditionals rather than attempting the corresponding simplifications using the standard probability formulas for conditional independence.

4 Summary

In order to adequately represent and manipulate explicitly conditional statements such as "A given B" the familiar Boolean algebra of propositions or events must be extended to ordered pairs of such propositions or events. This is quite analogous to the requirement to extend integers to order pairs in order to adequately represent fractions and allow division. The resulting system of Boolean fractions includes the original propositions and also allows the non-trivial assignment of conditional probabilities to these Boolean fractions. Boolean fractions are truth functional in the sense that their truth status is completely determined by the truth or falsity of the two Boolean components of the fraction. But since there are two components, the truth status of a Boolean fraction has three possibilities – one when the condition (denominator) is false and two more when the denominator is true. Just as all integer fractions with a zero denominator are "undefined", so too are all Boolean fractions with a false condition undefined or "inapplicable". When the condition is true then the truth status of a Boolean fraction is determined by the truth of the numerator. The four extended operations (or, and, not, and given) on the Boolean fractions reduce to ordinary Boolean operations when the denominators are equivalent. Just as with integer fractions, the system of Boolean fractions has some new properties but loses others that are true in the Boolean algebra of propositions or events.

A conditional statement is not an implication or a deduction; it is rather a statement in a given context. Deduction of one conditional by another can still be defined in terms of the (extended) operations, as is often done in Boolean algebra. Due to the two components of a conditional there is a question of what is being implied when one conditional implies another. It turns out that several plausible implications between conditionals can be reduced to ordinary implications between the Boolean components of the two conditionals. The applicability, truth, non-falsity or inapplicability of one conditional can imply the corresponding property in the second conditional. Any two or more of these four elementary implications can be combined to form a more stringent implication. With respect to any one of these implications, a set of conditionals will generally imply a larger set, and it is now possible to compute the set of all deductions generated by some initial set of conditionals, as illustrated by three examples in this paper.

While computations can be done in principle, in practice the complexity of partial and uncertain conditional information precludes the possibility of solving for all possible probability distributions that satisfy the partial constraints. What is feasible and already successfully implemented in the program SPIRIT is to compute the distribution with maximum information entropy. However, the amount of confidence that can be associated with the probabilities assigned by this "most likely", maximum entropy distribution is still an open question. Finally the idea of applying conditional independence directly to conditional events lacks a clear theoretical basis.

References

1. P.G. Calabrese, "An Algebraic Synthesis of the Foundations of Logic and Probability", *Information Sciences*, 42, (1987) 187-237.
2. D. Lewis, "Probabilities of Conditionals and Conditional Probabilities", *The Philos. Rev.* 85(3) (1976), 297-315.
3. P.G. Calabrese, "A Theory of Conditional Information with Applications", In *Special Issue on Conditional Event Algebra, IEEE Transactions on Systems, Man, and Cybernetics*, Vol. 24, No. 12 (1994) 1676-1684.
4. G. Schay, "An Algebra of Conditional Events", *Journal of Mathematical Analysis and Applications 24* (1968), 334-344.
5. Z. Domotor "Probabilistic Relation Structures and their Applications" (Technical Report 144), Stanford University, Institute for Mathematical Studies in the Social Sciences, Stanford, California (1969).
6. P. G. Calabrese, "Deduction with uncertain conditionals", *Information Sciences* 147, 2002, 143-191.
7. I.R. Goodman, H.T. Nguyen, E.A. Walker, Conditional Inference and Logic for Intelligent Systems: A Theory of Measure-Free Conditioning, North-Holland, 1991.
8. T. Hailperin, *Sentential Probability Logic*, Lehigh Univ. Press (1996).
9. E.W. Adams, "Probability and the Logic of Conditionals," in: *Aspects of Inductive Logic*, P. Suppes & J. Hintikka, eds., Amsterdam: North-Holland (1966) 265-316.
10. E.W. Adams, "On the Logic of High Probability", *Journal of Philosophical Logic* 15 (1986) 225-279.
11. P.G. Calabrese, "Reasoning with Uncertainty Using Conditional Logic and Probability," in: *Proceeding of the First International Symposium on Uncertainty Modeling and Analysis*, IEEE Computer Society (1990) 682-688.
12. P.G. Calabrese, "Deduction and Inference Using Conditional Logic and Probability", in: I. R. Goodman, M. M. Gupta, H. T. Nguyen and G. S. Rogers, eds., *Conditional Logic in Expert Systems*, North-Holland, Amsterdam (1991) 71-100.

13. E.W. Adams, *A primer of Probability Logic*, CSLI Publications (1998).
14. B. De Finetti, "La logique de la probabilite," in *Actes du Congres International de Philosophie Scientifique* Paris: Hermann Editeurs, IV, (1936) 1-8.
15. B. De Finetti, *Probability, Induction and Statistics*, Wiley, New York (1972).
16. P.G. Calabrese, "An extension of the fundamental theorem of Boolean algebra to conditional propositions", Part I (1-18) of "Conditional event algebras and conditional probability logics" by P. G. Calabrese and I. R. Goodman, *Probabilistic Methods in Expert Systems*, Romano Scozzafava ed., Proceedings of the International Workshop, Rome, Italy, Societá Italiana di Statistica, (1993) 1-35.
17. E.A. Walker, "Stone Algebras, Conditional Events, and Three valued Logic", *IEEE Transactions of Systems, Man and Cybernetics*, Vol. 24, Number 12 (Dec. 1994) 1699-1707.
18. B. Sobocinski, "Axiomatization of a partial system of the three-valued calculus of propositions", *J. Computing Systems* 1 (1) 23-55.
19. N. Rescher, *Many-Valued Logics*, McGraw-Hill (1969).
20. H. Pospesel, *Arguments: Deductive Logic Exercises*, Prentice Hall (1971) , p.27, #78.
21. J.E. Shore, "Axiomatic Derivation of the Principle of Maximum Entropy and the Principle of Minimum Cross-Entropy", *IEEE Transactions on Information Theory*, IT-26, 1 (Jan. 1980), 26-37.
22. W. Rödder, C.-H. Meyer, "Coherent Knowledge Processing at Maximum Entropy by SPIRIT", *Proceedings of the Twelfth Conference On Uncertainty in Artificial Intelligence*, Portland, Oregon, USA, (1996) 470-476.
23. W. Rödder, "Conditional logic and the Principle of Entropy", *Artificial Intelligence* 117 (Feb. 2000) 83-106
24. E.T. Jaynes, "Concentration of Distributions at Entropy Maxima", reprinted in R. D. Rosenkrantz (ed.), *E.T. Jaynes: Papers on Probability, Statistics and Statistical Physics*, Reidel, Dordrecht, 1983.
25. S. Amari, Differential-Geometrical Methods in Statistics, Springer-Verlag, 1985.
26. A. Caticha, "Maximum Entropy, Fluctuations and Priors", MaxEnt 2000, *the 20th International Workshop on Bayesian Inference and Maximum Entropy Methods*, July 8-13, 2000, Gif-sur-Yvette, France.
27. S. Amari & T.S. Han, "Statistical Inference Under Multiterminal Rate Restrictions: A Differential Geometric Approach", Information Theory, *IEEE Transactions on*, Volume: 35 Issue: 2, (March 1989) 217 –227.
28. C.R. Rao, "Information and the Accuracy Attainable in the Estimation of Statistical Parameters", *Bulletin Calcutta Mathematical Society*, Vol. 35, (1945) 199-210.
29. D. Dubois & H. Prade, "Conditional Objects as Non-monotonic Consequence Relations", *IEEE Transactions on Systems, Man, and Cybernetics*, Vol. **24**, No. 12 (1994) 1724-1740.
30. E. Adams, *The Logic of Conditionals*, D. Reidel, Dordrecht, Holland, 1975.
31. I.R. Goodman & H.T. Nguyen, "Mathematical foundations of conditionals and their probability assignments", *Int. J. Uncertainty, Fuzziness & Knowledge-Based Systems* 3(3), 1995, 247-339.
32. I.R. Goodman, R.P. Mahler & H.T. Nguyen, *Mathematics of Data Fusion*, Kluwer Academic Press, Dordrecht, Holland, 1997.
33. D. Bamber & I.R. Goodman, "New uses of second order probability techniques in estimating critical probabilities in command & control decision-making", *Proc. 2000 Command & Control Research & Technology Symposium*, June 26-28, 2000, Naval Postgraduate School, Monterey, CA. URL = http://www.dodccrp.org/2000CCRTS/cd/html/pdf_papers/Track_4/124.pdf/

34. D. Bamber & I.R. Goodman, "Reasoning with assertions of high conditional probability: entailment with universal near surety", *Proc. Second Int. Symp. On Imprecise Probabilities and Their Applications (ISIPTA '01)*, (G. De Cooman, T.L. Fine & T. Seidenfeld, eds.) June 26-29, 2001, Cornell University, Ithica, NY, 17-26.

35. I.R. Goodman, D. Bamber & H.T. Nguyen, "New relations between Adams-Calabrese and Product Space conditional event algebras with applications to second order Bayesian inference", *Proc. Workshop on Conditionals, Information & Inference* (G. Kern-Isberner & W. Rödder, eds.), May 13-15, 2002, Hagen, Germany, 149-163.

36. C. Robert, "An entropy concentration theorem: application in artificial intelligence and descriptive statistics", *Journal of Applied Probability* 27, Sept 1990, 303-331.

37. W. Rödder, "On the measurability of knowledge acquisition and query processing", *International Journal of Approximate Reasoning* 33, 2003, 203-218.

Acceptance, Conditionals, and Belief Revision

Didier Dubois, Hélène Fargier, and Henri Prade

IRIT-CNRS, Université Paul Sabatier,
31062 , Toulouse, France
{dubois, fargier, prade}@irit.fr

Abstract. This paper bridges the gap between comparative belief structures, such as those induced by probability measures, and logical representations of accepted beliefs. We add, to natural properties of comparative belief relations, some conditions that ensure that accepted beliefs form a deductively closed set. It is shown that the beliefs accepted by an agent in all contexts can always be described by a family of conditionals. These results are closely connected to the nonmonotonic 'preferential' inference system of Kraus, Lehmann and Magidor and the works of Friedman and Halpern on their so-called plausibility functions. Acceptance relations are also another way of approaching the theory of belief change after the works of Gärdenfors and colleagues.

1 Introduction

There is an old controversy in the framework of Artificial Intelligence between probabilistic and other numerical representations of uncertainty on the one hand, and the symbolic setting of logical reasoning methods on the other hand. The emergence and relative success of some numerical reasoning formalisms based on probability (especially Bayesian networks) or fuzzy sets has led AI to accept numerical representations as complementary to symbolic ones. However the basic issue underlying the controversy, that is to say, whether or not the two forms of reasoning (logical and numerical) are at all compatible at the formal level has not been widely addressed. Borrowing from Cohen [10], the question is to study the common ground between the probable and the provable.

Namely suppose the beliefs of an agent are represented simultaneously by a measure of confidence (such as a probability function for instance) on a set of states, and by a logical knowledge base that accounts for the propositions believed to be true according to this measure of confidence. One approach to bridging the gap between numerical and symbolic reasoning theories is to define the notion of belief according to a confidence measure. For a proposition p to be believed, a minimal requirement is that the agent believes p more than its negation. Of course, this is not a sufficient property for p to be an accepted belief. An event that is slightly more probable than its complement will not generally be accepted as true. Other natural requirements for an accepted belief are that first, any consequence of an accepted belief be an accepted belief and next, that the conjunction of two accepted beliefs be an accepted belief. In other words, a set of accepted beliefs should be deductively closed. This view of acceptance is shared by philosophers like Cohen [11]:

G. Kern-Isberner, W. Rödder, and F. Kulmann (Eds.): WCII 2002, LNAI 3301, pp. 38–58, 2005.
© Springer-Verlag Berlin Heidelberg 2005

« ...to accept that p is to have or adopt a policy of deeming, positing or postulating that p – that is, of going along with this proposition... as a premiss in some or all contexts for one's own or others' proofs, argumentations, deliberations, etc. »

While « belief that p ... is a disposition to feel it true that p ». Confidence relations that model acceptance thus model a very strong kind of belief, one that an agent is ready to take for granted in the course of deliberations. Then, two questions are worth investigating [14]:

- to what extent are inferences drawn from a measure of confidence and inferences drawn from the logical knowledge base coherent?
- upon arrival of new information, to what extent can the beliefs obtained from the measure of confidence after conditioning and the revised knowledge base be in agreement as well?

This question has been considered in the past only indirectly and with respect to particular uncertainty theories, basically probability theory by Kyburg [32, 33]. This paper addresses the problem in the more general setting of comparative belief structures, that is, relations comparing events, in terms of relative likelihood, certainty, plausibility and the like. Such relations can be induced by numerical representations of belief such as probability [23], possibility [36, 13], and the like. The assumption of deductive closure for sets of accepted beliefs alone severely restricts the type of comparative belief structure that can be envisaged for reasoning under uncertainty. They are called acceptance relations.

This paper shows that acceptance relations precisely lead to a notion of conditional first syntactically characterised by Kraus, Lehmann and Magidor [31] in their nonmonotonic inference system called preferential, and nothing else. Our work is also closely related to the works of Friedman and Halpern [24] that describe the semantics of nonmonotonic inference in terms of plausibility functions valued on a partially ordered set. The originality of the confidence relation approach to plausible inference is that, instead of starting from intuitive postulates on syntactic objects (like Lehmann and colleagues), our basic concepts are on the one hand the confidence relation, that is thought of as a natural tool for describing an agent's uncertain knowledge, and the notion of accepted belief on the other hand. This point of view enables plausible (nonmonotonic) reasoning to be cast in a general framework of uncertain reasoning that includes probabilistic reasoning. The analogy between nonmonotonic inference and probabilistic reasoning has already been pointed out. It has been stressed by Pearl [39] and Lehmann and Magidor [34] that System P has semantics in terms of infinitesimal probabilities, and comes close to Adams' conditional logic [1]. Paris [38] has viewed maximum entropy inference as a kind of default probabilistic inference. The relationship between maxent probabilistic reasoning and nonmonotonic inference is further explored by Kern-Isberner [29].

In probabilistic reasoning, the confidence relation stems from a probability measure or a family thereof. A set of generic rules is then encoded as a set of conditional probabilities characterizing a family of probability measures [38]. The most popular approach in AI currently uses a single probability measure, and the set of conditional probabilities defines a Bayesian network [39]. A Bayesian network basically represents generic knowledge, and so does a confidence relation. This network is built either from expert domain knowledge, or from statistical data via learning techniques. Probabilistic inference with a Bayesian network consists in calculating the (statistical) conditional probability of a conclusion, where the conditioning event encodes the available observations. The obtained conditional probability value is interpreted as the degree of belief of the conclusion in the current situation, assuming that this situation is a regular one in the context described by the observations. This procedure is very similar to the derivation of a plausible conclusion by conditioning an acceptance relation, or deducing a conditional from a conditional rule base. The condition part of the derived conditional again encodes the available observations. The derived default rule is valid « generally ». Its conclusion is considered as an accepted belief in the current situation, assuming that this situation is not an exceptional one in the context described by the observations.

Of course, there are also noticeable differences between probabilistic reasoning and ordinal plausible inference:

1. Plausible inference does not quantify belief.
2. Plausible reasoning considers the most plausible situations and neglects others, while probability theory performs reasoning in the average.
3. Lastly, probabilistic reasoning is generally considered not compatible with the notion of accepted belief.

Indeed, the conjunction of two highly probable events may fail to be highly probable and may even turn out to be very improbable. However, the arbitrary conjunction of accepted beliefs is still an accepted belief (this is because we assume that the agent considers accepted beliefs as tentatively true). This is the source of the lottery paradox that has been proposed as a counterexample to the use of classical deduction on accepted beliefs and to nonmonotonic reasoning at large (Kyburg [32, 33], Poole [41]).

Suppose $n > 1,000,000$ lottery tickets are sold and there is one winner ticket. So the probability $P(player\ i\ loses) > 0.99\ 999$. That is, one should believe that player i loses and the set of accepted beliefs contains all propositions of the form « *player i loses* » for all i. If accepted beliefs are deductively closed, the agent should conclude that all players lose. However, since there is one winning ticket, $Prob(all\ players\ lose) = 0$, for any value of n. That is, the proposition « *one player wins* » should be believed. So, the deductive closure assumption for accepted beliefs leads to inconsistency. Such accepted beliefs cannot match with high probability events, whatever « high » means. Deductive reasoning with accepted beliefs looks counterintuitive. This example seems to kill any attempt to exploit logical approaches in the computation of accepted beliefs. Yet, our claim in this paper is that, contrary to what the lottery paradox would suggest, there does exist some consistency between probability theory and symbolic approaches to plausible reasoning. The solution to the paradox lies in the existence of probability functions that agree with classical deduction from accepted beliefs. Such probabilities, for which the set $\{A, P(A) > 0.5\}$ of beliefs is deductively closed, exist and are laid bare in section 3: big-stepped (or lexicographic) probabilities on finite sets. They are probability measures for which some states of the world appear to be much more probable than others, in any context.

Another by-product of our setting is a connection between accepted beliefs and the AGM theory of belief revision, after Alchourron *et al.* [2] and Gärdenfors [25]. This paper also indirectly shows, relying on previous results by Dubois and Prade [19] that the range of compatibility between the deductively closed belief representations and uncertain reasoning basically reduces to possibility theory [21]. A direct proof of the representation of acceptance relations in terms of comparative possibility theory, as well as results on numerical set functions compatible with the postulates of acceptance relations can be found in a full-length report [15].

The paper is organised as follows: Section 2 introduces the postulates governing the notion of accepted beliefs and defines acceptance relations. Some examples of acceptance relations are discussed, especially the so-called possibility relations [13, 17, 21], each being characterized by a complete preorder of states. Section 3 describes comparative probabilities that are acceptance relations. Section 4 presents a semantics of non-monotonic conditionals in terms of acceptance relations. Section 5 relates acceptance orderings to belief revision.

2 Acceptance Relations

This section presents a general framework for confidence relations that may account for both probabilistic and logical reasoning about beliefs, focusing on the latter. Let S be a finite set of elements called « states ». States encode descriptions of possible situations, states of affairs, etc. Subsets A, B, C, of states are called events. $A \cup B$ and A^c denote the union of A and B, and the complementation of A, respectively. The intersection $A \cap B$ will be shortened into AB when necessary. The finiteness assumption is made for the sake of simplicity, and is natural when the description of situations is achieved by means of a formal language such as propositional logic. Then, S is the set of interpretations of the language. It comes down to understanding propositions as ultimately true or false. Here we consider a syntax-free environment, from a logical point of view. In particular, a proposition is understood as the subset of states where it is true.

The epistemic state of an agent about the normal state of the world is supposed to be modeled by a reflexive relation $\geq_L$ among events. Stating $A \geq_L B$ means « the agent has at least as much confidence in event A as in event B», and subscript L stands for « likelihood ». From this confidence relation, other associated relations are defined in the usual way [42]:

- $A >_L B$ iff $(A \geq_L B)$ and not $(B \geq_L A)$ (strict preference);
- $A \neq_L B$ iff neither $(A \geq_L B)$ nor $(B \geq_L A)$ (incomparability);
- $A \equiv_L B$ iff $(A \geq_L B)$ and $(B \geq_L A)$ (indifference).

Let us set the minimal properties required for a confidence relation. The strict part of a confidence relation is thus naturally assumed to be transitive.

- **Quasi-transitivity** (QT): $A >_L B$ and $B >_L C$ imply $A >_L C$.

But neither the incomparability relation nor the indifference relation it induces is necessarily so. Indifference between two events may be due to imperfect discernment about confidence levels and cannot always be chained. Moreover, if $A \subseteq B$, then A implies B. So it should be that A and B are comparable, and that the agent's confidence in A cannot be strictly greater than in B.

- **Monotony with respect to inclusion** *(MI):* $A \subseteq B$ *implies* $B \geq_L A$.

This property of monotony is not sufficient to ensure that the strict part of $\geq_L$ is coherent with classical deduction. The so-called *orderly axiom* of Halpern [28] must be added:

- **Orderly axiom** *(O): if* $A \subseteq A'$ *and* $B' \subseteq B$ *then* $A >_L B$ *implies* $A' >_L B'$

It should be emphasised that axiom O does not imply MI, because due to incompleteness of $\geq_L$ it may occur that $A \subset B$ and $A \neq_L B$. In general MI does not imply O because if $A >_L B$, $B \geq_L C$ then $A >_L C$ does not follow. For instance, if the confidence relation $\geq_L$ is not transitive, we may have that $A >_L B$, $B \equiv_L C$ and $A \neq_L C$. But it should be noticed that O is recovered from MI if full transitivity of $\geq_L$ is assumed.

The pair $(2^S, \geq_L)$ is called a *comparative belief structure* and a relation $\geq_L$ satisfying O, QT, and MI is called a *confidence relation.* Well-known confidence relations, such as comparative probabilities, satisfy a stability property with respect to adding or deleting elements common to any two sets. A confidence relation $\geq_L$ is preadditive iff

$$\text{whenever } A \cap (B \cup C) = \varnothing, A \cup B \geq_L A \cup C \text{ iff } B \geq_L C.$$

Denoting $\geq^T$ the relation dual to relation $\geq$, such that $A \geq^T B$ iff $B^c \geq A^c$, it can be checked that preadditive relations are self-dual, i.e., they are such that $A \geq B$ iff $A \geq^T B$. Comparative probabilities (Fishburn [23]) are complete and transitive, non-trivial $(S >_\Pi \varnothing)$ preadditive confidence relations. Probability functions induce such type of preadditive confidence preorders, but not all such relations can be represented by probability functions (Kraft *et al.*, [30]).

Another type of confidence relation was introduced by Lewis [36] in the setting of a modal logic of counterfactuals, and independently rediscovered by Dubois [13] and Grove [27]: the possibility relation $\geq_\Pi$; it is a non-trivial $(S >_\Pi \varnothing)$ transitive confidence relation satisfying the *disjunctive stability* property:

$$\forall A, B \geq_\Pi C \text{ implies } A \cup B \geq_\Pi A \cup C.$$

Possibility relations are not preadditive nor even self-dual. The dual set-relation defined $A \geq_N B$ iff $B^c \geq_\Pi A^c$ is called a necessity relation [13, 17] and epistemic entrenchment by Gärdenfors [25]. It is a non-trivial $(S >_\Pi \varnothing)$ transitive confidence relation satisfying the *conjunctive stability* property:

$$\forall A, B \geq_N C \text{ implies } A \cap B \geq_N A \cap C.$$

Conjunctive stability cannot be compared with the preadditivity property. The duality between possibility and necessity relations is a translation of the duality between possibility and necessity modalities in modal logic [22]. Contrary to comparative probabilities, a possibility relation $\geq_\Pi$ and its dual necessity relation $\geq_N$ are fully characterised by a complete preorder $\geq_\pi$ on the set of states S expressing relative plausibility. It is the restriction of the possibility relation to singletons [13]:

$$s \geq_\pi s' \text{ iff } \{s\} \geq_\Pi \{s'\}.$$

The set of states is partitioned via $\geq_\pi$ into a totally ordered set of m clusters of equally possible states $S = E_1 \cup E_2 \cup ... \cup E_m$ by means of the equivalence relation $\equiv_\Pi$ induced by the symmetric part of $\geq_\pi$. This set of equivalence classes is a set of events, that is linearly ordered by the strict part of the possibility relation

$$E_1 >_\Pi E_2 >_\Pi ... >_\Pi E_m.$$

The set E_1 contains the most possible states that will be considered as the normal ones. The set E_m contains the least possible states and, by convention, these states are not fully impossible ($E_m >_\Pi \varnothing$). The family $\{E_1, E_2... E_m\}$ is called a well-ordered partition (wop) by Spohn [44]. The possibility and necessity relations on non-empty sets are recovered as follows [13]:

$$\forall A \neq \varnothing, \ \forall B \neq \varnothing, \ A \geq_\Pi B \text{ iff } \exists s \in A, \ \forall s \in B, \ s \geq_\pi s';$$

$$\forall A \neq \varnothing, \ \forall B \neq \varnothing, \ A \geq_N B \text{ iff } \exists s \in B^c, \ \forall s \in A^c, \ s \geq_\pi s'.$$

Possibilistic belief focuses on most normal states. $A \geq_\Pi B$ reads « A is at least as plausible (for the agent) as B » in the sense that the most normal state where A is true is at least as plausible as the most normal state where B is true. $A \geq_N B$ reads: « A is at least as certain (entrenched for the agent) as B ». In particular, $A >_N \varnothing$ whenever A is true in all normal states ($E_1 \subseteq A$). Linear possibility relations are those induced by a linear ordering of states (the E_i's are singletons).

The weakest and most natural way to define what it means for a proposition to be believed in the sense of a confidence relation is to consider that A is a belief if the agent is more confident in A than in its negation (modelled by the complement of the set A):

Definition 1: Let $\geq_L$ be a confidence relation. A proposition A is called a belief induced by $\geq_L$ if and only $A >_L A^c$. The set of beliefs according to $\geq_L$ is thus: $A_L = \{A: A >_L A^c\}$.

Note that, using duality, $A >_L A^c$ is equivalent to $A >_L^T A^c$. The set A_L generalises the definition of a belief set that is implicit in Gärdenfors theory of revision using the epistemic entrenchment relation, i.e. $\{A: A >_N \varnothing\}$. Indeed, for necessity relations, $A >_N A^c$ is equivalent to $A >_N \varnothing$ since it cannot be that $A >_N \varnothing$ and $A^c >_N \varnothing$ for any event A. The belief set A_L is always deduced from the confidence relation, but neither the original confidence relation nor even its strict part can be re-built from this belief set alone. In order to remain compatible with deductive inference, a set of *accepted* beliefs A_L must be deductively closed. It requires that any consequence of an accepted belief be an accepted belief and that the conjunction of two accepted beliefs be an accepted belief. In terms of the confidence relation, it leads to the following additional postulates [20]:

-   **Consequence stability** *(CS): if $A \subseteq B$ and $A >_L A^c$ then $B >_L B^c$*
-   **Conjunction stability** (AND): *if $A >_L A^c$ and $B >_L B^c$ then $A \cap B >_L A^c \cup B^c$.*

Actually, CS is an obvious consequence of the Orderly axiom. Possibility and necessity relation satisfy both properties and yield deductively closed sets of beliefs. But, in general, the logical closure of the set of beliefs induced by any confidence measure is not guaranteed. For instance, the set $A_P = \{A, P(A) > 0.5\}$ of beliefs induced by a probability measure P is generally not deductively closed. This phenomenon remains, changing the threshold 0.5 into any larger value (Snow, [43]). This remark is the basis of the lottery paradox mentioned in the introduction (one can believe that any given player in a one-winner lottery game will lose with arbitrary high probability, all the more so as players are numerous, but one cannot believe that all of them will lose).

Due to incomplete knowledge, a belief is only tentative and may be questioned by the arrival of new information. Conditioning a confidence relation on a set C representing a piece of evidence about the current situation comes down to restricting the confidence relation to subsets of C. The agent's confidence in A is said to be at least as high as the confidence in B in the context C if and only if $A \cap C \geq_L B \cap C$. The set of beliefs in context C induced by a confidence relation $\geq_L$ is thus defined as

$$A_L(C) = \{A : A \cap C >_L A^c \cap C\}. \qquad (*)$$

This proposal is a natural way of revising a set of current beliefs A_L about a given situation, on the basis of a confidence relation and a new information item C about the current situation. Clearly, $A_L = A_L(S)$ can be viewed as the set of prior beliefs when no evidence is yet available. Note that revising the confidence relation itself is another problem not dealt with here. In this paper we only consider the change of current beliefs about a particular situation when prior generic knowledge is encoded by the confidence relation.

It should also be noticed that here, $A_L(\varnothing) = \varnothing$. This is contrary to classical logic tradition, which assumes $A_L(\varnothing) = 2^S$. We consider that, in the presence of contradictory information on the current situation, an agent cannot entertain beliefs. This is also true if $C \equiv_L \varnothing$, which clearly entails $A_L(C) = \varnothing$. If the agent is supposed to entertain beliefs in each non-contradictory context, one is thus led to adopt a non-dogmatism postulate for the confidence relation:

- **Non-dogmatism** *(ND)*: $C >_L \varnothing$ iff $C \neq \varnothing$.

Now, the revision of *accepted* beliefs should yield new accepted beliefs. Hence $A_L(C)$ should be deductively closed. It leads to stating conditional versions of CS and the AND properties [20]:

- **Conditional consequence stability** (CCS): *if $A \in A_L(C)$ and $A \subseteq B$ then $B \in A_L(C)$.* It also reads: *If $A \subseteq B$ and $C \cap A >_L C \cap A^c$ then $C \cap B >_L C \cap B^c$;*

- **Conditional conjunctive stability** (CAND): *if $A \in A_L(C)$ and $B \in A_L(C)$ then $A \cap B \in A_L(C)$.* It also reads:

If $C \cap A >_L C \cap A^c$ and $C \cap B >_L C \cap B^c$ then $C \cap A \cap B >_L C \cap (A^c \cup B^c)$.

Notice that CCS and CAND reduce to the CS and AND rules when $C = S$. It is obvious that the properties that make the set of accepted beliefs according to a confidence relation a deductively closed set only involve the strict part of the confidence relation, moreover restricted to disjoint subsets. The remaining part of the confidence relation has no influence on the set of accepted beliefs. Moreover, it is not necessary to explicitly require CCS for the confidence relation, since the Orderly property clearly implies CCS. The latter is indeed the axiom O restricted to pairs of disjoint subsets (A, B) and $(A\#, B\#)$ such that $A \subseteq A\#$, $B\# \subseteq B$ and $A \cup B = A\# \cup B\#$. Conversely, due to O, $A > B \cup C$ implies $A \cup C > B$, which trivially implies CCS. We can now define a formal ordinal framework that captures the concept of acceptance [14]:

Definition 2: An acceptance relation $\geq_L$ on 2^S is a non-dogmatic confidence relation that satisfies the CAND property.

It is easy to see that CAND can be written as follows:

Lemma 1: For disjoint subsets A, B, C, D, CAND is equivalent to

$$If\ A \cup B >_L C \cup D\ and\ A \cup C >_L B \cup D\ then\ A >_L B \cup C \cup D. \tag{SN}$$

Proof: Let $B' = A \cap C \cap B^C$, $C' = A^C \cap C \cap B$, $A' = A \cap B \cap C$, $D' = C \cap A^C \cap B^C$. Then notice that $C \cap A = A' \cup B'$, $C \cap A^C = C' \cup D'$ and $C \cap B = A' \cup C'$ and $C \cap B^C = B' \cup D'$. Then CAND is another way of writing (SN).

As pointed out by Dubois and Prade [20] (also Friedman and Halpern [24]) the following basic result can be obtained, which yields yet another form of the CAND axiom:

Theorem 1: For any relation $>$ on disjoint events that satisfies O, CAND is equivalent to the « negligibility property »: $\forall A, B, C$ three disjoint events,

$$If\ A \cup B > C\ and\ A \cup C > B\ then\ A > B \cup C. \tag{NEG}$$

Proof: First O and NEG imply CAND: indeed, from $A \cup C > B \cup D$ and O, we get $A \cup C \cup D > B$. Moreover, both $A \cup C \cup D > B$ and $A \cup B > C \cup D$ are of the form $A \cup X > B$ and $A \cup B > X$ with A, B, X disjoint. So, by NEG, $A > B \cup C \cup D$, which is (SN). Conversely CAND implies NEG, letting $D = \varnothing$ in (SN).

The name « negligibility property » can be explained as follows: for three disjoint events A, B, C, when $A >_L C$ and $A >_L B$, it means that each of B and C is less plausible than A. However, whatever their respective plausibilities, their disjunction is always less plausible than A since, from O, $A \cup B >_L C$ and $A \cup C >_L B$, and by NEG, $A >_L B \cup C$. So $A >_L C$ means that the confidence in C is negligible in front of the one of A. If $>_L$ derives from a probability function, the property NEG certainly does not hold in general. However, for a proposition to be an accepted belief, one expects it to be much more believed than its negation. So axiom NEG is not as counterintuitive as it might look, from the standpoint of acceptance.

Possibility relations are typical examples of transitive acceptance relations. It is easy to see that the strict part $>_\Pi$ of a non dogmatic possibility relation satisfies the negligibility axiom NEG because obviously, $max(a, b) > c$ and $max(a, c) > b$ implies $a > max(b, c)$. It suggests a much stronger property than NEG called « qualitative-ness » by Halpern [28]. A relation $>$ on 2^S is called « qualitative » when

$$\forall A, B, C, \text{ if } A \cup B > C \text{ and } A \cup C > B \text{ then } A > B \cup C. \tag{QUAL}$$

QUAL is much stronger than negligibility since it applies to any sets A, B, C (not necessarily disjoint). Any reflexive orderly qualitative relation is an acceptance relation. The only example of a reflexive and complete qualitative preorder relation is a possibility relation since QUAL is actually equivalent to the characteristic axiom of possibility relations. Note that necessity relations are not acceptance relations because, even if they satisfy NEG, the property $C >_N \varnothing$ iff $C \neq \varnothing$ is not valid. Generally, if $>_L$ satisfies the non-dogmatism axiom its dual does not.

Another example of a acceptance relation $>_F$ is obtained by considering a family F of possibility relations, letting $A >_L B$ iff $A >_\Pi B$ for all $\geq_\Pi$ in F. The complement of such a relation, defined by letting $A \geq_L B$ iff $not(A >_L B)$, is complete but not transitive. Another acceptance relation defined by letting $A \geq_L B$ iff $A >_\Pi B$ for all $\geq_\Pi$ in F, or $A \equiv_\Pi B$ for all $\geq_\Pi$ in F is not complete but is transitive. Both these acceptance relations share the same strict part. Such families play an important role in the representation of general acceptance relations [15]. Namely, any acceptance relation $\geq_L$ can be replaced by a family F of possibility relations, in such a way that their strict parts $>_F$ and $>_L$ coincide on disjoint events, hence generate the same set of accepted beliefs.

Some acceptance relations are preadditive. For instance, let $\geq_\Pi$ be a possibility relation and consider the strict ordering $>_{\Pi L}$ comparing only the non-common part of events [14]:

$$A >_{\Pi L} B \text{ iff } A \cap B^c >_\Pi A^c \cap B; \ A \geq_{\Pi L} B \text{ iff } not \ (B >_{\Pi L} A).$$

Relation $\geq_{\Pi L}$ is an acceptance relation. Its strict part $>_{\Pi L}$ refines $>_\Pi$ and $>_N$ (each of $A >_\Pi B$ and $A >_N B$ implies $A >_{\Pi L} B$). If A and B are disjoint, $A >_{\Pi L} B$ is identical to $A >_\Pi B$, and then $A_\Pi(C) = A_{\Pi L}(C)$. But $\geq_{\Pi L}$ is generally not transitive: $A \cap B^c =_\Pi A^c \cap B$ and $B \cap C^c =_\Pi B^c \cap C$ do not imply $A \cap C^c =_\Pi C^c \cap B$. The same construction works with a family of possibility relations.

The relation $>_{\Pi L}$ is not qualitative. Consider for instance $S = \{s_1, s_2\}$ and a possibility relation such that $s_1 >_\pi s_2 >_\pi 0$. It holds that $\{s_1, s_2\} >_{\Pi L} \{s_1\} >_{\Pi L} \{s_2\} >_{\Pi L} \varnothing$. This relation does not satisfy (QUAL). Indeed: $\{s_1\} \cup \{s_1\} >_{\Pi L} \{s_2\}$ and $\{s_1\} \cup \{s_2\} >_{\Pi L} \{s_1\}$ but $\{s_1\} >_{\Pi L} \{s_1\} \cup \{s_2\}$ does not hold. However the relative position of $\{s_1, s_2\}$ w.r.t. $\{s_1\}$ (be it higher or at the same rank) is immaterial for characterising accepted beliefs.

3   Acceptance Relations and Comparative Probabilities

Probability measures generally do not induce acceptance relations. Comparative probabilities are preadditive confidence complete preorders (they are reflexive, transitive and complete relations [42]). They are more general than confidence relations induced by probabilities. It is thus interesting to characterise comparative probabilities modeling acceptance. For a preaddidive confidence complete preorder $\geq_L$, it holds, for any A, B, C such that $A \cap (B \cup C) = \varnothing$:

$$A \cup B \equiv_L A \cup C \text{ iff } B \equiv_L C, \text{ and } A \cup B >_L A \cup C \text{ iff } B >_L C.$$

In the context of transitive acceptance relations, it turns out that there is a strong incompatibility between the existence of equally plausible disjoint events and preadditivity. Indeed, acceptance complete preorders have the following property [14]:

Lemma 2. Let A, B and C be three disjoint events. If an acceptance relation $\geq_L$ is a complete preorder, then $A \equiv_L C >_L B$ implies $A \equiv_L A \cup B \equiv_L C \equiv_L C \cup B >_L B$.

Proof: $C >_L B$ imply, by O, $C \cup A >_L B$. Moreover, $C \equiv_L A$ also implies $A \cup B \geq_L C$. And $A \cup B >_L C$ is impossible (otherwise, by NEG, $A >_L C \cup B$ and thus $A >_L C$ by O). So, $A \cup B \equiv_L C$, i.e. by transitivity $A \cup B \equiv_L A$.

Lemma 2 makes it clear that when $A >_L B$, the plausibility of B is negligible when compared to the one of A since $A \equiv_L A \cup B$. However, under transitivity, preadditivity will severely constrain the possibility of expressing negligibility between events. Indeed, $B >_L \varnothing$ holds for acceptance relations if $B \neq \varnothing$. So $A \cup B >_L A$ is enforced by preadditivity for any two disjoints nonempty sets A and B. It contradicts the property $A \equiv_L A \cup B$ enforced by the presence of another disjoint set $C \equiv_L A$. More formally [14]:

Lemma 3. If $\geq_L$ is a preaddiative acceptance complete preorder, then for any three disjoint events A, B and C: $C \equiv_L A >_L B$ implies $B = \varnothing$.

Proof: Clearly, $A \neq \varnothing$. $C \equiv_L A > B$ implies $A \equiv_L A \cup B$ by Lemma 2. But preadditivity and non-dogmatism enforce strict preference $A \cup B >_L B$. This is compatible only if $B = \varnothing$.

Hence as soon as equivalence is allowed between disjoint events, preaddive acceptance complete preorders have to be very particular: no state can be less plausible than two equivalent states, except if it is impossible. In summary, assuming transitivity, preadditivity is little compatible with negligibility. The conflict between these two properties makes deductive closure hard to reconcile with the notion of partial belief graded on a numerical scale. Clearly the following result follows for comparative probabilities:

Theorem 2: If an acceptance relation $\geq_L$ is a comparative probability on S, and S has more than two elements, then, there is a permutation of the elements in S, such that $s_1 >_L s_2 >_L \ldots >_L s_{n-1} \geq_L s_n$.

Proof: From the above results, if $s_i >_L s_k$ and $s_j >_L s_k$ we have that either $s_i >_L s_j$ or $s_j >_L s_i$. Hence the only group of equally plausible states are the least plausible ones. Suppose now that there are more than one least plausible state. Assume $\{s_i, s_j\} \equiv_L$

$\{s_i\}$; then $\varnothing \equiv_L \{s_i\}$, which is impossible since the relation is not dogmatic. Hence $\{s_i, s_j\} >_L \{s_i\}$. Hence, in the set of least plausible states, events are strictly ordered in agreement with set-inclusion. Suppose there are more than two equally plausible states s_i, s_j, s_k. Then preadditivity implies $\{s_i, s_j\} >_L \{s_k\}$ and $\{s_i, s_k\} >_L \{s_j\}$, which due to (NEG) implies $\{s_i\} >_L \{s_k, s_j\}$, and this is impossible since, by assumption, states s_i, s_j, s_k are equally plausible. Hence there cannot be more than two equally plausible states, and these are the least plausible ones.

Note that the possibilistic likelihood relation $\geq_{\pi L}$ of section 2 is a preadditive acceptance relation, but it is not transitive, so it allows for equally plausible states and escapes Theorem 4.

The problem of finding probability measures that are acceptance functions can be easily solved on this basis. Necessarily the states are totally ordered (but for the two least probable ones). Moreover such comparative probabilities coincide with possibility relations on disjoint events A and B [5]: $P(A) > P(B)$ if and only if $A >_\Pi B$, that is $max_{i \in A} \ p_i > max_{i \in B} \ p_i$. A necessary and sufficient condition for this probability-possibility compatibility is that $\forall A \ \exists s \in A$ such that $P(\{s\}) > P(A \setminus \{s\})$. This leads to very special probability measures such that the probability of a state is much bigger than the probability of the next probable state [43], [5] – we call them *big-stepped* probabilities. They are lexicographic probabilities. It is clear that any comparative probability that is an acceptance function can be represented by any big-stepped probability P with $p_i = P(\{s_i\})$ such that:

$$p_1 > ... > p_{n-1} \geq p_n > 0; \ \forall i < r - 1, p_i > \Sigma_{j \ = \ i+1, ... n} \ p_j.$$

Example: S has five elements and consider $s_1/0.6$, $s_2/0.3$, $s_3/0.06/$, $s_4/0.03$, $s_5/0.01$.

Note that in the lottery paradox described in the introduction, it is implicitly assumed that all players have equal chance of winning. The underlying probability measure is uniform. Hence there is no regularity at all in this game: no particular occurrence is typical and randomness prevails. It is thus unlikely that an agent can come up with a consistent set of beliefs about the lottery game. So, in the situation described in the example, deriving accepted beliefs about who loses, and reasoning from such beliefs is not advisable indeed.

However, big-stepped probabilities are the total opposite of uniformly distributed ones, since the probability of any state is larger than the sum of the probabilities of less probable states. So, the paradox disappears if the underlying phenomenon on which the agent entertains accepted beliefs is ruled by a big-stepped probability since $\{A, P(A \ /C) > 0.5\}$ remains logically consistent and deductively closed. It suggests that big-stepped probabilities (and plausible reasoning based on acceptance relations) model an agent reasoning in front of phenomena that have typical features, where some non-trivial events are much more frequent than other ones. We suggest that such domains, where a body of default knowledge exists, can be statistically modelled by big-stepped probabilities on a meaningful partition of the sample space. Default reasoning based on acceptance should then be restricted to such situations in order to escape the lottery paradox.

4 Relationships with Non-monotonic Inference

The above framework is close to the one proposed by Friedman and Halpern [24]. They account for the preferential approach of Kraus *et al.* [31] to nonmonotonic reasoning (system P) by means of so-called « plausibility measures », which are set-functions valued in a partially ordered set, basically what we call confidence relations. Let $A \rightarrow B$ be a conditional assertion relating two propositions, and stating that if A holds then generally B holds too. We neglect the syntax of propositions. A conditional assertion should be understood as: in the context A, the agent accepts B. Basic properties of such conditionals, viewed as nonmonotonic inference consequence relationships have been advocated by Kraus *et al.* [31] (referred to as KLM properties in the following):

- **Reflexivity:** $A \rightarrow A$
- **Right weakening:** $A \rightarrow B$ and $B \subseteq C$ imply $A \rightarrow C$
- **AND:** $A \rightarrow B$ and $A \rightarrow C$ imply $A \rightarrow B \cap C$
- **OR:** $A \rightarrow C$ and $B \rightarrow C$ imply $A \cup B \rightarrow C$
- **Cautious monotony** *(CM):* $A \rightarrow B$ and $A \rightarrow C$ imply $A \cap B \rightarrow C$
- **Cut:** $A \rightarrow B$ and $A \cap B \rightarrow C$ imply $A \rightarrow C$

The above postulates embody the notion of plausible inference in the presence of incomplete information. Namely, they describe the properties of deduction under the assumption that the state of the world is as normal as can be. The crucial rules are Cautious Monotony and the Cut. Cautious Monotony claims that if A holds, and if the normal course of things is that B and C hold in this situation, then knowing that B and A hold should not lead us to situations that are exceptional for A: C should still normally hold. The Cut is the converse rule: If C usually holds in the presence of A and B then, if situations where A and B hold are normal ones among those where A holds (so that A normally entails B), one should take it for granted that A normally entails C as well. The other above properties are not specific to plausible inference: OR enables disjunctive information to be handled without resorting to cases. The Right Weakening rule, when combined with AND, just ensures that the set of nonmonotonic consequences is deductively closed in every context. Reflexivity sounds natural but can be challenged for the contradiction ($A = \varnothing$). These basic properties can be used to form the syntactic inference rules of a logic of plausible inference. It has close connections with measure-free conditionals of Calabrese [9], or Goodman and Nguyen [26], which provide a three-valued logic semantics to conditional assertions (Dubois and Prade [18]).

A conditional knowledge base is a set Δ of conditionals $A \rightarrow B$ built from subsets of S. A strict plausibility order $>$ on disjoint events is induced by a set of conditional assertions by interpreting $A \rightarrow B$ as the statement that the joint event $A \cap B$ is more plausible than $A \cap B^c$:

$$A \cap B > A \cap B^c \ iff \ A \to B \in \Delta \tag{**}$$

Conversely, the relation $A > B$ between disjoint sets corresponds to the conditional $A \cup B \to B^c$. In this context, the properties of nonmonotonic preferential inference can be written as properties of the confidence order >:

- Or: $A \cap C > A \cap C^c$ and $B \cap C > B \cap C^c$ imply $(A \cup B) \cap C > (A \cup B) \cap C^c$
- RW : if $B \subseteq C, A \cap B > A \cap B^c$ implies $A \cap C > A \cap C^c$
- CM: $A \cap B > A \cap B^c$ and $A \cap C > A \cap C^c$ imply $A \cap B \cap C > A \cap B \cap C^c$
- Cut: $A \cap B > A \cap B^c$ and $A \cap B \cap C > A \cap B \cap C^c$ imply $A \cap C > A \cap C^c$.

The AND and the RW axioms are exactly the CAND and the CCS axioms of acceptance relations. Kraus *et al.* [31] also assume the reflexivity axiom $A \to A$, which is hard to accept for $A = \varnothing$ in the present framework since it means $\varnothing > \varnothing$ and violates the irreflexivity requirement for relation >. $A \to A$ makes sense for $A \neq \varnothing$, and we shall consider the restricted reflexivity condition:

$$RR: A \to A, \ \forall A \neq \varnothing$$

Note that $A \to \varnothing$ is inconsistent with acceptance relations (otherwise, $\varnothing > A$). Hence the consistency preservation condition:

$$CP: A \to \varnothing \ never \ holds.$$

The aim of this section is to show, without the reflexivity axiom, that relations induced by conditional assertions in the above sense are strict parts of acceptance relations and that it reciprocally turns out that the entailment relation induced by acceptance relations is precisely preferential inference, but for the reflexivity $A \to A$. The main point is to prove that the KLM axioms enforce the negligibility axiom and the transitivity and orderly properties of >.

Lemma 4: if Δ is a conditional knowledge base closed under RR, AND, OR, RW, CM, and CP, then the confidence relation it defines via (**) is the restriction, to disjoint sets, of the strict part of an acceptance relation.

Proof: • Non Dogmatism: $A > \varnothing$ for $A \neq \varnothing$ is $A \to S$, which is obvious by RR.

- Axiom O: Let A, B, C be disjoint sets, A and C being not empty. Note that the OR axiom can be written for disjoint sets A, B, C, D: $A > B$ and $C > D$ imply $A \cup C > B \cup D$. Assume $A > B$, note that $C > \varnothing$. Hence by OR, $A \cup C > B$. Now, assume A, B, C be disjoint sets, A being not empty, and $A > B \cup C$. Write it $(B \cup A) \cap (B \cup C)^c > (B \cup C) \cap (B \cup C)$. By RW, $(B \cup A) \cap C^c > (B \cup C) \cap C$, i.e., $A \cup B > C$. Now apply CM to $A > B \cup C$ and $A \cup B > C$ and get $A \cap (A \cup B) > (A \cup B) \cap (B \cup C)$; that is, $A > B$.
- Negligibility: This is obvious since the AND axiom of $\to$ directly yields the CAND property for >, which added to O, yields NEG (see Theorem 1, section 2).
- Transitivity: Consider disjoint sets A, B, C, and assume $A > B, B > C$. Then by O, $A \cup C > B$, and $A \cup B > C$. Now by NEG, $A > B \cup C$, and by O, $A > C$ follows.

Theorem 3: The strict part of any acceptance relation satisfies CAND, RW, CM, CUT, OR, CP, RR. The conditional assertions satisfy system P except for $A \rightarrow \varnothing$.

Proof: it is obvious for CP, RR.

- CM: $AB > AB^c$ means $ABC \cup ABC^c > A B^cC \cup AB^cC^c$ (1)

   $AC > AC^c$ means $ABC \cup AB^cC > ABC^c \cup AB^cC^c$ (2)

hence, applying CAND to (1) and (2): $ABC > ABC^c \cup AB^cC^c \cup AB^cC$. Thus (by O): $ABC > ABC^c$

- RW reads: $AB > AB^t$ implies $A(B \cup D) > AB^tD^t$, a consequence of O.
- OR: $AC > AC^c$ means $ABC \cup AB^cC > ABC^c \cup AB^cC^c$;

by (O) : $ABC \cup AB^cC > ABC^c$ and $ABC \cup AB^cC > AB^cC^c$;

$BC > BC^c$ means $ABC \cup A^cBC > ABC^c \cup A^cBC^c$;

by (O) : $ABC \cup A^cBC > ABC^c$ and $ABC \cup A^cBC > A^cBC^c$;

Thus by (O): $ABC \cup A^cBC \cup AB^cC > ABC^c$

and by (O) and (NEG) : $ABC \cup A^cBC \cup AB^cC > A^cBC^c \cup AB^cC^c$;

finally, by (O) and (NEG) : $ABC \cup A^cBC \cup AB^cC > ABC^c \cup A^cBC^c \cup AB^cC^c$,

i.e. $(A \cup B)C > (A \cup B)C^c$.

- CUT: $AB > AB^c$ means $ABC \cup ABC^c > AB^cC \cup AB^cC^c$, and assume $ABC > ABC^c$; by (O) and (NEG): $ABC > AB^cC \cup AB^cC^c \cup ABC^c$. Thus, by (O), $ABC \cup AB^cC > AB^cC^c \cup ABC^c$ i.e., $AC > AC^c$.

The relation on disjoint events induced by a conditional knowledge base via system P can thus be extended into an acceptance relation. The main difference between Friedman's and Halpern's axiomatics and ours comes from a different understanding of system P. Friedman and Halpern [24] admit reflexivity of $\rightarrow$ and interpret $A \rightarrow B$ as $(A \cap B > A \cap B^c$ or $A \equiv \varnothing)$ after the suggestion of Makinson and Gärdenfors [37]. Our simpler interpretation of the conditional assertion $A \rightarrow B$ as $A \cap B > A \cap B^c$ allows us not to use one axiom they need (namely, that $A \equiv B \equiv \varnothing$ implies $A \cup B \equiv \varnothing$), since it is vacuous here for non-dogmatic confidence relations. We never assume that the inference $A \rightarrow B$ holds when A is empty. The non-dogmatism assumption $(A > \varnothing)$ provides for the reflexivity axiom when $A \neq \varnothing$.

Suppose that $A > B$ iff $A >_\Pi B$ for all $\geq_\Pi$ in F, and $A \geq B$ iff $not(A > B)$. Then, $A \cup B > A \equiv B \equiv \varnothing$ for some A and B occurs if a possibility 0 is assigned to A by some of the possibility relations, a possibility 0 is assigned to B by the other ones: hence, it may hold that $A \cup B > \varnothing$ although $A \equiv B \equiv \varnothing$ when $>$ is represented by a family of possibility relations.

Given a conditional knowledge base $\Delta = \{A_i \rightarrow B_i, i = 1, n\}$, its preferential closure Δ^P is obtained by applying to it the KLM properties as inference rules. The above results clearly show that the relation on events induced by Δ^P is an acceptance relation $\geq_L$, and that the set $\{A, C \rightarrow A \in \Delta^P\}$ coincides with the belief set $A_L(C)$.

The relation defined by a consistent base of defaults can be represented by a non empty family of possibility measures F_Δ such that [19]: $A \rightarrow B$ iff $A \cap B > A \cap B^c$ iff

$\forall \Pi \in F_\Delta, A \cap B >_\Pi A \cap B^c$. We have seen in section 2 that the relation $>$ defined in this way by means of a family of possibility relations is the strict part of an acceptance relation. Benferhat *et al.* [5] proved that a family of possibility relations representing preferential inference can always be chosen as families of possibility relations induced by linear plausibility orderings of states, or equivalently families of big-stepped probabilities.

Lehmann and Magidor [34] have considered an additional property a non-monotonic inference relation might satisfy:

- **Rational monotony:** $A \to C$ and $\neg(A \to B^c)$ imply $A \cap B \to C$.

In the above, $\neg(A \to B^c)$ means that it is not the case that B generally does not hold, in situations where A holds. Indeed if B^c is expected then it might well be an exceptional A-situation, where C is no longer normal. In terms of confidence relation this axiom reads:

$A \cap C >_L A \cap C^c$ and not $(A \cap B^c >_L A \cap B)$ imply $A \cap B \cap C >_L A \cap B \cap C^c$.

Adding rational monotony to the basic KLM properties, the nonmonotonic inference $A \to B$ can always be modelled as $A \cap B >_\Pi A \cap B^c$ for a possibility ordering $\geq_\Pi$ because it forces the underlying confidence relation $\geq$ to be a complete preorder. See for instance Benferhat *et al.*, [4]. This is also the basis of the so-called « rational closure » [34]. What these previous results prove is that RM is equivalent to the transitivity of $\geq$ for an acceptance relation. A conditional knowledge base Δ is thus equivalent to a set of constraints of the form $A > B$ restricting a family of possibility relations. Selecting the least specific possibility relation corresponds to the computation of the rational closure of Δ after Lehmann and Magidor [34], or the most compact ranking according to Pearl [40]. The actual computation of this rational closure of Δ can be carried out by finding the well-ordered partition (wop, for short) induced by this set of constraints by means of a ranking algorithm, several versions of which have appeared in the literature in possibilistic logic [3], also system Z [40]. A wop $PART = \{E_1, E_2, ... E_k\}$ of S can be built from any set of constraints $G(>)$ of the form $A > B$ such that $A \cap B = \emptyset$ (where $>$ is any irreflexive relation):

$T = S, G = G(>), PART = \emptyset$. Do until $G = \emptyset$:

1) Let $E = T \setminus \cup \{A, \exists B \subseteq T, B > A \in G\}$;
2) If $E \neq \emptyset$, add E to PART; $T \leftarrow T \setminus E$. Otherwise stop: inconsistency.
3) $GG = \{(A, B) \in G, A > B, E \cap A = \emptyset\}$.
4) If $GG \neq G, G \leftarrow GG$. Otherwise stop: inconsistency.

The algorithm finds $E_1, E_2, ... E_k$ in a row, that form a well-ordered partition induced by the constraints in G, if this set of constraints is not violating the axioms of acceptance. Indeed, it can be shown that if G is a partial acceptance relation, then, no inconsistency is found in step 2 nor in step 4 of the algorithm. If $\geq_\Pi$ is the obtained possibility relation, then the rational closure of Δ is $\{A \to B : A \cap B >_\Pi A \cap B^c\}$.

5 Revising Accepted Beliefs Versus Revising an Acceptance Relation

At this point, it can be asked whether, given an acceptance relation $\geq_L$, the change operation that turns the belief set A_L into the belief set $A_L(C)$ when the information stating that proposition C is true, will satisfy the main postulates of belief revision (after Gärdenfors [25]). A set of accepted beliefs $A_L(C)$ being deductively closed, is characterised by a subset of states $B(C)$ called its *kernel*, such that $A_L(C) = \{A, B(C) \subseteq A\}$. It is thus possible to describe a belief change operation as a mapping from $2^S \times 2^S$ to 2^S, changing a pair *(B, C)* of subsets of states into another subset of states $B(C)$. This approach is in some sense easier to understand than the general logical setting used by Gärdenfors, where a syntactic construction is proposed with a syntax-independence axiom. Interpreting any sentence ψ in a propositional language as its set of models, the translation of the postulates characterising revision operations * then goes as follows:

- **Postulate 1**: For any sentence ψ, and any belief set K, $K*\psi$ is a belief set. It means that for any pair of subsets *(B, C)* of S, $B(C)$ is a subset of S, where B is the set of models of K and C the set of models of ψ, $B(C)$ the set of models of $K*\psi$.
- **Postulate 2**: $K*\psi$ contains ψ. This is the success postulate, that reads $B(C) \subseteq C$.
- **Postulate 3**: $K*\psi \subseteq Cons(K \cup \{\psi\})$ where the latter is the belief set obtained by simply adding ψ to K and taking the logical closure (also called the expansion operation). In set-theoretic terms it reads: $B \cap C \subseteq B(C)$.
- **Postulate 4**: if $\neg\psi \notin K$ then $Cons(K \cup \{\psi\}) \subseteq K*\psi$. In other words, if the new information is coherent with the old belief set then it should be simply absorbed by it. It reads: if $B \not\subset C^c$, then $B(C) \subseteq B \cap C$.
- **Postulate 5:** If ψ is a logical contradiction then $K*\psi$ contains the whole language. It means that $B(\varnothing) = \varnothing$.
- **Postulate 6**: If ψ and ϕ are logically equivalent then $K*\psi = K*\phi$. This is the claim for syntax invariance. It is obviously satisfied in the set-theoretic setting, and it is the reason why the set-theoretic description of the revision theory makes sense.
- **Postulate 7**: $K*(\psi \wedge \phi) \subseteq Cons(K*\psi \cup \{\phi\})$.
  It means that $B(C) \cap D \subseteq B(C \cap D)$.
- **Postulate 8**: if $\neg\phi \notin K*\psi$ then $Cons(K*\psi \cup \{\phi\}) \subseteq K*(\psi \wedge \phi)$.
  It reads: if $B(C) \not\subset D^c$ then $B(C \cap D) \subseteq B(C) \cap D$.

It is obvious that Postulates 1 and 2 are valid for sets of accepted beliefs. However, postulate 5 is not assumed. On the contrary, $B(\varnothing)$ does not exist, because the set $A_L(\varnothing)$ is empty. So, we restrict ourselves to non contradictory input information, and Postulate 1 is to be restricted as well to pairs of subsets *(B, C)* of S, with $C \neq \varnothing$. It makes sense because the new information is supposed to be a piece of observed evidence on which the confidence relation will focus. It is impossible to focus on the empty set. If $B \cap C = \varnothing$ then postulate 3 is obviously verified. Similarly postulate 7 trivially holds if $B(C) \cap D = \varnothing$. More generally, these postulates do hold in the acceptance framework:

Proposition: Let $\geq_L$ be an acceptance ordering. Then $A_L(C \cap D) \subseteq Cons\{A \cap D, A \in A_L(C)\}$.

Proof: The proposition says that for each A, $A \cap C \cap D >_L A^c \cap C \cap D$ implies that for some E, with $E \cap C >_L E^c \cap C$, $E \cap D \subseteq A$. The only non trivial case is if $E \cap D \neq \emptyset$, $\forall E \in A_L(C)$ (otherwise, $Cons\{E \cap D, E \in A_L(C)\} = 2^S$). Using O, $A \cap C \cap D >_L A^c \cap C \cap D$ implies $(A \cup D^c) \cap C >_L A^c \cap C \cap D$, and indeed, $(A \cup D^c) \cap D \subseteq A$.

This proposition proves that postulate 7 is valid in the acceptance framework, and postulate 3 as well since the latter comes down to letting $C = S$ in the above proof.

Conjoining the effects of postulates 3 and 4, it is requested that if $B \not\subseteq C^c$, then $B(C) = B \cap C$. This property does not hold in the acceptance relation setting, where postulate 4 fails. Indeed, $B \not\subseteq C^c$ is equivalent to $C^c \notin A_L$ and means that $C^c >_L C$ does not hold. Let us use the family of *linear* possibility relations F representing $>_L$ (following [5, 15])). $C^c >_L C$ does not hold iff $C >_{\Pi L} C^c$ for a linear possibility relation $>_{\Pi L} \in F$. However we may have that $C^c >_{\Pi} C$ does hold for other possibility relations $>_{\Pi} \in F$. Postulate 4 claims that if that $C^c >_L C$ does not hold, it should be that $A \cap C >_L A^c \cap C$, whenever $A >_L A^c$. However, if $C^c >_{\Pi} C$ for some possibility relation $>_{\Pi} \in F$, it may be that $A^c \cap C >_{\Pi} A \cap C$, even if $A >_{\Pi} A^c$. In that case, $A \cap C >_L A^c \cap C$ fails even if $A >_L A^c$ while $C^c >_L C$ does not hold. Postulate 4 is thus not presupposed in the acceptance relation framework, hence postulate 8 fails too. However, if an acceptance relation is transitive, then, it corresponds to a unique possibility relation and postulates 4 and 8 will hold as well (since $C \geq_{\Pi} C^c$ and $A >_{\Pi} A^c$ imply $A \cap C >_{\Pi} A^c \cap C$). These postulates are clearly related to Rational Monotony.

The direct link between acceptance functions and revision postulates provided here is somewhat parallel to the comparison between revision and non-monotonic inference first studied by Makinson and Gärdenfors [37]. However, our paper considers the AGM theory of revision as only being concerned with the revision of the *current* beliefs of an agent pertaining to the *present* situation. It is the ordinal counterpart to the notion of focusing an uncertainty measure on the proper reference class pointed at by the available factual evidence [16]. The AGM theory is not concerned with the revision of the generic knowledge of the agent regarding what is normal and what is not. The acceptance function setting is in some sense more general than the belief revision setting not only because less postulates are assumed, but because it lays bare the existence of these two kinds of revision problems: the revision of the accepted beliefs on the basis of new observations, and the revision of the acceptance relation itself, due to the arrival of new pieces of knowledge (like a new default rule). The AGM theory seems to deal with the first problem only.

Moreover, the acceptance relation framework also makes it clear that $A_L(C)$ is not computed from A_L, but it can solely induced by the input information C and the acceptance relation, a point that is not crystal-clear when adopting the notations of the AGM theory.

Some scholars have wondered what becomes of the epistemic entrenchment relation after an AGM belief revision step, a problem left open by the AGM theory. Interpreting the latter in the acceptance relation framework suggests that it should remain unchanged. It has been argued that belief revision cannot be iterated, because the epistemic entrenchment relation that underlies the revision operation is lost and it prevents a further revision step from being performed. This is questionable: in the AGM theory, if, after revising A_L into $A_L(C)$, the agent receives a new piece of information D consistent with C, the belief set becomes $A_L(C \cap D)$. This is very natural in our framework. The fact that the order of arrival of inputs is immaterial in this case points out that acceptance relations pertain to a static world, and that $\{C, D\}$ is just the evidence gathered at some point about this static world. *By assumption*, observations C and D *cannot* be contradictory in the acceptance relation framework, otherwise one of the observations C or D is wrong, which kills the success postulate. This feature is common to plausible and probabilistic reasoning : in the latter, successive conditioning of a probability distribution P by C and D comes down to conditioning by the conjunction $C \cap D$, which cannot be a contradiction.

One may object that the success postulate makes no epistemic commitment to the input information C being in fact true in the domain. However, in our view, the success postulate means that the agent believes in observations made on the current situation. So, receiving, in a static world, a subsequent input D contradicting C implies that one of C or D must be false in the domain, even if the agent cannot prove that each individual observation is true, since a reasonable agent will not believe in contradictions. In particular, the problem of handling a sequence of contradictory inputs of the form C and C^c (Lehmann [35]; Darwishe and Pearl [12]), makes no sense in the acceptance framework.

The issue of revising the acceptance relation is different, and beyond the scope of this paper. The problem has been addressed for complete preorderings on states by some authors like Spohn [44], Boutilier and Goldszmidt [8], Williams [45] Darwishe and Pearl [12], and Benferhat *et al.* [6].

6 Conclusion

Our results confirm a significant formal incompatibility between, on the one hand, logical representations of what logic-based AI calls beliefs, viewed here as accepted propositions, for which deductive closure is allowed, and, on the other hand, partial belief in the sense of probability theory. These results are a severe impediment to a generalised view of theory revision based on relations on formulae other than epistemic entrenchment, derived from uncertainty theories, simply because the assumption of closed belief sets is devastating in this respect. The paper indicates that the notion of deductive closure is not fully adapted to the modelling of partial belief, not only because this notion disregards syntactic aspects and presupposes logical omniscience, but also because the closure under conjunction may be counter-intuitive when reasoning with partial belief, as already revealed in the lottery paradox. However, our discussion of the lottery paradox suggests that nonmonotonic reasoning with accepted beliefs is coherent with probabilistic reasoning in situations where typical phenomena can be observed and default statements make sense because some states of nature are much more probable that others. This is modelled by big-stepped probabilities. On the contrary, the lottery paradox especially applies to random situations where no typical trend prevails. So, the significant mathematical incompatibility between probabilistic reasoning and plausible reasoning nevertheless goes along with a strong similarity between them. They just apply to different types of situations.

While we cannot expect to find sample spaces where statistical big-stepped probability functions pop up right away, one may think that for phenomena that have regularities, there may exist partitions of the sample space that form conceptually meaningful clusters of states for the agent and that can be ordered via a big -stepped probability. If this conjecture is valid, it points out a potential link between non-monotonic reasoning and statistical data, in a knowledge discovery perspective. An interesting problem along this line is as follows: Given statistical data on a sample space and a language, find the « best » linguistically meaningful partition(s) of the sample space, on which big-stepped probabilities are induced and default rules can be extracted (See [7] for preliminary results). The difference between other rule extraction techniques and the one suggested here, is that, in our view, the presence of exceptions is acknowledged in the very definition of symbolic rules, without having to keep track of the proportion of such exceptions.

In order to model symbolic reasoning methods that come closer to uncertain reasoning with partial beliefs, weaker types of « deductive closures » might be considered for this purpose, as for instance unions of standard deductively closed sets of propositions (that may be globally inconsistent). This type of closure is encountered in argument-based reasoning under inconsistency. Tolerating inconsistency is indeed incompatible with standard deductive closure. It turns out that most confidence functions (and noticeably probabilities) synthesise partially conflicting pieces of information while possibility measures do not (as witnessed by their nested structure). It may explain why the latter seem to be the only simple ones that accounts for the concept of accepted beliefs in the sense of propositions one can reason about as if they were true. Finally, further work is needed to lay bare numerical uncertainty functions other than probabilities and possibilities that induce acceptance relations.

References

1. E.W. Adams, *The Logic of Conditionals*. Dordrecht: D. Reidel (1975).
2. C.E.P. Alchourrón, P. Gärdenfors, D. Makinson, On the logic of theory change: Partial meet functions for contraction and revision. *J. of Symbolic Logic*, 50 (1985) 513-530.
3. S. Benferhat, D. Dubois, H. Prade, Representing default rules in possibilistic logic. *Proc. of the 3rd Inter. Conf. on Principles of Knowledge Representation and Reasoning* (KR'92), Cambridge, MA, (1992) 673-684.
4. S. Benferhat, D. Dubois, H. Prade Nonmonotonic reasoning, conditional objects and possibility theory. *Artificial Intelligence*, 92 (1997) 259-276.
5. S. Benferhat, D. Dubois, H. Prade Possibilistic and standard probabilistic semantics of conditional knowledge. *Journal of Logic and Computation,* 9 (1999) 873-895.

6. S. Benferhat, D. Dubois, O. Papini A sequential reversible belief revision method based on polynomials.: *Proceedings National American AI conference* (AAAI-99), Orlando, Floride (USA) AAAI Press/The MIT Press, (1999) 733-738.
7. S. Benferhat, D. Dubois, S. Lagrue, H. Prade A big-stepped probability approach for discovering default rules. *Proc. of the Ninth Int. Conf. Information Processing and Management of Uncertainty in Knowledge-based Systems* (IPMU 2002), Annecy, France, (2002) 283-289. (a revised version is to appear in the *Int. J. Uncertainty, Fuzziness and Knowledge-based Systems*).
8. C. Boutilier M. Goldszmidt Revision by conditionals beliefs. In: G. Crocco, L. Fariñas del Cerro, A. Herzig, (eds.) *Conditionals. From Philosophy to Computer Sciences*, Oxford University Press, Oxford, UK, (1995) 267-300.
9. P. Calabrese An algebraic synthesis of the foundations of logic and probability. *Information Sciences*, 42 (1987) 187-237.
10. L.J. Cohen, *The Probable and the Provable*. Clarendon Press Oxford (1977).
11. L. J. Cohen, Belief and acceptance. *Mind*, XCVIII, No. 391 (1989) 367-390.
12. A. Darwishe, J. Pearl On the logic of iterated belief revision. *Artificial Intelligence*, 89, 1997 (1997) 1-29.
13. D. Dubois Belief structures, possibility theory and decomposable confidence measures on finite sets. *Computers and Artificial Intelligence* (Bratislava), 5(5) (1986) 403-416.
14. D. Dubois, H. Fargier, H. Prade Comparative uncertainty, belief functions and accepted beliefs. *Proc. of 14th Conf. on Uncertainty in Artificial Intelligence* (UAI-98), Madison, Wi., (1998) 113-120
15. D. Dubois, H. Fargier, H. Prade, Ordinal and numerical representations of acceptance. Research report 02-20R, IRIT, Toulouse (2002).
16. D. Dubois, S. Moral and H. Prade, Belief change rules in ordinal and numerical uncertainty theories. In D. Dubois H. Prade, (eds.), *Belief Change*, Handbook on Defeasible Reasoning and Uncertainty Management Systems, Vol. 3: Kluwer Academic Publ., Dordrecht, The Netherlands, (1998) 311-392.
17. D. Dubois, H. Prade, Epistemic entrenchment and possibilistic logic. *Artificial Intelligence* 50 (1991) 223-239.
18. D. Dubois, H. Prade, Conditional objects as non-monotonic consequence relationships. *IEEE Trans. on Systems, Man and Cybernetics*, 24 (1994) 1724-1740.
19. D. Dubois, H. Prade, Conditional objects, possibility theory and default rules. In: G. Crocco, L. Farinas del Cerro, A. Herzig (eds.), *Conditionals: From Philosophy to Computer Science* (Oxford University Press, UK, (1995) 311-346.
20. D. Dubois, H. Prade, Numerical representation of acceptance. *Proc. of the 11th Conf. on Uncertainty in Artificial Intelligence,* Montréal, Quebec (1995) 149-156
21. D. Dubois, H. Prade, Possibility theory: qualitative and quantitative aspects. P. Smets, editor, *Quantified Representations of Uncertainty and Imprecision*. Handbook on Defeasible Reasoning and Uncertainty Management Systems, Vol. 1: Kluwer Academic Publ., Dordrecht, The Netherlands (1998) 169-226
22. L.Fariñas del Cerro, A. Herzig A modal analysis of possibility theory. In Ph. Jorrand, J. Kelemen, (eds.) *Fundamentals of Artificial Intelligence Research* (FAIR'91) Lecture Notes in Computer Sciences, Vol. 535, Springer Verlag, Berlin (1991) 11-18.
23. P. Fishburn, The axioms of subjective probability. *Statistical Science,* 1 (1986) 335-358.
24. N. Friedman, J. Halpern, Plausibility measures and default reasoning. *Proc of the 13th National Conf. on Artificial Intelligence* (AAAI'96), Portland, OR, (1996) 1297-1304.
25. P. Gärdenfors, *Knowledge in Flux*. MIT press, Cambridge, MA. (1988).

26. I.R. Goodman, H.T. Nguyen, E.A. Walker, *Conditional Inference and Logic for Intelligent Systems: a Theory of Measure-Free Conditioning*. North-Holland, Amsterdam, The Netherlands (1991).

27. A. Grove, Two modellings for theory change. *J. Philos. Logic*, 17 (1988) 157-170.

28. J. Halpern Defining relative likelihood in partially-ordered preferential structures. *J. AI Research*, 7 (1997) 1-24.

29. G. Kern-Isberner, *Conditionals in Nonmonotonic Reasoning and Belief Revision*. Lecture Notes in Artificial Intelligence, Vol. 2087, Springer Verlag, Berlin (2001).

30. C.H. Kraft, J.W Pratt and A. Seidenberg, Intuitive probability on finite sets. *Ann. Math. Stat.* 30 (1959) 408-419.

31. K. Kraus, D. Lehmann, M. Magidor, Nonmonotonic reasoning, preferential models and cumulative logics. *Artificial Intelligence*, 44 (1990) 167-207.

32. H. E. Kyburg, Probability and the Logic of Rational Belief. Wesleyan University Press. Middletown, Ct. (1961).

33. H. E. Kyburg, Knowledge. In: *Uncertainty in Artificial intelligence*, vol 2, J. F. Lemmer and L. N. Kanal eds, Elsevier, Amsterdam (1988) 263-272.

34. D. Lehmann, M. Magidor, What does a conditional knowledge base entail? *Artificial Intelligence*, 55(1) (1992) 1-60.

35. D. Lehmann Belief revision, revised. *Proc. of the 14th Inter. Joint Conf. on Artificial Intelligence* (IJCAI'95), Montreal, Quebec, (1995) 1534-1540.

36. D. L. Lewis *Counterfactuals*. Basil Blackwell, Oxford, UK (1973).

37. D. Makinson and P. Gärdenfors Relations between the logic of theory change and nonmonotonic reasoning. In: *The Logic of Theory Change* (A. Fürmann, M. Morreau, Eds), LNAI 465, Sprnger Verlag, (1991) 185-205.

38. J. Paris, *The Uncertain Reasoner's Companion*. Cambridge University Press, Cambridge, UK (1994).

39. J. Pearl, *Probabilistic Reasoning Intelligent Systems: Networks of Plausible Inference*, Morgan Kaufmann, San Mateo, CA. (1988).

40. J. Pearl System Z: a natural ordering of defaults with tractable applications to default reasoning. *Proc. of the 3rd Conf. on the Theoretical Aspects of Reasoning About Knowledge (TARK'90)*, Morgan and Kaufmann, San Mateo, CA. (1990) 121-135.

41. D. Poole The effect of knowledge on belief: conditioning, specificity and the lottery paradox in defaut reasoning. *Artificial Intelligence*, 49 (1991) 281-307

42. M. Roubens, P. Vincke, *Preference Modelling*. Lecture Notes in Economics and Mathematical Systems, Vol. 250, Springer Verlag, Berlin (1985).

43. P. Snow Diverse confidence levels in a probabilistic semantics for conditional logics. *Artificial Intelligence*, 113 (1999) 269-279.

44. W. Spohn Ordinal conditional functions: A dynamic theory of epistemic states. In: W.L. Harper, B. Skyrms, (eds.), *Causation in Decision, Belief Change, and Statistics*, Vol. 2, D. Reidel, Dordrecht (1988) 105-134.

45. M.A. Williams Transmutations of knowledge systems. In: J. Doyle, E. Sandewall, P. Torasso (eds.), *Proc. of the 4th Inter. Conf. on Principles of Knowledge Representation and Reasoning (KR'94)*, Bonn, Germany, (1994) 619-629

Getting the Point of Conditionals: An Argumentative Approach to the Psychological Interpretation of Conditional Premises

Jean-François Bonnefon and Denis J. Hilton

Dynamiques Socio-Cognitives et Vie Politique, Université Toulouse-2,
5 allées Antonio Machado, 31058 Toulouse Cedex 9, France
{bonnefon, hilton}@univ-tlse2.fr

Abstract. Processes that govern the interpretation of conditional statements by lay reasoners are considered a key-issue by nearly all reasoning psychologists. An argumentative approach to interpretation is outlined, based on the idea that one has to retrieve the intention of the speaker to interpret a statement, and an argumentative based typology of conditionals is offered. Some empirical support to the approach is provided, as well as some theoretical support from an evolutionary perspective.

1 Introduction

Whatever the theoretical divergences between psychologists interested in conditional reasoning, there is reasonable agreement on at least one point: We will never obtain a satisfying account of conditional reasoning without some idea on the way human reasoners *interpret* conditional statements before reasoning *per se* takes place. That is, we need to specify the pragmatic processes that turn a conditional statement into the mental premise that will serve as the input of reasoning proper. While there is agreement on the importance of these pragmatic processes, it is less clear what the word "pragmatic" is taken for. Yet, broadly speaking, it seems that in the context of reasoning psychology, "pragmatics" refers to the way reasoners use their background knowledge to enrich, restrict, or transform the premise set they are given by means of conversational implicatures.

While we firmly believe this kind of pragmatic processes to be indeed at work in reasoning in general and in conditional reasoning in particular (see [1], [2], [3]), we will suggest in the present paper that our understanding of conditional reasoning would benefit from a broader conception of the pragmatics of reasoning, a conception that would include considerations about *speaker's intention* and illocutionary uptake.

In Section 2, we outline our proposal, and suggest that in reasoning situations, considerations of illocutionary force can be handled within an argumentative framework, which allows us to offer a very simple typology of conditional statements. Section 3 is devoted to recent experimental observations and the way we can account for those using this typology. Lastly, Section 4 summarises our contribution and offer some evolutionary perspective on the idea that "domain-general" reasoning is of argumen-

G. Kern-Isberner, W. Rödder, and F. Kulmann (Eds.): WCII 2002, LNAI 3301, pp. 59–64, 2005.

tative nature, and that interpretation processes are rooted in the activity that consists in extracting arguments from the discourse of others.

2 To Interpret a Conditional Statement Is to Get Its Point

We begin by suggesting that in order to interpret a conditional statement, one has to retrieve the claim it is making. We then offer a typology of conditionals based on the argumentative conventions they are associated to.

2.1 The Intention of the Speaker in a Reasoning Situation

In our daily lives, a large part of those conditional premises we use to reason come from conversations, or situations that we treat like conversations. (As argued in [4], "many of our everyday inferences are made from premises that we read or hear from various sources".) Even when a situation does not (or is not supposed to) have any salient conversational feature, as it is the case with most of experimental reasoning tasks, lay people treat the premises they are given as if they were contributions to a conversation [1].

As contributions to a conversation, conditional statements have *points*. That is, the speaker asserting a conditional has the intention of achieving some effect by mean of this assertion. The "illocutionary force" of a conditional statement refers to the effect its assertion is intended to achieve. Correct uptake on the addressee's side demands the retrieval of this illocutionary force: That is, in order to interpret a conditional statement, a reasoner has to understand what the intention of the speaker who asserted it was.

Generally speaking, conversation is so ubiquitous that a speaker may have any of a large number of possible intentions – e.g., to inform, to threaten, to apologize, etc. But reasoning situations are more specific in that they greatly narrow the array of possible intentions on the speaker's side: When it is manifest that the speaker expects the addressee to reason (i.e., to derive a conclusion) from a statement, the addressee can assume the intention to be *argumentative*. That is, when a statement is meant to be used in reasoning, its contribution to the conversation takes the form of an *incitation to a conclusion*. Let us call the "claim" of a statement the conclusion it is an incitation for.

To illustrate this specificity of reasoning situations, let us consider the two following examples of conditional statements:

1. If you're fond of Asian food, I know of a wonderful Vietnamese restaurant not far from your place.
2. If you use adult websites at work, you'll be fired.

Addressee of statement (1) is not expected to engage into any conditional reasoning. The contribution of statement (1) to a conversation would most probably be to inform the addressee of the existence of such a restaurant, the if-clause being merely mentioning some condition for the then-clause to be conversationally relevant [5]. On the other hand, statement (2) clearly is an incitation to a conclusion: It is claiming that the addressee had better not visit adult websites at work, a conclusion that the addressee has to calculate for him(her)self. We have stressed the importance of retrieving the intention of the speaker in order to interpret a conditional statement. We have suggested that in reasoning situations, this intention can be assumed to be of argumentative nature. This greatly simplifies the reasoner's interpretative task: All (s)he has to do to get the point of the conditional is to get its *claim*, that is, to understand what is the conclusion the speaker is arguing for or against. The task is made even simpler by the restricted set of conventional argumentative uses of conditional statements. A tentative typology of those conventional uses is offered in the next section.

2.2 An Argumentation-Based Typology of Conditionals

We can basically make two broad kinds of points: points about beliefs (theoretical reasoning) or points about actions (practical reasoning). Similarly, it seems there are two main types of conditional arguments: arguments pointing to the truthfulness of a belief and arguments pointing to the appropriateness of an action.

Epistemic conditionals (dealing with beliefs) have received most of the attention of reasoning researchers. In particular, much has been said about what we consider to be their first subcategory, *basic epistemic conditionals*. Examples of such basic epistemic conditionals are numerous and sometimes extremely famous, as "if it is a bird, then it flies". Such a conditional claims that its consequent is true, provided that its antecedent is the case.

A second subcategory of epistemic conditionals is what we have dubbed [3] the *preconditional statement*. A preconditional statement is of the form "if (prerequisite of the conclusion) then (conclusion)". The claim of this statement is actually that the prerequisite is not the case (and therefore that the conclusion cannot be the case, more on this in Section 3).

Practical conditionals (dealing with actions) come in two main types: the *consequential conditional* and the *conditional instruction*. The conditional instruction is of the form "if (event) then (action)". Its analysis in terms of speaker's intention is rather more complex than those of other conditionals, see [6] for a detailed exposition. The consequential conditional is of the form "if (action) then (outcome)". Its claim is that the action is either appropriate or not, depending on the valence of the outcome: If the outcome is desirable (viz. undesirable), the action is appropriate (viz. inappropriate).[1]

Note that two out of those four types of arguments are self-sufficient, in that sense that they do not need any further premise to make their claim: While the basic epistemic conditional and the conditional instruction need some additional categorical premise to allow a conclusion, consequential conditionals and preconditionals make a claim on their own. Section 3 will focus on experiments involving those two interesting categories, which lead to inferential behaviours that would be somewhat puzzling from a strict deductive point of view.

---

[1] Social contracts, Inducements/Deterrents, and Means-end statements are other contenders for the practical conditional category. The possibility that they may all be consequential conditionals is considered in [7].

It should be clear now what we believe reasoners to do when they interpret a conditional premise: Reasoners retrieve the claim of the statement by relying on the argumentative conventions associated to the type they identified the conditional with. Remember the example: "If you use adult websites at work, you'll be fired." This conditional is immediately identified as a consequential conditional with an undesirable outcome: It is therefore interpreted as making the claim that using adult websites at work is an inappropriate action.

From our interpretational framework, we can derive predictions about what inferences reasoners will derive when confronted to simple problems involving various types of conditionals, which is the focus of the next section.

3 Experimental Observations

To illustrate the predictions that can be derived from our interpretational framework, we briefly present some data relevant to preconditionals (Section 3.1) and consequential conditionals (Section 3.2).

3.1 Preconditionals

The consideration of the "preconditional" subcategory of conditional statements was actually driven by experimental observations. Byrne [8] was the first to present subject with problems of the form: "If A then C; if (precondition of C) then C"; A." She then observed that less than 40% of reasoners endorsed the conclusion C from these premises. For example, when presented with the following problem, less than 35% of participants endorsed the conclusion "Mary will study late at the library":

"If Mary has an essay to write then she will study late at the library,
If the library stays open late then she will study late at the library,
Mary has an essay to write."

From a deductive point of view, reasoners should derive the conclusion C – but according to our argumentative approach to interpretation, there is an explanation why they do not. Reasoners first retrieve the claims of the various arguments they can isolate in the premise set. Here the first (epistemic conditional) and third (categorical) premises of the set constitute an argument supporting the conclusion C. But the second premise is a preconditional whose claim is that the precondition for C might not be satisfied: it is thus an argument *against* the conclusion C.[2] Thus reasoners interpret the premise set as containing one argument in favour of C and one argument against it: It is then no longer surprising that the endorsement rate of C falls well below 100%.

Experimental support to this explanation can be found in [3]: It is reported there that 60 students were presented with various problems of the form above. They rated on a 7-point scale (from "no chance to be true" to "certainly true") the confidence they had in the conclusion C, and said if, in their opinion, the second conditional was meant to convey the idea that chances were for the precondition not to be satisfied. Almost 80% of participants judged that the preconditional statement was indeed

---

[2] The rationale for this claim is detailed in [2] and [3].

meant to convey the idea that chances were for the precondition not to be satisfied. More importantly, participants judging so were also the ones that gave the conclusion C a low confidence rating, whereas participant not judging so granted C a normal confidence rating (i.e., comparable to the rating they would have given in the absence of the preconditional).

3.2 Consequential Conditionals

Just like preconditionals, consequential conditionals can make a claim on their own, without the need for a categorical premise. This property paves the way for interesting problems – consider the following premise set:

"If Marie's TV is broken, she will have it fixed,
If Marie has her TV fixed, she will not be able to pay the electricity bill,
Marie's TV is broken."

Will Marie have her TV fixed? Deductively speaking, she will, and she will not be able to pay the bill. But our argumentative interpretational framework makes another prediction.

The first and third premises of the set are a standard epistemic argument in favour of the conclusion "Marie will have her TV fixed". Yet, the second conditional is of consequential nature, with an undesirable outcome: its claim is thus that the action considered in its *if*-part (having the TV fixed) is inappropriate. Reasoners are then left with one argument in favour of the conclusion and one argument against it. Endorsement rate of the conclusion should therefore fall again well below 100%, and so was observed in [7]. Endorsement rate was never higher than 40% when the outcome of the consequential was clearly negative. In another experiment, endorsement rates were intermediate (around 55%) when the outcome was only mildly undesirable.

4 Conclusion

We have argued that for a reasoner to interpret a conditional statement is to get the point the conditional is making, that is, to retrieve the claim the speaker intended to make. This task is made simple by the fact that there are only a few possible conditional argumentative conventions: We listed four, divided into two main types (epistemic and practical). This typology can account for a variety of experimental phenomena.

At the core of our proposal is the idea that interpretation processes are strongly determined by social activity: We interpret premises of a conditional reasoning problem as we would consider the arguments of one that would try to convince us either of the truth of a belief or of the appropriateness of an action. As a final word, we would like to stress that this postulate can receive theoretical support from evolutionary psychology.

Evolutionary psychologists find it implausible that the human mind would have evolved some domain-general reasoning ability [9]. Such an ability would probably suffer from fatal flaws compared to domain and task-specific mechanisms (in terms of speed and accuracy). Yet, Sperber [10] [11] has suggested that one plausible evolved general ability would be the logical-argumentative skills needed to persuade others and not to be too easily persuaded by others. It is argued in [11] that: "It is generally taken for granted that the logical and inferential vocabulary is – and presumably emerged – as tools for reflection and reasoning. From an evolutionary point of view, this is not particularly plausible. The hypothesis that they emerged as tools of persuasion may be easier to defend." Accordingly, we believe the words *if* and *then* to be rooted in persuasion rather than in philosophical logic. Hence, it should not be surprising that lay reasoners interpret conditionals as pieces of persuasive argumentation rather than pieces of logical reasoning.

References

1. Hilton, D.J.: The social context of reasoning: Conversational inference and rational judgement. Psychological Bulletin 118 (2) (1993) 248-271
2. Bonnefon, J.F., & Hilton, D.J.: Formalizing human uncertain reasoning with default rules: A psychological conundrum and a pragmatic suggestion. In S. Benferhat, & P. Besnard (Eds) Proceedings of the 6th European Conference on Symbolic and Quantitative Approaches to Reasoning with Uncertainty, ECSQARU'2001. Berlin: Springer Verlag (2001) 628-634
3. Bonnefon, J.F., & Hilton, D.J.: The suppression of Modus Ponens as a case of pragmatic preconditional reasoning. Thinking & Reasoning 8 (1) (2002) 21-40
4. Stevenson, R.J., & Over, D.E.: Reasoning from uncertain premises: Effects of expertise and conversational context. Thinking & Reasoning 7 (4) (2001) 367-390
5. DeRose, K., & Grandy, R.E.: Conditional assertions and "biscuit" conditionals. Noûs 33 (3) (1999) 405-420
6. Hilton, D.J., Bonnefon, J.F., & Kemmelmeier, M.: Pragmatics at work: Formulation and interpretation of conditional instructions. In J.D. Moore, & K. Stenning (Eds), Proceedings of the 23rd Annual Conference of the Cognitive Science Society, Hillsdale: Lawrence Erlbaum (2001) 417-422
7. Bonnefon, J.F., & Hilton, D.J.: Consequential conditionals: Invited and suppressed inferences from valued outcomes. Journal of Experimental Psychology: Learning, Memory, & Cognition (in press).
8. Byrne, R.M.J.: Suppressing valid inferences with conditionals. Cognition 31 (1989) 61-83
9. Cosmides, L.: The logic of social exchange: Has natural selection shaped how humans reason? Studies with Wason Selection Task. Cognition 31 (1989) 187-276
10. Sperber, D.: Metarepresentations in an evolutionary perspective. In D. Sperber (ed.) Metarepresentations: A Multidisciplinary Perspective. Oxford University Press (2000) 117-137
11. Sperber, D.: An evolutionary perspective on testimony and argumentation. Philosophical topics 29 (2001) 401-413.

Projective Default Epistemology

- A First Look -

Emil Weydert

University of Luxembourg
emil.weydert@ist.lu

Abstract. We show how to extract iterated revision strategies for complex evidence, e.g. epistemic conditionals, from reasonable default inference notions. This approach allows a more extensive cross-fertilization between default reasoning and belief revision. To illustrate it, we use powerful default formalisms exploiting canonical ranking constructions, like system JZ, for specifying new ranking-based revision mechanisms. They extend Spohn's ranking revision to sets of conditionals and satisfy basic rationality principles.

1 Introduction

It is well-known that there are formal links between default reasoning, i.e. non-monotonic inference from defaults, and belief revision, i.e. the dynamics of epistemic states [8]. However, previous work has mostly focused on the connections between monotonic logics of default conditionals (object-level), or consequence relations (meta-level), and single-step revision in the tradition of AGM, e.g. exploiting epistemic proritizations. In fact, both are generally based on similar plausibility orderings over worlds or propositions. But this is only the tip of the iceberg – it hardly addresses the core issues, iterated belief change and proper default inference. As illustrated by the existence of many competing approaches, the lack of consensus, and the scarcity of substantial rationality principles, these areas – and a fortiori their mutual relationships – are much less well understood.

The overall goal of this paper is therefore to establish and explore new links between these major cognitive tasks. More concretely, we want to introduce and investigate a general strategy for extracting a reasonable iterated revision method from any sufficiently well-behaved default formalism. This approach should pave the way for

- specifying new, natural and better iterated revision formalisms,
- gaining deeper insights into the relation between these epistemic tasks,
- exploiting the cross-border perspective in order to evaluate existing and to derive fresh rationality postulates for both sides.

First, we are going to introduce resp. recall general frameworks for belief revision and default reasoning, thereby delimiting our playground. We shall then develop our new account for belief change which characterizes epistemic states by default

G. Kern-Isberner, W. Rödder, and F. Kulmann (Eds.): WCII 2002, LNAI 3301, pp. 65–85, 2005.
© Springer-Verlag Berlin Heidelberg 2005

knowledge bases and uses a suitable default consequence notion to determine the admissible beliefs. On top of this, it realizes revision by simply adding appropriate conditionals and propositions interpreting the epistemic content of the incoming evidence. This may be seen as a more sophisticated implementation of Rott's notion of vertical revision [11].

In particular, we will investigate revision strategies derivable from well-behaved (e.g. inheritance to exceptional subpropositions) quasi-probabilistic default formalisms based on minimal ranking constructions, like system JZ/JLZ [18] [19]. These default inference concepts are also of interest because they are themselves rooted in epistemic ranking revision techniques [13] [14]. This cross-fertilization allows us to extend Spohn-style revision from single propositions to multiple epistemic conditionals constraining the choice of the revised state. The resulting default revision procedures are not only intuitively transparent and appealing (as long as you accept the default formalism), but they also satisfy some relevant rationality desiderata.

Although the perspectives are quite promising, we want to stress that our present implementation of the vertical revision philosophy is best seen as a preliminary account whose main role is to provide a first flavour of some relevant issues.

2   Formal Background

First, we fix the basic logical framework. Let $\mathcal{L}$ be a language of propositional logic with an infinite set of variables Va, $\mathcal{W} = 2^{Va}$ the corresponding collection of propositional valuations or $\mathcal{L}$-structures, and $\models\, \subseteq \mathcal{W} \times \mathcal{L}$ the classical satisfaction relation. $(\mathcal{L}, \models)$ constitutes our standard background logic. For $\Sigma \cup \{\varphi\} \subseteq \mathcal{L}$, we set $[\varphi] = \{w \in \mathcal{W} \mid w \models \varphi\}$, $[\Sigma] = \cap\{[\psi] \mid \psi \in \Sigma\}$, and $\Sigma \vdash \varphi$ iff $[\Sigma] \subseteq [\varphi]$. Let $\mathcal{B}_{\mathcal{L}} = \{[\varphi] \mid \varphi \in \mathcal{L}\}$ be the corresponding boolean propositional system.

In this paper, we heavily exploit the ranking measure framework [15], which offers a powerful and intuitively appealing plausibility semantics for default reasoning and belief revision. Ranking measures are coarse-grained quasi-probabilistic valuations meant to express the degree of surprise or disbelief of propositions. More specifically, their ranking range carries an additive structure which supports reasonable notions of conditionalization and independence (e.g. important for handling plausibility networks). They include Spohn's discrete-valued κ-functions, originally used to model iterated revision of graded full belief [13] [14] before being successfully introduced into default reasoning [10], as well as Dubois and Prade's dense-valued possibility functions [4].

Interestingly, for some purposes, e.g. the formally correct implementation of entropy maximization at the ranking level, or the specification of system JZ (which we are going to exploit), neither Spohn's integer-valued, nor the rational-valued possibilistic approach are fine-grained enough. The reason is that their valuation structures fail to verify divisibility (e.g. no roots for some possibility values). But they admit divisible extensions. Therefore, we are going to focus on divisible ranking measures, namely $\kappa\pi$-measures.

Definition 21 ($\kappa\pi$-measures).
Let $\mathcal{B}$ be a boolean set system. $R : \mathcal{B} \to \mathcal{V}$ is called a $\kappa\pi$-measure iff

- $\mathcal{V} = (V, 0, \infty, +, <)$ *shares the first-order-theory of* $([0, \infty], 0, \infty, +, <)^1$,
- $R(\mathcal{W}) = 0$, $R(\emptyset) = \infty$,
- $R(A \cup B) = \min\{R(A), R(B)\}$,
- $R(\cup_i A_i) = \infty$ *if* $\cup_i A_i \in \mathcal{B}$ *and* $R(A_i) = \infty$ *for all i.*

R_0 *with* $R_0(A) = 0$ *for* $A \neq \emptyset$ *is the uniform $\kappa\pi$-measure. The conditional $\kappa\pi$-measure* $R(.|.): \mathcal{B} \times \mathcal{B} \to \mathcal{V}$ *associated with $R(.)$ is given by*

- $R(B|A) = R(B \cap A) - R(A)$ *if* $R(A) \neq \infty$, *otherwise* $R(B|A) = \infty$.

R is called real/rational iff $\mathcal{V}$ is the additive structure of the positive reals/rationals with ∞. Let $\mathcal{KP}$ be the set of real $\kappa\pi$-measures over $\mathcal{B}$, and $R([\varphi]) = R(\varphi)$.

$\kappa\pi$-measures bridge the gap between the powerful probabilistic and the practical preferential perspective. Their features can be explained by the order of magnitude interpretation, which links the $\kappa\pi$-value $R(A) = r$ to the nonstandard probability $P(A) = \varepsilon^r$, for some arbitrary but fixed infinitesimal $0 < \varepsilon$.

A useful tool for specifying ranking transformations, like ranking revision, is the shifting function, which implements one half of Spohn's conditionalization procedure. For practical reasons, we define shifting over "non-normalized $\kappa\pi$-measures" $R^{\pm} : \mathcal{B} \to \mathcal{V}$, i.e. verifying all the $\kappa\pi$-conditions except $R^{\pm}(\mathcal{W}) = 0$. Let $\mathcal{KP}^{\pm}$ be the set of all such real-valued $R^{\pm}$ over $\mathcal{B}$.

Definition 22 (Shifting).
Let $R^{\pm} \in \mathcal{KP}^{\pm}$, $r \in [-\infty, \infty]$, and $A \in \mathcal{B}$. Shifting of length r maps $R^{\pm}$ to $R^{\pm} + rA \in \mathcal{KP}^{\pm}$ such that for all $B \in \mathcal{B}$,

- $(R^{\pm} + rA)(B) = \min\{R^{\pm}(B \cap A) + \max\{-R(A), r\}, R^{\pm}(B \cap -A)\}$.

In other words, we obtain $R^{\pm} + rA$ by uniformly shifting the A-worlds by the amount r – upwards if $r \geq 0$, downwards if $r \leq 0$ – but at most until one of them touches the bottom 0.

3   Revision Framework

Belief revision is the process of modifying epistemic attitudes in the light of new evidence. For the sake of expressiveness, we may assume that beliefs concern the actual history of the world – not only its current state. Following this perspective, belief revision subsumes what has been called belief update, i.e. the epistemic account of a changing world. Although we are here not going to address nonmonotonic reasoning about action and change explicitly, our approach is well suited to cope with this issue. We are now going to sketch our general framework

---

[1] In other words, $\mathcal{V}$ is the positive half of an ordered commutative group extended by an absorptive maximum ∞, i.e. for all $x \in V$, $x \leq x + \infty = \infty + x = \infty$.

for modeling revision. To make our epistemic spaces slightly more realistic, we do not only consider epistemic states, inputs, and transition functions, but also some further ingredients to refine the specification. Note that "dynamic" here refers to changing beliefs and varying inputs (about the history of the external world).

Definition 31 (Dynamic epistemic system).
A dynamic epistemic system is a tuple $(\mathcal{E}, \mathcal{E}_0, \mathcal{I}, ev, \star, \mathcal{B}, \mathcal{ES}, Bel)$ *with*

- *a collection* $\mathcal{E} \neq \emptyset$ *of epistemic states,*
- *a subset* $\mathcal{E}_0 \subseteq \mathcal{E}$ *of initial states (original priors, e.g. states of ignorance),*
- *a set* $\mathcal{I} \neq \emptyset$ *of admissible epistemic inputs (possible "observations"),*
- *an input evaluation function* $ev : \mathcal{E} \times \mathcal{I} \to 2^{\mathcal{E}}$,
- *a revision function* $\star : \mathcal{E} \times \mathcal{I} \to \mathcal{E}$ *with* $e \star i \in ev(e, i)$,
- *a boolean propositional system* $\mathcal{B}$ *with* $\emptyset, \mathcal{W} \in \mathcal{B} \subseteq 2^{\mathcal{W}}$,
- *a family* $\mathcal{ES}$ *of epistemic specification functions,*
- *a belief set function* $Bel : \mathcal{E} \to 2^{\mathcal{B}}$ *in* $\mathcal{ES}$ *s.t.* $Bel[e]$ *is upwards* $\subseteq$*-closed.*

The first three concepts describe the basic cognitive entities, the following two specify the belief change process, and the remaining three determine the interpretation framework characterizing the epistemic attitudes. More concretely, we stipulate a two-phase revision strategy where we first interpret the cognitive input – in the context of a belief state e – as an epistemic constraint fixing a set of candidate states $ev(e, i)$, from which we then pick up the most suitable one, which becomes our official revised state $e \star i$. We may see the initial input evaluation step as the conservative "monotonic", and the proper revision step as the adventurous "nonmonotonic" part of evidence integration. The task of the epistemic specification functions – the prototypical example being the belief function – is to query those cognitive facets of the epistemic states which are considered relevant in a given application context, e.g. an agent model. Because here we are just looking for a flexible reference framework, we do not discuss more specific desiderata at this point. But we may illustrate these concepts by three different types of dynamic epistemic systems, namely probabilistic conditioning, a more syntax-oriented form of prioritized base revision, and ranking revision accounts inspired by Spohn.

Standard probabilistic revision
There are two variants, using a single distinguished, resp. all positive priors.

- $\mathcal{E} = \{P : \mathcal{B} \to [0, 1] \mid P \text{ probability measure}\}$,
- $\mathcal{E}_0^{uni} = \{P_0\}$ (canonical uniform/symmetric prior), resp.

  $\mathcal{E}_0^{bay} = \{P \in \mathcal{E} \mid P(A) > 0 \text{ for } A \neq \emptyset\}$ (arbitrary non-degenerate priors),
- $\mathcal{I} = \mathcal{B}$,
- $ev(P, A) = \{P^* \in \mathcal{E} \mid P^*(A) = 1\}$ if $P(A) \neq 0$, else $ev(P, A) = \mathcal{E}$,
- $P \star A = P_A \in ev(P, A)$ with $P_A(B) = P(B|A)$ if $P(A) \neq 0$, else $P_A = P$,
- $\mathcal{B} = \{[\varphi] \mid \varphi \in \mathcal{L}\}$,
- $\mathcal{ES} = (Bel_{\geq a} \mid a > 0.5)$ where $Bel_{\geq a}[P] = \{A \in \mathcal{B} \mid P(A) \geq a\}$,
- $Bel[P] = Bel_{\geq 1}[P]$.

Linear base revision
This is a computationally attractive prioritized syntactic approach [9].

- $\mathcal{E} = \{(\varphi_0, \ldots, \varphi_n) \mid \varphi_i \in \mathcal{L}, n \in Nat\}$,
- $\mathcal{E}_0 = \{()\}$,
- $\mathcal{I} = \mathcal{L}$,
- $ev(\vec{\varphi}, \psi) = \{(\vec{\varphi}, \psi)\}$,
- $\vec{\varphi} \star \psi = (\vec{\varphi}, \psi)$,
- $\mathcal{B} = \{[\varphi] \mid \varphi \in \mathcal{L}\}$,
- $\mathcal{ES} = (Bel)$,
- $Bel[\varphi] = \{[\psi] \mid \lfloor \varphi_{i_1} \wedge \ldots \wedge \varphi_{i_m} \rfloor \sqsubseteq \lfloor \psi \rfloor\}$ where $(\varphi_{i_1}, \ldots, \varphi_{i_m})$ is the maximally consistent right-prioritized subsequence of $\vec{\varphi}$,

  e.g. $Bel[(B, A, \neg A)] = \{[\psi] \mid B \wedge \neg A \vdash \psi\}$.

Spohn-style ranking revision
There exist three variants with slightly different revision functions.

- $\mathcal{E} = \mathcal{KP}$,
- $\mathcal{E}_0 = \{R_0\}$,
- $\mathcal{I} = \mathcal{L}$,
- $ev(R, \varphi) = \{R^* \in \mathcal{E} \mid R^*(\varphi) \geq 1\}$ if $R(\varphi) \neq \infty$, else $ev(R, \varphi) = \mathcal{KP}$,
- $R \star_{sp} \varphi,\ R \star_{rsp} \varphi,\ R \star_{msp} \varphi \in ev(R, \varphi)$ (see below),
- $\mathcal{B} = \{[\varphi] \mid \varphi \in \mathcal{L}\}$,
- $\mathcal{ES} = (Bel, Know)$,
- $Bel[R] = \{[\varphi] \in \mathcal{B} \mid R(\neg\varphi) > 0\}$,

  $Know[R] = \{[\varphi] \in \mathcal{B} \mid R(\neg\varphi) = \infty\}$.

Here the epistemic states are the real-valued $\kappa\pi$-measures, the canonical initial prior is the uniform $\kappa\pi$-measure, the inputs are the $\mathcal{L}$-formulas (to stay close to traditional belief revision), the propositional system consists of the $\mathcal{L}$-model sets, and the epistemic specification functions for belief resp. knowledge collect those propositions whose complement is surprising resp. impossible. Each evidence item defines a ranking constraint to be satisfied (if possible) by the revised state $R \star \varphi$. By default, we attribute at least belief strength 1 to incoming epistemically consistent evidence. Note that because of density, $]0, \infty]$ has no minimal element.

There are three major ranking-based revision strategies (with strength parameters) in the tradition of Spohn, which we can specify with the help of shifting functions. If $R(\varphi) = \infty$, we reject the input and set $R \star \varphi = R$. So, let us suppose $R(\varphi) \neq \infty$ and $a > 0$. Note that, reminiscent of the Levi-identity, we always start with a contraction step $R \to R - R(\varphi)[\varphi]$ to neutralize belief in $\neg\varphi$.

Classical Spohn revision: pushes $R(\neg\varphi)$ to a.
- $R \star_{sp}^a \varphi = R - R(\varphi)[\varphi] + a[\neg\varphi]$.

Rigid Spohn revision: adds a to $R(\neg\varphi)$.
- $R \star_{rsp}^a \varphi = R - R(\varphi)[\varphi] + (R(\neg\varphi) + a)[\neg\varphi]$.

Minimal Spohn revision: pushes $R(\neg\varphi)$ to a, if necessary.
- $R \star_{msp}^a \varphi = R - R(\varphi)[\varphi] + max\{R(\neg\varphi), a\}[\neg\varphi]$.

If $R(\neg\varphi) \leq a$, we have $(R \star_{sp}^a \varphi)(\neg\varphi) = a$, $(R \star_{rsp}^a \varphi)(\neg\varphi) = a + R(\neg\varphi)$, and $(R \star_{msp}^a \varphi)(\neg\varphi) = a$. If $R(\neg\varphi) > a$, we still get $(R \star_{sp}^a \varphi)(\neg\varphi) = a$, as well as $(R \star_{rsp}^a \varphi)(\neg\varphi) = a + R(\neg\varphi)$, but also $(R \star_{msp}^a \varphi)(\neg\varphi) = R(\neg\varphi) > a$. That is, only $\star_{msp}^a$ preserves the ranking state if the input constraint already holds. If no parameter is explicitly given, w.l.o.g. we set $a = 1$. This gives us the revision functions for the above Spohn-style dynamic epistemic systems.

- $\star_{sp} = \star_{sp}^1$, $\quad \star_{rsp} = \star_{rsp}^1$, $\quad \star_{msp} = \star_{sp}^1$.

Classical Spohn revision, also known as (A, a)-conditionalization, ignores potentially useful information, namely the original belief strength of A. Rigid Spohn revision, the main approach discussed by Darwiche and Pearl [3], blindly increases the belief strength of A by a constant. Minimal Spohn revision, which we have introduced in the context of default reasoning [18], requires the least shifting efforts and seems to be the most sensible implementation of the minimal change principle in this context. In particular, it is a special instance of $\kappa\pi$-MCE, cross-entropy minimization - a well-behaved probabilistic revision procedure minimizing information loss [12] - at the ranking level [18]. Therefore, minimal Spohn revision is of particular interest and serves as a reference formalism when looking for more general approaches.

4   Default Framework

To exploit default reasoning for iterated belief revision, we are first going to introduce a sufficiently general framework for default formalisms and formulate reasonable basic principles within this context. Defaults are conditional expressions (implications or rules) of the form $\varphi \Rightarrow \psi$ expressing normal relationships and telling us that φ plausibly implies ψ. Let $\mathcal{L}(\Rightarrow) = \{\varphi \Rightarrow \psi \mid \varphi, \psi \in \mathcal{L}\}$. Problems may arise if there are defaults with conflicting conclusions or facts violating them. The role of default inference relations $\vdash$ is to manage these conflicts and to derive plausible consequences from usually finite collections of specific facts, e.g. $\Sigma \subseteq \mathcal{L}$, and generic defaults, e.g. $\Delta \subseteq \mathcal{L}(\Rightarrow)$. We write

- $\Sigma \cup \Delta \vdash \psi$ iff $\Sigma \vdash_\Delta \psi$ iff $\psi \in C_\vdash^\Delta(\Sigma)$.

The literature exhibits an impressive number of default consequence notions but there is no real consensus about which one might be most appropriate in which context. In fact, real progress has often been blocked by historical choices ignoring later insights, clashes of intuitions, and the resulting proliferation of systems with a narrow, benchmark-oriented perspective. However, even if the area is still characterized by methodological gaps, conflicting concepts, and a certain amount of confusion, there are nevertheless some fundamental principles which we may want to impose for any kind of default reasoning serving belief revision. The first requirement encompasses the most basic KLM-postulates [7]. Let $Cn(\Phi) = C_\vdash^\emptyset(\Phi)$.

Definition 41 (Logicality - LOG).
A default inference $\vdash\!\!\!\sim$ verifies logicality iff for all finite $\Sigma \subseteq \mathcal{L}$, $\Delta, \Delta' \subseteq \mathcal{L}(\Delta)$,

- $Cn(\Sigma) \subseteq C_{\vdash\!\!\!\sim}^{\Delta}(\Sigma)$ *(supraclassicality)*,
- $Cn(C_{\vdash\!\!\!\sim}^{\Delta}(\Sigma)) = C_{\vdash\!\!\!\sim}^{\Delta}(\Sigma)$ *(right weakening, conjunction)*,
- $Cn(\Sigma) = Cn(\Sigma')$ *implies* $C_{\vdash\!\!\!\sim}^{\Delta}(\Sigma) = C_{\vdash\!\!\!\sim}^{\Delta}(\Sigma')$ *(left logical equivalence)*.

Logicality enforces a minimal commitment to classical logic. Examples of nonmonotonic inference concepts violating it are consistency-restoration (relaxation) and probabilistic threshold reasoning. However, in the context of qualitative default reasoning, they shouldn't be controversial. Of course, they ignore the nature of the generic default knowledge in Δ. The most elementary requirement w.r.t. the default content may be the following weak form of left logical equivalence for defaults. It guarantees that the results only depend on the semantic content of the facts, the default antecedents and consequents.

Definition 42 (Basic local logical invariance for defaults - BLLI).
$\vdash\!\!\!\sim$ satisfies basic local logical invariance iff for all finite $\Sigma \subseteq \mathcal{L}$, $\Delta = \{\varphi_i \Rightarrow \psi_i \mid i \leq n\}$ and $\Delta' = \{\varphi_i' \Rightarrow \psi_i' \mid i \leq n\}$,

- $\varphi_i \dashv\vdash \varphi_i'$, $\psi_i \dashv\vdash \psi_i'$ *for all $i \leq n$ implies* $\vdash\!\!\!\sim_\Delta = \vdash\!\!\!\sim_{\Delta'}$.

The above invariance conditions are special instances (set $\pi = id_{\mathcal{B}}$) of a more general, but quite natural principle which requires that the inferential behaviour is governed by the semantical structure of the propositional components. Technically, this amounts to ask for invariance under partial boolean isomorphisms within $\mathcal{B}$.

Definition 43 (Representation independence - RI).
Let $\pi : \mathcal{B}_0 \rightarrow \mathcal{B}_1 \subseteq \mathcal{B}$ be an isomorphism between two boolean subsystems $\mathcal{B}_0, \mathcal{B}_1 \subseteq \mathcal{B}$, and $\pi' : \mathcal{L} \rightarrow \mathcal{L}$ a compatible representation map with $[\pi'(\varphi)] = \pi([\varphi])$, $\Sigma^{\pi'} = \{\pi'(\varphi) \mid \varphi \in \Sigma\}$, and $\Delta^{\pi'} = \{\pi'(\varphi) \Rightarrow \pi'(\psi) \mid \varphi \Rightarrow \psi \in \Delta\}$.

- $\Sigma \cup \Delta \vdash\!\!\!\sim \psi$ *iff* $\Sigma^{\pi'} \cup \Delta^{\pi'} \vdash\!\!\!\sim \pi'(\psi)$.

The next requirement establishes a simple but fundamental link between the default conditional and default inference notions.

Definition 44 (Defeasible modus ponens - WDMP, DMP).
$\vdash\!\!\!\sim$ verifies the weak defeasible modus ponens iff

- **WDMP:** $\{\varphi\} \cup \{\varphi \Rightarrow \psi\} \vdash\!\!\!\sim \psi$ *for all consistent $\varphi, \psi \in \mathcal{L}$.*

$\vdash\!\!\!\sim$ verifies the strong defeasible modus ponens iff for $\varphi, \psi \in \mathcal{L}$, $\Delta \subseteq \mathcal{L}(\Rightarrow)$,

- **DMP:** $\{\varphi\} \cup \{\varphi \Rightarrow \psi\} \cup \Delta \vdash\!\!\!\sim \psi$.

Whereas WDMP is hard to criticize, it is also almost useless. The situation is different for DMP. Although the principle appears to be very reasonable, it is nevertheless a bit elusive. In particular, it implements a simple form of specificity reasoning, just consider $\Delta = \{\varphi \vee \varphi' \Rightarrow \neg\psi\}$. Thus, Reiter's default logic fails to verify DMP, even if we assume that φ, ψ are consistent.

- $\{\varphi\} \cup \{\varphi : \psi/\psi, \mathbf{T} : \neg\psi/\neg\psi\} \not\hspace{-0.3em}\sim \psi.$

Nonetheless, DMP is very important for applying default reasoning to belief revision. Our basic desiderata are now as follows (BLLI follows from RI).

Definition 45 (Regularity).
A default inference relation $\sim$ is regular iff it satisfies LOG, RI, DMP.

Of course, there are other interesting principles, but they are less popular and/or more difficult to realize. The most famous ones are the KLM postulates, which characterize the preferential resp. rational consequence relations [7]. Let $\Delta^\rightarrow = \{\varphi \rightarrow \psi \mid \varphi \Rightarrow \psi \in \Delta\}$.

Definition 46 (Preferentiality - PREF, Rationality - RAT).
$\sim$ *is preferential iff it verifies logicality and, for all finite Δ with consistent $\Delta^\rightarrow$, $\sim_\Delta$ satisfies reasoning by cases, cut, and cautious monotony. $\sim$ is rational if $\sim_\Delta$ also verifies rational monotony.*

Another desirable property is that additional consistent information expressed in a new vocabulary should not affect the conclusions expressible in the original language.

Definition 47 (Strong irrelevance - IRR).
Let $\mathcal{L}_1, \mathcal{L}_2 \subseteq \mathcal{L}$ be two sublanguages over disjoint variable sets. $\sim$ verifies IRR iff for all finite $\Sigma_i \subseteq \mathcal{L}_i$, $\Delta_i \subseteq \mathcal{L}_i(\Rightarrow)$ with $\Sigma_2 \cup \Delta_2 \not\hspace{-0.3em}\sim \mathbf{F}$, and $\psi \in \mathcal{L}_1$,

- $\Sigma_1 \cup \Sigma_2 \cup \Delta_1 \cup \Delta_2 \sim \psi$ *iff* $\Sigma_1 \cup \Delta_1 \sim \psi.$

5   Projective Default Revision

In the context of our frameworks for belief revision and default reasoning, we are now going to investigate when and how we may exploit default inference relations for specifying adequate revision strategies. This approach looks particularly promising for the handling of complex evidence, i.e. finite sets of default conditionals representing epistemic constraints. Our first goal is to build a reasonably well-behaved default revision system $(\mathcal{E}, \mathcal{E}_0, \mathcal{I}, ev, \star, \mathcal{B}, (Bel), Bel)$ from a given regular default consequence relation $\sim$. Our next task is then to realize further desiderata by strengthening the requirements for the default reasoning component.

The idea to link nonmonotonic reasoning and belief revision has been around for some time [8]. On the semantic side, this earlier work has been mainly concerned with the relation between monotonic logics of default conditionals and single-step revision, which both exploit similar preferred world semantics. On the syntactic side, maximal consistency techniques for prioritized knowledge bases have seen applications in both areas [1]. Our project is more ambitious – we consider iterated revision for conditionals – and also more general – we exploit a broader range of default formalisms on a more abstract level. This should help to pave the way for hand-tailored revision notions and cross-fertilization spinoffs in both domains.

One approach is to identify the belief set with the collection of nonmonotonic consequences derivable from a knowledge base representing the belief state. In our formal environment this means that we associate with each epistemic state e an underlying default knowledge base $\Sigma_e \cup \Delta_e$ which defeasibly infers exactly the beliefs backed by e using a suitable given regular $\vdash\!\!\!\sim$. This immediately fixes the main epistemic specification function.

- **Belief set function:** $Bel[\Sigma_e \cup \Delta_e] = \{[\psi] \in \mathcal{B} \mid \Sigma_e \cup \Delta_e \vdash\!\!\!\sim \psi\}.$

The regularity of $\vdash\!\!\!\sim$ ensures that the belief set is closed under the classical background logic, that is, $Bel[e]$ is a filter. Note that a single epistemic state may support several belief set notions resulting from different, more or less speculative $\vdash\!\!\!\sim$. The simplest approach identifies e with $\Sigma_e \cup \Delta_e$ and characterizes epistemic states by default knowledge bases. In fact, because of regularity, we do not have to distinguish between $\Sigma_e \cup \Delta_e$ which are locally logically equivalent, i.e. with the same $[\Sigma_e]$ and $\{([\varphi], [\psi]) \mid \varphi \Rightarrow \psi \in \Delta_e\}$. The elements of Σ_e may represent explicit basic or background facts whereas the conditionals from Δ_e may encode plausibility judgments guiding decisions and revisions. For practical reasons, we restrict ourselves to finite sets, which is not a very serious restriction. The only a priori requirement would be consistency w.r.t. $\vdash\!\!\!\sim$. We start with a first, preliminary version, which we are going to refine in the following.

- **Belief states (1):** $\mathcal{E} = \{\Sigma \cup \Delta \subseteq \mathcal{L} \cup \mathcal{L}(\Rightarrow) \mid \Sigma, \Delta \text{ finite}, \Sigma \cup \Delta \not\vdash\!\!\!\sim \mathbf{F}\}.$

The maximally ignorant state, the generic prior, is here of course the empty set.

- **Initial prior:** $\mathcal{E}_0 = \{\emptyset\}.$

To represent the evidence, we use finite sets of default conditionals.

- **Epistemic inputs (1):** $\mathcal{I} = \{\Delta \subseteq \mathcal{L}(\Rightarrow) \mid \Delta \text{ finite}\}.$

What does this mean in practice? In fact, one possibility is to think of these epistemic inputs as precompiled evidence, as the result of a preceding more or less sophisticated interpretation and deliberation process. For instance, such a preliminary step could translate

- parallel external – usually defeasible – observations of some $\varphi_0, \ldots, \varphi_n$ into the default set $\Delta_0 = \{\mathbf{T} \Rightarrow \varphi_i \mid i \leq n\}$, or
- parallel internal – usually indefeasible – introspective insights $\psi_0, \ldots, \psi_n$ into the strict necessity set $\Delta_0 = \{\neg\psi_i \Rightarrow \mathbf{F} \mid i \leq n\}$.

In fact, we could see the meaning of default evidence as given by its expected revision impact, fixed by the the input evaluation and the revision function.

Let $\Sigma \cup \Delta$ be some prior epistemic state, Δ_0 a set of conditionals representing the incoming evidence, and $\Sigma^* \cup \Delta^* = (\Sigma \cup \Delta) \star (\Sigma_0 \cup \Delta_0)$ the revised state. For the sake of simplicity, we are going to follow the tradition of the qualified AGM success postulate and – as long as there is no conflict with epistemic necessities – integrate the new information without further questioning, implicitly stipulating an anterior deliberation and acceptance phase. However, our approach is also applicable to non-prioritized revision or fusion. But what does success in the present context mean? The DMP principle suggests that, if we accept the default evidence Δ_0, the inference behaviour of $\Sigma^* \cup \Delta^*$ should reflect the defaults from Δ_0. That is, we ask for

- $\varphi \Rightarrow \psi \in \Delta_0$ implies $\Sigma^* \cup \Delta^* \cup \{\varphi\} \hspace{0.5em}\vdash\hspace{-0.9em}\sim\hspace{0.3em} \psi$ (*projection principle*).

In particular, if $\mathbf{T} \Rightarrow \psi \in \Delta_0$, we should get $\psi \in Bel[\Sigma^* \cup \Delta^*]$. The only reason to reject default evidence would be a fundamental incoherence with the prior state, i.e. if the naive application of the basic revision mechanism would risk to produce an inconsistent state. For instance, in classical probability theory we have to disallow conditioning on null-events. For us, coherence requires at least that the plausible consequences of the default evidence are not considered impossible within the prior epistemic context. This should be enough to avoid inconsistent states, i.e. $\Sigma^* \cup \Delta^* \hspace{0.5em}\vdash\hspace{-0.9em}\sim\hspace{0.3em} \mathbf{F}$.

Definition 51 (Coherence).
Δ_0 *is coherent with* $\Sigma \cup \Delta$ *iff* $\Sigma \cup \{\delta_0\} \cup \Delta \hspace{0.5em}\nvdash\hspace{-0.9em}\sim\hspace{0.3em} \mathbf{F}$, *where* $\delta_0 = \wedge \Delta_0^{\rightarrow}$.

All this suggests the following straightforward input evaluation function for characterizing revision candidates based on the hard "monotonic content" of the default evidence.

- **Evaluation function:** Let $Imp_{\sim}(\Sigma^*, \Delta^*) = \{\varphi \Rightarrow \psi \mid \Sigma^* \cup \{\varphi\} \hspace{0.5em}\vdash\hspace{-0.9em}\sim_{\Delta^*}\hspace{0.3em} \psi\}$. If $\Sigma \cup \Delta, \Delta_0$ coherent, $ev(\Sigma \cup \Delta, \Delta_0) = \{\Sigma^* \cup \Delta^* \mid \Delta_0 \subseteq Imp_{\sim}(\Sigma^*, \Delta^*)\}$, else, $ev(\Sigma \cup \Delta, \Delta_0) = \mathcal{E}$.

What remains to be done is to pick up the most plausible candidate $\Sigma^* \cup \Delta^*$ from $ev(\Sigma \cup \Delta, \Delta_0)$, i.e. to specify the revision function $\star$ itself. Obviously, setting $\Sigma^* \cup \Delta^* = \Sigma \cup \Delta \cup \Delta_0$ is not enough because – similar to standard conditionalization – this would trivialize any iterated revision sequence including conflicting evidence items (e.g. $\emptyset \star \{\mathbf{T} \Rightarrow \varphi\} \star \{\mathbf{T} \Rightarrow \neg\varphi\}$). However, the general idea to generate $\Sigma^* \cup \Delta^*$ by simply adding appropriate finite knowledge chunks $\Phi^+ \cup \Psi^+$ to $\Sigma \cup \Delta$, with $\Phi^+ \subseteq \mathcal{L}$ and $\Psi^+ \subseteq \mathcal{L}(\Rightarrow)$, still looks quite attractive. In the terminology of Rott [11], this amounts to combine weak horizontal (simple transformations of the belief base) and strong vertical (exploitation of nonmonotonic inference) elements.

Given Σ, Δ, Δ_0, what could be an appropriate choice for $\Phi^+ \cup \Psi^+$? First, we observe that such a simple expansion procedure can only work if the knowledge in Σ is unrevisable. The idea is now – in a first approximation – to introduce for each incoming default evidence $\varphi \Rightarrow \psi$ a kind of virtual background context statement X and to put X into Φ^+ and $\varphi \wedge X \Rightarrow \psi$ into Ψ^+. These input contexts X are meant to represent hard but non-specific knowledge. All we know specifically about them is derived from the conditional relationships they support. They allow a kind of virtual conditioning and their main role is to anchor the default evidence in the revised state.

For practical purposes, we are going to split Va into (the usual) concrete world-description variables and abstract input-context variables X_i ($i \in Nat$). This defines two sub-languages $\mathcal{L}_{obj}$ and $\mathcal{L}_{inp}$ of $\mathcal{L}$. Furthermore, we redefine or restrict $\mathcal{E}$ and $\mathcal{I}$ accordingly.

- **Belief states:** % $\mathcal{E} = \{\Sigma \cup \Delta \subseteq \mathcal{L}_{inp} \cup \mathcal{L}(\Rightarrow) \mid \Sigma, \Delta \text{ finite}, \Sigma \cup \Delta \not\vdash \mathbf{F}\}$,
- **Epistemic inputs:** $\mathcal{I} = \{\Delta \subseteq \mathcal{L}_{obj}(\Rightarrow) \mid \Delta \text{ finite}\}$.

How do we determine $\Sigma^* \cup \Delta^* = (\Sigma \cup \Delta) \star \Delta_0$? First, we choose a fresh input-context variable X_i, w.l.o.g. the one with the smallest index among those not occurring in $\Sigma \cup \Delta \cup \Delta_0$, and set $\Sigma^* = \Sigma \cup \Phi^+$ where $\Phi^+ = \{X_i\}$. To allow $\Sigma^* \cup \Delta^* \in ev(\Sigma \cup \Delta, \Delta_0)$, we specify Ψ^+ by relativizing the conditional input Δ_0 to the conjunction of Σ^*. Note that we have to go beyond our preliminary stipulations from above. In fact, is not enough to relativize Δ_0 to X_i alone because this violates success for iterated revision with conflicting evidence. For instance, we would get the following undesirable result.

- $\emptyset \star \{\mathbf{T} \Rightarrow \neg\varphi\} \star \{\mathbf{T} \Rightarrow \varphi\} = (\{X_0\} \cup \{X_0 \Rightarrow \neg\varphi\}) \star \{\mathbf{T} \Rightarrow \varphi\} =$
  $\{X_0, X_1\} \cup \{X_0 \Rightarrow \neg\varphi\} \cup \{X_1 \Rightarrow \varphi\} \not\vdash \varphi, \neg\varphi$ (by symmetry).

Let $\Delta^\alpha = \{\varphi \wedge \alpha \Rightarrow \psi \mid \varphi \Rightarrow \psi \in \Delta\}$ and $\sigma^* = \wedge\Sigma^*$. If $\Sigma^* \cup \Delta_0^{\sigma^*} \cup \Delta \not\vdash \mathbf{F}$, we set $\Delta^* = \Delta \cup \Psi^+$ where $\Psi^+ = \Delta_0^{\sigma^*}$. If not, we stipulate $\Psi^+ = \emptyset$. For the strong inference notions $\vdash$ we are going to consider in the present paper, this alternative type of coherence prerequisite is equivalent to the one we have given above. However, this is not necessarily true for arbitrary $\vdash$. Now we use this default projection strategy to define our revision function.

Definition 52 (Projective default revision).
Projective default revision $\star_{pr}$ is given by

- $(\Sigma \cup \Delta) \star_{pr} \Delta_0 = \Sigma \cup \{X\} \cup \Delta \cup \Delta_0^{\sigma \wedge X}$.

where $X \in \mathcal{L}_{inp}$ is the new variable with the smallest index and $\sigma = \wedge\Sigma$.

In this way, we can associate with each regular default inference notion $\vdash$ a corresponding default revision system. Whether the resulting revision mechanism is reasonable mainly depends on the qualities of $\vdash$. Although regularity is already quite demanding – DMP requires weak specificity, which excludes Reiter's default logic – we need further assumptions to meet some standard requirements for belief revision. Consider for instance the preservation principle, which we may express as usual by translating propositional revision with φ into default revision with $\mathbf{T} \Rightarrow \varphi$.

- $\neg\varphi \notin Bel[\Sigma \cup \Delta]$ implies $Bel[\Sigma \cup \Delta] \subseteq Bel[(\Sigma \cup \Delta) \star_{pr} \{\mathbf{T} \Rightarrow \varphi\}]$.

Within the default revision framework, this gives us

- $\Sigma \not\vdash_\Delta \neg\varphi$ and $\Sigma \vdash_\Delta \psi$ implies $\Sigma \cup \{X\} \cup \Delta \cup \{\sigma \wedge X \Rightarrow \varphi\} \vdash \psi$.

With moderate assumptions about the interference of X with the inferential relationships, it can be shown that preservation presupposes rational monotony, i.e. $\Sigma \not\vdash_\Delta \neg\varphi$ and $\Sigma \vdash_\Delta \psi$ entails $\Sigma \cup \{\varphi\} \vdash_\Delta \psi$, which is a rather strong demand. This suggests that our discussion of default revision should focus on rational regular default inference notions, preferably verifying IRR to handle the input-context variables appropriately. Thus, we have gotten a very general approach for generating new revision formalisms, but the question is whether the subclass of those with the desirable features may not be too small? The answer is that this restriction has only a modest impact because there are different strategies – the most rudimentary one being rational closure – which allow us to extend a preferential default inference notion to a suitable rational one.

6   Ranking Construction Entailment

To grasp the potential of the projective default revision paradigm, we are going to investigate the behaviour and evaluate the power of those revision formalisms resulting from reasonable ranking-based default entailment notions, like system JZ. Their semantics is well-motivated and well-understood, and it sanctions the relevant inference desiderata. In particular, this approach also allows us to extend minimal Spohn revision to multiple conditional evidence. So, let us start with a short sketch of this type of default reasoning. For further information, we refer the reader to [18] [19]. First, we want to recall the (strong) monotonic semantics for default conditionals within the $\kappa\pi$-measure framework. For the sake of clarity, and w.l.o.g., we focus on real-valued $\kappa\pi$-measure models.

Definition 61 ($\kappa\pi$-measure semantics).
The satisfaction relation $\models_{\kappa\pi} \subseteq \mathcal{KP} \times \mathcal{L}(\Rightarrow)$ is given by

- $R \models_{\kappa\pi} \varphi \Rightarrow \psi$ *iff* $R(\varphi \wedge \psi) + 1 \leq R(\varphi \wedge \neg\psi)$.

For $\Delta \subseteq \mathcal{L}(\Rightarrow)$, let $Mod_{\kappa\pi}(\Delta) = \{R \in \mathcal{KP} \mid R \models_{\kappa\pi} \Delta\}$. The monotonic entailment notion $\vdash_{\kappa\pi}$ is given by $\Delta \vdash_{\kappa\pi} \delta$ iff $Mod_{\kappa\pi}(\Delta) \subseteq Mod_{\kappa\pi}(\delta) = Mod_{\kappa\pi}(\{\delta\})$.

We use the threshold semantics because it simplifies the definition of default formalisms by guaranteeing the existence of minima. Note that we could replace 1 by any threshold value $\alpha \neq 0, \infty$ given that they are exchangeable by automorphisms of the valuation structure. But we could also reformulate our definitions so as to fit the weaker truth condition $R(\varphi \wedge \psi) < R(\varphi \wedge \neg\psi)$ (stipulating $\infty < \infty$).

Within the ranking measure framework, we can use appropriate ranking choice functions $\mathcal{F}$ to specify default formalisms. They associate with each finite default base Δ a set of distinguished $\kappa\pi$-models $\mathcal{F}(\Delta) \subseteq Mod_{\kappa\pi}(\Delta)$ which determines the plausible consequences in the context of Δ. There are two basic conditions we want to impose. First of all, any proper semantic approach should be invariant under the substitution of individual $\kappa\pi$-semantically equivalent defaults (local semanticality), which constitutes a strengthening of BLLI. Let $Mod_{\kappa\pi}^{sg}(\Delta) = \{Mod_{\kappa\pi}(\delta) \mid \delta \in \Delta\}$. Furthermore, a $\kappa\pi$-consistent default set should never support an inconsistency, i.e. it should admit at least one distinguished $\kappa\pi$-model. Because we sometimes want to exploit the internal structure of a default knowledge base, a necessary condition for some inheritance features, we have to drop invariance w.r.t. $\kappa\pi$-semantically equivalent default bases (global semanticality), i.e.

- $Mod_{\kappa\pi}(\Delta) = Mod_{\kappa\pi}(\Delta')$ may fail to imply $\mathcal{F}(\Delta) = \mathcal{F}(\Delta')$.

Definition 62 (Ranking choice).
$\mathcal{F}$ is called a ranking choice function iff for each finite $\Delta \subseteq \mathcal{L}(\Rightarrow)$,

- $\mathcal{F}(\Delta) \subseteq \mathcal{KP}(\Delta)$ (choice),
- $Mod_{\kappa\pi}(\Delta) \neq \emptyset$ implies $\mathcal{F}(\Delta) \neq \emptyset$ (nonmonotonic consistency),
- $Mod^{sg}_{\kappa\pi}(\Delta) = Mod^{sg}_{\kappa\pi}(\Delta')$ implies $\mathcal{F}(\Delta) = \mathcal{F}(\Delta')$ (local semanticality).

Each ranking choice function $\mathcal{F}$ defines a default inference notion $\vdash^{\mathcal{F}}$.

Definition 63 (Default ranking entailment).
Let $\mathcal{F}$ be a ranking choice function and $\sigma = \wedge\Sigma$.

- $\Sigma \vdash^{\mathcal{F}}_{\Delta} \psi$ iff for all $R \in \mathcal{F}(\Delta)$, $R(\neg\psi|\sigma) > 0$.

There are two reasons for choosing > 0. First, to ensure rational monotony if there is a single preferred $\kappa\pi$-model, and secondly, to maximize the default consequences. It is easy to see that the $\vdash^{\mathcal{F}}$ always verify DMP, BLLI, LOG, and PREF. In fact, the $\vdash^{\mathcal{F}}_{\Delta}$ exactly describe the finitary preferential inference notions over $\mathcal{L}$, with $\mathcal{F}(\Delta) = Mod_{\kappa\pi}(\Delta)$ corresponding to system P.

A simple strategy for generating powerful ranking-based default inference relations is the ranking construction paradigm [16] [17] [18] [19], which tells us to focus on those $\kappa\pi$-measures which are constructible by iterated parametrized minimal Spohn revision with $\varphi \rightarrow \psi$, or equivalently, by uniformly shifting upwards the $[\varphi \wedge \neg\psi]$-worlds, for suitable $\varphi \Rightarrow \psi \in \Delta$.

Definition 64 (Ranking constructibility).
$R \in \mathcal{KP}$ is constructible over $\Delta = \{\varphi_i \Rightarrow \psi_i \mid i \leq n\}$, or $R \in Constr(\Delta)$, iff

- $R = R_0 + a_0[\varphi_0 \wedge \neg\psi_0] + \ldots + a_n[\varphi_n \wedge \neg\psi_n]$ for some $a_i \in [0, \infty]$.

In practice, this amounts to impose a context-dependent, possibly vanishing penalty for the violation of each default and to construct a $\kappa\pi$-measure by adding these penalties for each world and computing the ranks of propositions as ususal. This gives us a whole class of well-behaved default inference relations.

Definition 65 (Constructible inference).
A default inference notion $\vdash^{\mathcal{F}}$ is said to be constructible iff

- $\mathcal{F}(\Delta) \subseteq Mod_{\kappa\pi}(\Delta) \cap Constr(\Delta)$.

The weakest constructible default inference relation is system J [17].

- $\vdash^{J} = \vdash^{\mathcal{F}_J}$: $\mathcal{F}_J(\Delta) = Mod_{\kappa\pi}(\Delta) \cap Constr(\Delta)$.

It satisfies many relevant desiderata, including defeasible chaining, inheritance to exceptional subpropositions, as well as strong irrelevance (IRR). It is also quite robust w.r.t. intuitively minor changes of the default knowledge base. But the price to pay is a lack of inferential commitment, the rejection of some reasonable conclusions, a modesty which may not always fit the agent's needs when a decision is required. Thus we should also look for more speculative inference strategies. One possible strengthening is to focus on the minimally constructible $\kappa\pi$-models of a given Δ, i.e. on those which do not allow a further shrinking of the shifting lengths and implement a kind of local shifting and surprise minimization.

Definition 66 (Minimal constructibility).
$R^* = R_0 + \Sigma_{i\leq n}a_i[\varphi_i \wedge \neg\psi_i]$ *is a minimally constructible $\kappa\pi$-model of and over* Δ *iff*

- *for all* $R_0 + \Sigma_{i\leq n} x_i[\varphi_i \wedge \neg\psi_i] \models_{\kappa\pi} \Delta$ *with* $\vec{x} \leq \vec{a}$, *we have* $\vec{x} = \vec{a}$.

However, this is not enough to guarantee canonicity because there are Δ with infinitely many minimally constructible models. Let us therefore turn to the maximally speculative approaches, those based on the construction of a canonical preferred ranking model, i.e. with $\mathcal{F}_{JX}(\Delta) = \{JX[\Delta]\}$ for consistent Δ. A major goal has been to find a minimally constructible counterpart of system Z [10] (or Z^+) which shares its semantic attractiveness but not its inheritance weaknesses. We have developed two well-behaved approaches meeting these demands, namely system JZ [18] and system JLZ [19]. Here we are going to exploit the former, which is inspired by three complementary (rough) construction guidelines.

- **Minimizing surprise:** *Prefer ranking constructions making more propositions less surprising.* For instance, $1[a \wedge b]+1[\neg a \wedge b]+3[\neg a \wedge \neg b]$ is preferable to $1[a \wedge b] + 2[\neg a \wedge b] + 2[\neg a \wedge \neg b]$ because the set of propositions reaching rank 2 in the first construction, i.e. $\{[\neg a \wedge \neg b]\}$, is a strict subset of those achieving this in the second one, i.e. $\{[\neg a \wedge \neg b], [\neg a \wedge b]\}$.
- **Minimizing shifting:** *Minimize the length of shifting moves aimed at pushing propositions to a given rank.* For instance, $1[a]+1[b]+1[a \vee b]$ is preferable to $2[a] + 2[b] + 0[a \vee b]$ because the second construction uses longer shifting moves, although both push $[a], [b], [a \vee b]$ to rank 2.
- **Minimizing redundancy:** *Reject proper unjustified shifting moves* $a_i[\varphi_i \wedge \neg\psi_i]$ *where* $a_i > 0$ *and* $JX[\Delta](\varphi_i \wedge \psi_i) + 1 < JX[\Delta](\varphi_i \wedge \neg\psi_i)$. In [18] [19], there is an example which shows that this requirement goes beyond minimal constructibility.

To obtain an intelligible specification of the JZ-construction algorithm, we introduce some auxiliary concepts. The first one is relative normality maximization, which helps to maximize plausibility, or minimize surprise, at each construction step. Let us call R compatible with Δ iff $R(\delta) = 0$ for $\delta = \wedge\Delta^{\rightarrow}$. Let $R \leq R'$ iff for all $A \in \mathcal{B}$, $R(A) \leq R'(A)$.

Definition 67 (Relative normality maximization).
If R is compatible with Δ, the relative normality maximization model of Δ over R exists and is given by

- $NM[R, \Delta] = Min\{R' \in Mod_{\kappa\pi}(\Delta) \mid R \leq R'\}$.

$NM[R, \Delta]$ is the unique least surprising model of Δ above R. Next, we introduce a lexicographic preference relation $\prec_{lsh}$ for comparing the local shifting efforts. Let $\vec{x} = (x_i \mid i \leq m)$ and $\vec{y} = (y_i \mid i \leq m)$ be tuples from $[0, \infty]$ representing the lengths of shifting moves. The idea is to prefer those (partial) constructions with less longer shifting moves. Let $S^r_{\vec{x}} = \{j \mid x_j = r\}$.

- $\vec{x} \prec_{lsh} \vec{y}$ iff for the largest r with $S^r_{\vec{x}} \neq S^r_{\vec{y}}$, $S^r_{\vec{x}} \subset S^r_{\vec{y}}$.

For instance, $(2, 5, 3, 2, 2) \prec_{lsh} (3, 5, 3, 1, 0)$ because $(\{1\}, \{2\}, \{0, 3, 4\}, \emptyset, \emptyset) \subset_{lex}$ $(\{1\}, \{0, 2\}, \emptyset, \{3\}, \{4\})$, the tuples corresponding to $(S^5_{\vec{z}}, \ldots, S^0_{\vec{z}})$. To implement minimal shifting, we pick up the $\prec_{lsh}$-minimal local constructions shifting a chosen collection of A_i $(i \leq m)$ above a given rank $\alpha \in [0, \infty]$, starting at some prior R, possibly resulting from previous construction steps.

Definition 68 (Lexicographic shifting minimization).
Let R, A_i and α be as before. We say that the construction $\Sigma_{i \leq m} a_i A_i$ is (lexicographically) shifting minimal w.r.t. reaching rank α from R iff $\vec{a}$ is $\prec_{lsh}$-minimal s.t.

- $(R + \Sigma_{i \leq m} a_i A_i)(A_j) \geq \alpha$ for all $j \leq m$.

If $\alpha < \infty$, we obtain a single minimal solution, for $\alpha = \infty$, we may stipulate $\vec{a} = \vec{\infty}$. This is useful for realizing our canonical construction. Concerning uniqueness, we observe that if $\vec{a}$ and $\vec{x}$ verify the above inequality, then so does $(\vec{a} + \vec{x})/2$. However, if $\vec{a} \neq \vec{x}$, we have $(\vec{a} + \vec{x})/2 \prec_{lsh} max_{\prec_{lsh}}\{\vec{a}, \vec{x}\}$. Hence, there cannot be two $\prec_{lsh}$-minima. Concerning existence, i.e. the construction of a minimum, we first note that for $a_0 = \ldots = a_m = \alpha$, we have $(R + \Sigma_{i \leq m} a_i A_i)(A_j) \geq \alpha$ for all $j \leq m$. Then, all we have to do is to uniformly and jointly decrease the individual a_j until $(R + \Sigma_{i \leq m} a_i A_i)(A_j) = \alpha$ or $a_j = 0$ (see also [19]).

We are now ready to specify the JZ-construction algorithm in a transparent way. We use an inductive bottom-up construction starting at the uniform prior R_0 and proceeding from more plausible to less plausible levels. More specifically, we construct the distinguished $\kappa\pi$-model rank by rank, trying to approximate normality maximization while minimizing the shifting efforts at each rank. This means building two increasing sequences of real $\kappa\pi$-measures $(R_i)_{i \leq h}$ and $(R^*_i)_{i \leq h}$, with $R_i \leq R_{i+1}, R^*_i \leq R^*_{i+1}$ and $R_i \leq R^*_i$, which will eventually converge to a single preferred constructible $\kappa\pi$-model $JZ[\Delta] = R_h = R^*_h$ of Δ. The major ingredients are

- R_j : $\kappa\pi$-measure resulting from the current partial construction,
- R^*_j : current approximating $\kappa\pi$-model of Δ, i.e. $NM[R_j, \Delta]$,
- s_j : current target rank,
- I : index set of $Prop(\Delta) = \{[\varphi_i \wedge \neg\psi_i] \mid i \leq n\} = \{[\varphi \wedge \neg\psi] \mid \varphi \Rightarrow \psi \in \Delta\}$,
- I_j : index set of the shiftable propositions considered at rank j.

JZ-procedure: Let Δ be consistent.
Induction start: $j = 0$.

- $s_0 = 0$, $I_0 = \emptyset$,
- $R_0 = \Sigma_{i \in I} 0[\varphi_i \wedge \neg\psi_i]$, $R_0^* = NM[R_0, \Delta]$.

Induction step: $j \to j+1$.

- s_{j+1} smallest $s > s_j$ of the form $s = R_j^*(\varphi_i \wedge \neg\psi_i)$,
- $I_{j+1} = \{i \in I \mid R_j^*(\varphi_i \wedge \neg\psi_i) = s_{j+1}\} \subseteq I - (I_0 \cup \ldots \cup I_j)$,
- $R_{j+1} = R_j + \Sigma_{i \in I_{j+1}} a_i[\varphi_i \wedge \neg\psi_i]$ where $\vec{a}$ specifies the lexicographically shifting minimal construction reaching s_{j+1} from $R_j + \Sigma_{i \notin I_{\leq j+1}} \infty[\varphi_i \wedge \neg\psi_i]$, i.e. while ignoring the shiftable propositions with $R_j^*(\varphi_i \wedge \neg\psi_i) > s_{j+1}$,
- $R_{j+1}^* = NM[R_{j+1}, \Delta]$.

Induction stop: $j = h$.

- If $I_j = \emptyset$, then $JZ[\Delta] = R_{j+1} = R_j$.

Definition 69 (System JZ).
If Δ is consistent, JZ-entailment is given by $\mathcal{F}_{JZ}(\Delta) = \{JZ[\Delta]\}$.

As documented in [18], system JZ exhibits nice inferential features. It follows from the construction that $JZ[\Delta]$ is minimally constructible w.r.t. Δ.

7   JX-Revision

We are now going to explore specific ranking revision strategies derived from projective default revision to evaluate the potential of this approach. Accordingly, we shall focus on well-behaved default entailment notions $\vdash^{JX}$ (generic notation) based on a canonical, minimally constructible distinguished $\kappa\pi$-model $JX[\Delta]$, like system JZ. Among others, this guarantees the validity of the AGM-postulates for single-step revision (modulo technical details). The obvious first move is to stipulate $R_0 \star \Delta = JX[\Delta]$. The more difficult question is how to revise arbitrary R, or at least those which are epistemically accessible, i.e. of the form $R = R_0 \star \Delta_1 \star \ldots \star \Delta_n$. Our starting idea is to represent the prior $R : \mathcal{B}_{obj} \to [0, \infty]$ by a finite default knowledge base $\Sigma_R \cup \Delta_R \subseteq \mathcal{L}_{inp} \cup \mathcal{L}_{obj}(\Rightarrow)$ verifying

- $R = JX[\Delta_R](.|\sigma_R)$,

where $\sigma_R = \wedge \Sigma_R$. Σ_R is meant to represent the hard concrete reference context determining the actual plausibility assumptions in the context of the generic ones expressed by Δ_R. Of course there are many $\Sigma \cup \Delta \subseteq \mathcal{L}_{inp} \cup \mathcal{L}_{obj}(\Rightarrow)$ with $R = JX[\Delta](.|\sigma)$. So we need a non-trivial choice function to pick up Σ_R and Δ_R. Of course, the existence of such representations also depends on the expressiveness of the default language. If a $\kappa\pi$-measure R with range V_0 is characterizable over a finite subsystem $\mathcal{B}_0 \subseteq \mathcal{B}$, then it is always representable by default conditionals

$\Rightarrow^a$ with strength parameter $a \in V_0$ and truth condition $R(\varphi \wedge \psi) + a \leq R(\varphi \wedge \neg \psi)$. Alternatively, we may restrict ourselves to those constructible ranking states reachable by iterated revision from R_0.

Using default projection and the above default base representation, we can now try to reformulate the default revision problem for $\kappa \pi$-measures over $\mathcal{B}_{obj}$ into one for default knowledge bases $\Sigma_R \cup \Delta_R$. Let R be a prior and Δ_0 be default evidence with $R(\delta_0) \neq \infty$ for $\delta_0 = \wedge \Delta_0^{\rightarrow}$ (i.e. Δ_0 coherent with $\Sigma_R \cup \Delta_R$). Our proposal for defining a ranking revision function $\star_{jx}$ to exploit system JX and default projection is then as follows.

Definition 71 (JX-revision).
For $R = JX[\Delta_R](.|\sigma_R)$ and $\Sigma^* \cup \Delta^* = (\Sigma_R \cup \Delta_R) \star_{pr} \Delta_0$, we set

- $R \star_{jx} \Delta_0 = JX[\Delta^*](.|\sigma^*)$ *(over $\mathcal{B}_{obj}$).*

To ensure well-definedness, we must guarantee that neither the choice of a new input context X, nor the choice of a representation $\Sigma_R \cup \Delta_R$ for R have any influence on the ranking revision result.

Theorem 72 (Projection invariance).
Let $\mathcal{L}' \subseteq \mathcal{L}$ be a sublanguage with $\Sigma \subseteq \mathcal{L}', \Delta, \Delta_0 \subseteq \mathcal{L}'(\Rightarrow)$, and let $X, Y \in \mathcal{L}_{inp}$ be logically independent from the formulas in $\mathcal{L}'$. Then, for $\varphi \in \mathcal{L}_{obj}$,

- $JX[\Delta \cup \Delta_0^{\sigma \wedge X}](\varphi | \sigma \wedge X) = JX[\Delta \cup \Delta_0^{\sigma \wedge Y}](\varphi | \sigma \wedge Y)$.

This immediately follows from the existence of a suitable Boolean automorphism exchanging $[X]$ and $[Y]$ while keeping $\mathcal{B}'$ stable, and the representation independence of JX-entailment notions. Furthermore, for any representation $\Sigma_R \cup \Delta_R$ of R, JX-revision always produces the same $\kappa \pi$-measure over the relevant language fragment $\mathcal{L}_{obj}$.

Theorem 73 (Representation invariance).
Let $R = JX[\Delta_R](.|\sigma_R) = JX[\Psi_R](.|\phi_R)$, $\Sigma^* \cup \Delta^* = (\Sigma_R \cup \Delta_R) \star_{pr} \Delta_0$, and $\Phi^* \cup \Psi^* = (\Phi_R \cup \Psi_R) \star_{pr} \Delta_0$. Then, for $\varphi \in \mathcal{L}_{obj}$,

- $JX[\Delta^*](\varphi | \sigma^*) = JX[\Psi^*](\varphi | \phi^*)$.

Here we can exploit the logical independence of X from the anterior formulas and especially the fact that the ranking construction within X only depends on the result, but not on the shifting details of the preceding JX-construction of R. We observe that we also have the possibility to pack together a whole sequence of revision steps in a single ranking construction and conditionalization step. Here we may use similar arguments as before.

Theorem 74 (Aggregation).
Let $\Delta_0, \ldots, \Delta_n$ be a sequence of default sets and $X_1, \ldots, X_n$ be different new propositional variables in $\mathcal{L}_{inp}$. Then we have over $\mathcal{B}_{obj}$,

- $R_0 \star_{jx} \Delta_0 \ldots \star_{jx} \Delta_n = JX[\Delta_0 \cup \Delta_1^{X_1} \ldots \cup \Delta_n^{X_1 \wedge \ldots \wedge X_n}](.|X_1 \wedge \ldots \wedge X_n)$.

It is easy to see that the JX-revision functions are quite natural extensions of minimal Spohn revision to finite sets of default conditionals Δ interpreted as epistemic constraints.

Theorem 75 (Spohn extension).
For all minimally constructible default entailment notions $\mathrel{\vert\!\sim}^{JX}$,

- $R \star_{msp} \varphi = R \star_{jx} \{\mathbf{T} \Rightarrow \varphi\}$.

8   Principles and Illustrations

For reasons of space, and given the importance of system JZ, we are going to use JZ-revision to explore the power of the default projection paradigm for iterated revision. We start with a very simple technical example to illustrate the procedure. Let $\Delta_1 = \{\mathbf{T} \Rightarrow \neg\varphi, \mathbf{T} \Rightarrow \neg\psi\}$ and $\Delta_2 = \{\mathbf{T} \Rightarrow \varphi, \mathbf{T} \Rightarrow \psi\}$, where $\varphi, \psi \in \mathcal{L}_{obj}$ and $X, Y \in \mathcal{L}_{inp}$ are logically independent. The following equalities are valid over $\mathcal{B}_{obj}$ and produce rather reasonable revision outcomes.

- $R_1 = R_0 \star_{jz} \Delta_1 = JZ[\{X \Rightarrow \neg\varphi, X \Rightarrow \neg\psi\}](.|X) =$
  $(R_0 + 1[X \wedge \varphi] + 1[X \wedge \psi])(.|X) = R_0 + 1[\varphi] + 1[\psi].$

  $R_2 = R_1 \star_{jz} \Delta_2 = JZ[\Delta_1 \cup \{X \Rightarrow \varphi, X \Rightarrow \psi\}](.|X) =$
  $(R_0 + 1[\varphi] + 1[\psi] + 2[X \wedge \neg\varphi] + 2[X \wedge \neg\psi])(.|X) = R_0 + 1[\neg\varphi] + 1[\neg\psi].$

- $R_0 \star_{jz} \Delta_1 \star_{jz} \Delta_2 =$
  $JZ[\{X \Rightarrow \neg\varphi, X \Rightarrow \neg\psi\} \cup \{X \wedge Y \Rightarrow \varphi, X \wedge Y \Rightarrow \psi\}](.|X \wedge Y) =$
  $(R_0 + 1[X \wedge \varphi] + 1[X \wedge \psi] + 2[X \wedge Y \wedge \neg\varphi] + 2[X \wedge Y \wedge \neg\psi])(.|X \wedge Y) = R_2.$

The next examples illustrate the sensitivity of JZ-revision to the ordering and grouping of default evidence. Let φ, ψ be as above.

- $R_0 \star_{jz} \{T \Rightarrow \varphi, \varphi \Rightarrow \psi\} \star_{jz} \{\psi \Rightarrow \neg\varphi\} =$
  $R_0 \star_{jz} \{T \Rightarrow \varphi\} \star_{jz} \{\varphi \Rightarrow \psi\} \star_{jz} \{\psi \Rightarrow \neg\varphi\} =$
  $R_0 \star_{jz} \{\varphi \Rightarrow \psi\} \star_{jz} \{T \Rightarrow \varphi\} \star_{jz} \{\psi \Rightarrow \neg\varphi\} = R_0 + 1[\varphi \wedge \psi].$

- $R_0 \star_{jz} \{T \Rightarrow \varphi\} \star_{jz} \{\psi \Rightarrow \neg\varphi\} \star_{jz} \{\varphi \Rightarrow \psi\} =$
  $R_0 \star_{jz} \{T \Rightarrow \varphi\} \star_{jz} \{\varphi \Rightarrow \psi, \psi \Rightarrow \neg\varphi\} = R_0 + 1[\varphi] + 1[\varphi \wedge \neg\psi].$

We haven't yet seen any intuitive benchmark examples for default set revision in the literature. Although the discusssion and testing of concrete patterns has helped to disqualify formalisms, they are poor guides when it comes to construct better proposals. Among the reasons are their singular character and their possible implicit dependence from world knowledge. Therefore, the present paper focuses on general principles. They are usually better suited to evaluate the overall qualities of a revision methodology.

We start with a ranking version of the Shore/Johnson axioms for discrete probabilistic revision which characterize cross-entropy minimization [12]. However, it is important to note that in our coarse-grained ranking context, these natural requirements are a bit less powerful and do no longer provide a canonical characterization.

Qualitative SJ-principles (QSJ): Let $\delta = \wedge \Delta^{\rightarrow}$ and $R(\delta) \neq \infty$.

1. **Identity.** $R \star \Delta = R$ iff $R \models_{\kappa\pi} \Delta$.
2. **Invariance.** $(R \star \Delta)^{\pi'} = R^{\pi'} \star \Delta^{\pi'}$ where $R^{\pi'} = R \circ \pi'^{-1}$, for any boolean automorphism π and compatible representation map π'.
3. **System independence.** $(R_1 \times R_2) \star (\Delta_1 \cup \Delta_2) = (R_1 \star \Delta_1)^+ + (R_2 \star \Delta_2)^+$, where $\Delta_i \subseteq \mathcal{L}_i(\Rightarrow)$, the $\mathcal{L}_i$ have disjoint Va_i, $dom(R_i) = \mathcal{B}_i$, the product $\kappa\pi$-measures are defined using the analogy with probability theory, and R_i^+ is the canonical uniform extension of R_i to $dom(R_1 \times R_2)$.
4. **Subset independence.** $R\star(\Delta_0^{\eta_0} \cup \ldots \cup \Delta_n^{\eta_n} \cup \Delta)(.|\eta_i) = R(.|\eta_i) \star \Delta_i^{\eta_i}$, where the $[\eta_i]$ form a partition over $\mathcal{B}$ s.t. $NM[R, \Delta](\eta_i) \neq \infty$ (i.e. $R(\eta_i) \neq \infty$ and $\Delta \not\models_{\kappa\pi} \mathbf{F}$), the defaults in Δ refer to disjunctions of η_i, and $R(\delta_i|\eta_i) \neq \infty$ for $\delta_i = \wedge \Delta_i^{\rightarrow}$.

Theorem 81 (QSJ for JZ-revision).
$\star_{jz}$ verifies identity, invariance, system and subset independence.

The first two postulates are obvious. The third one follows from the use of additive constructions which do not interfere with each other. For subset independence we observe that, by additive shifting and given the assumptions about Δ, the defaults from Δ affect each $[\eta_i]$ uniformly, without trivializing it, whereas those from Δ_i are only active within $[\eta_i]$. Furthermore, the JZ-construction for each Δ_i is independent from those for the other Δ_j and Δ. This entails the result. However, the construction for Δ may well depend on the local constructions for the Δ_i because proper revision with Δ_i may increase the rank of $[\eta_i]$.

Axioms for iterated conditional revision extending those for propositional revision [3] have also been proposed by Kern-Isberner [5] [6]. Because for single defaults, $\star_{jx}$-revision is reconstructible by Spohn-style revision, the reference formalism for choosing these principles, it is hardly surprising that her main axioms, CR1-CR7, are valid here as well (but would fail for projective default revision w.r.t. normality maximization approaches like system Z). Only the add-on postulate CR8 fails because of our minimal shifting requirement.

Theorem 82 (CR-postulates).
The $\star_{jx}$ verify CR1-CR7 for single conditional revision, but not CR8.

In addition, Kern-Isberner has suggested five principles CSR1 - CSR5 for conditional set revision. CSR1 encodes success, CSR2 is subsumed by identity - QSJ1, CSR3 imposes global semanticality, CSR4 constitutes a further strengthening of CSR3, and CSR5 reflects a feature of cross-entropy minimization. More formally,

- CSR3: $Mod_{\kappa\pi}(\Delta) = Mod_{\kappa\pi}(\Delta')$ implies $R \star \Delta = R \star \Delta'$,
- CSR4: $R \star \Delta \models_{\kappa\pi} \Delta'$ and $R \star \Delta' \models_{\kappa\pi} \Delta$ implies $R \star \Delta = R \star \Delta'$,
- CSR5: $R \star (\Delta \cup \Delta') = R \star \Delta \star (\Delta \cup \Delta')$.

If we assume $R(\delta) \neq \infty$, CSR1, CSR2 are valid for $\star_{jx}$. But CSR3 and therefore CSR4 fail because for independent φ, ψ, $R_0 \star_{jx} \{\mathbf{T} \Rightarrow \varphi, \mathbf{T} \Rightarrow \psi\} \neq R_0 \star_{jx} \{\mathbf{T} \Rightarrow \varphi \wedge \psi\}$, although both default sets are $\kappa\pi$-equivalent. In fact, a slight modification of the exceptional inheritance paradox argument [18] [19] shows that the violation of CSR3 is a direct consequence of the QSJ-principles. We just have to assume – rather modestly – that $R_0 \star \{\mathbf{T} \Rightarrow \varphi\} = R_0 \star \{\mathbf{T} \Rightarrow \neg\varphi\} \star \{\mathbf{T} \Rightarrow \varphi\}$, which blocks circumscriptive solutions pushing the exceptional parts of defaults, e.g. $[\varphi]$ for $\mathbf{T} \Rightarrow \neg\varphi$, to ∞. But $\star_{jz}$ satisfies CSR3^{loc}, a weaker, local version of CSR3 which requires also structural $\kappa\pi$-equivalence. Let $Mod_{\kappa\pi}^{sg}(\Delta) = \{Mod_{\kappa\pi}(\delta) \mid \delta \in \Delta\}$.

- CSR3^{loc}: $Mod_{\kappa\pi}^{sg}(\Delta) = Mod_{\kappa\pi}^{sg}(\Delta')$ implies $R \star \Delta = R \star \Delta'$.

On the other hand, $\star_{jz}$ rejects CSR5 because for independent φ, ψ,

- $R_0 \star_{jz} \{\mathbf{T} \Rightarrow \varphi, \mathbf{T} \Rightarrow \varphi \wedge \psi\} = R_0 + 1[\neg\varphi \vee \neg\psi] \neq$
  $R_0 \star_{jz} \{\mathbf{T} \Rightarrow \varphi\} \star_{jz} \{\mathbf{T} \Rightarrow \varphi, \mathbf{T} \Rightarrow \varphi \wedge \psi\} = R_0 + 1[\neg\varphi \vee \neg\psi] + 1[\neg\varphi]$.

9   Conclusion

In this paper, we have introduced projective default revision, a general mechanism for generating revision methods from default formalisms, which is applicable to multiple conditional evidence (epistemic constraints). This work, which is still a bit experimental, may pave the way for a broader exchange of techniques and results between proper default reasoning and iterated belief revision. Our current investigation has focused on well-behaved semantic accounts which are based on minimal canonical ranking constructions, in particular system JZ. We have seen that the resulting JX-revision strategies extend minimal Spohn revision and verify major desiderata compatible with implicit independence assumptions and the minimal information philosophy.

A promising area for future research may be the comparison of these approaches with $\kappa\pi$-MCE. This is the ranking-based counterpart of cross-entropy minimization which translates the ranking conditions into nonstandard probability constraints, determines the minimum-cross-entropy model, and transfers the result back into the ranking framework [16] [19]. The surprising behaviour of $\kappa\pi$-MCE and $\star_{jz}$ in some contexts suggests that we should also take a look at default revision strategies based on ranking constructions starting from nonuniform priors.

References

1. Benferhat, S.; Cayrol, C.; Dubois, D.; Lang, J.; Prade, H.: Inconsistency management and prioritized syntax-based entailment. In *Proc. of IJCAI 93*. Morgan Kaufmann, 1993.
2. Benferhat, S.; Saffiotti, A.; Smets, P.: Belief functions and default reasoning. In *Artificial Intelligence*, 122 : 1-69, 2000.

3. Darwiche, A.; Pearl, J.: On the logic of iterated belief revision. In *Artificial Intelligence*, 89:1-29, 1997.

4. Dubois, D.; Prade, H.: Possibility theory : qualitative and quantitative aspects. In *Quantified Representation of Uncertainty and Imprecision, Handbook on Defeasible Reasoning and Uncertainty Management Systems - Vol. 1*, P. Smets (ed.). Kluwer, 1998.

5. Kern-Isberner, G.: Postulates for conditional belief revision. In *Proc. of IJCAI 99*. Morgan Kaufmann, 1999.

6. Kern-Isberner, G.: *Conditionals in nonmonotonoic reasoning and belief revision, LNAI 2087*. Springer, 2001.

7. Kraus, S.; Lehmann, D.; Magidor, M.: Nonmonotonic reasoning, preferential models and cumulative logics. In *Artificial Intelligence*, 44:167-207, 1990.

8. Gaerdenfors, P.; Makinson, D.: Nonmonotonic Inference Based on Expectations. In *Artificial Intelligence*, 65(2):197-245, 1994.

9. Nebel, B.: Base Revision Operations and Schemes: Semantics, Representation, and Complexity. In *Proc. of ECAI 94*. John Wiley, 1994.

10. Pearl, J.: System Z: a natural ordering of defaults with tractable applications to nonmonotonic reasoning. In *Proc. of the Third Conference on Theoretical Aspects of Reasoning about Knowledge*. Morgan Kaufmann, 1990.

11. Rott, H.: *Change, Choice and Inference: A Study of Belief Revision and Nonmonotonic Reasoning*. Oxford Logic Guides, Vol. 42. Oxford University Press, Oxford, 2001.

12. Shore, J.E.; Johnson, R.W.: Axiomatic derivation of the principle of cross-entropy minimization. In *IEEE Transactions on Information Theory*, IT-26(1):26-37, 1980.

13. Spohn, W.: Ordinal conditional functions: a dynamic theory of epistemic states. In *Causation in Decision, Belief Change, and Statistics*, W.L.Harper, B.Skyrms (eds.). Kluwer, 1988.

14. Spohn, W.: A general non-probabilistic theory of inductive reasoning. In *Uncertainty in Artificial Intelligence 4*, R.D.Shachter et al. (eds.). North-Holland, Amsterdam, 1990.

15. Weydert, E.: General belief measures. In *Proc. of UAI 94*. Morgan Kaufmann, 1994.

16. Weydert, E.: Defaults and infinitesimals. Defeasible inference by non-archimdean entropy maximization. In *Proc. of UAI 95*. Morgan Kaufmann, 1995.

17. Weydert, E.: System J - Revision entailment. In *Proc. of FAPR 96*. Springer, 1996.

18. Weydert, E.: System JZ - How to build a canonical ranking model of a default knowledge base. In *Proc. of KR 98*. Morgan Kaufmann, 1998.

19. Weydert, E.: System JLZ - Rational default reasoning by minimal ranking constructions. In *Journal of Applied Logic*, Elsevier, 2003.

On the Logic of Iterated Non-prioritised Revision

Richard Booth

Department of Computer Science, University of Leipzig,
Augustusplatz 10/11, 04109 Leipzig, Germany
booth@informatik.uni-leipzig.de

Abstract. We look at iterated non-prioritised belief revision, using as a starting point a model of non-prioritised revision, similar to Makinson's *screened revision*, which assumes an agent keeps a set of *core* beliefs whose function is to block certain revision inputs. We study postulates for the iteration of this operation. These postulates generalise some of those which have previously been proposed for iterated AGM ("prioritised") revision, including those of Darwiche and Pearl. We then add a second type of revision operation which allows the core itself to be revised. Postulates for the iteration of this operator are also provided, as are rules governing mixed sequences of revisions consisting of both regular and core inputs. Finally we give a construction of both a regular and core revision operator based on an agent's *revision history*. This construction is shown to satisfy most of the postulates.

1 Introduction and Preliminaries

The most popular basic framework for the study of belief revision has been the one due to Alchourrón, Gärdenfors and Makinson (AGM) [1, 10]. This framework has been subjected in more recent years to several different extensions and refinements. Two of the most interesting of these have been the study of so-called *non-prioritised* revision [2, 12, 13, 19], i.e., revision in which the input sentence is not necessarily accepted, and of *iterated* revision [3, 5, 7, 8, 17, 21], i.e., the study of the behaviour of an agent's beliefs under a *sequence* of revision inputs. However, most of the extensions in the former group are concerned only with single-step revision. Similarly, most of the contributions to the area of iterated AGM revision are in the setting of normal, "prioritised" revision in which the input sentence is always accepted. However the question of *iterated non-prioritised revision* is certainly an interesting one, as can be seen from the following example.[1]

Example 1. Your six-year-old son comes home from school and tells you that today he had lunch at school with King Gustav. Given your expectations of the

---

[1] Based on an example given in [9–Ch. 7] to illustrate non-prioritised revision.

G. Kern-Isberner, W. Rödder, and F. Kulmann (Eds.): WCII 2002, LNAI 3301, pp. 86–107, 2005.

King's lunching habits, you dismiss this information as a product of your son's imagination, i.e., you reject this information. But then you switch on the TV news and see a report that King Gustav today made a surprise visit to a local school. Given *this* information, your son's story doesn't seem quite so incredible as it did. Do you now believe your son's information?

What this example seems to show is that information which is initially rejected (such as your son's information) may still have an influence on the results of *subsequent* revisions. In particular if subsequent information lends support to it, then this could cause a *re-evaluation* of the decision to reject, possibly even leading the input to be accepted *retrospectively*. The main purpose of this paper is to study patterns of iterated non-prioritised revision such as these.

We will use as a starting point one particular model of non-prioritised revision, the idea behind which first appeared behind Makinson's *screened revision* [19], and then again as a special case of Hansson et al.'s *credibility-limited revision* [13]. It is that an agent keeps, as a subset of his beliefs, a set of *core beliefs* which he considers "untouchable". This set of core beliefs then acts as the determiner as to whether a given revision input is accepted or not: if an input ϕ is consistent with the core beliefs then the agent accepts the input and revises his belief set by ϕ using a normal AGM revision operator. On the other hand if ϕ contradicts the core beliefs then the agent rejects ϕ rather than give up any of the core beliefs. In this case his belief set is left undisturbed. We will see that this quite simple model will already give us a flavour of some of the interesting issues at stake. For a start, to be able to iterate this operator we need to say not only what the new belief set is after a revision, but also what the new *core* belief set is.

The explicit inclusion in an agent's epistemic state of a second set of beliefs to represent the agent's core beliefs invites the question of what would happen if this set too were to be subject to revision by external inputs, just like the normal belief set. This question will also be taken up in this paper. Thus we will have *two* different types of revision operator existing side-by-side: the usual operators described above, which we shall call *regular revision* operators, and *core revision* operators. Again both single-step and iterated core revision will be looked at. We also look at the particularly interesting possibility of performing *mixed* sequences of revisions consisting of both regular and core revisions.

The plan of the paper is as follows. We start in Sect. 2 by briefly describing the revision operators of AGM and introducing our primitive notions of *epistemic state* and *epistemic frame*. Then, in Sect. 3, we look at regular revision. We consider postulates for both the single-step and the iterated case. The latter will involve adapting some well-known postulates from the literature on iterated AGM revision – principally those proposed by Darwiche and Pearl [8] – to our non-prioritised situation. In Sect. 4 we look at some postulates for single-step and iterated core revision. Some possible rules for mixed sequences of revision inputs will be looked at in Sect. 5. Then in Sect. 6, using a particular representation of an agent's epistemic state, we provide a construction of both a regular and a core revision operator. These operators are shown to display most of the behaviour described by our postulates. We conclude in Sect. 7.

1.1   Preliminaries

We assume a propositional language generated from finitely many propositional variables. Let L denote the set of sentences of this language. Cn denotes the classical logical consequence operator. We write $Cn(\theta)$ rather than $Cn(\{\theta\})$ for $\theta \in L$ and use L^+ to denote the set of all classically consistent sentences. Formally, a *belief set* will be any set of sentences $K \subseteq L$ which is (i) consistent, i.e., $Cn(K) \neq L$, and (ii) deductively closed, i.e., $K = Cn(K)$. We denote the set of all belief sets by $\mathcal{K}$. Given $K \in \mathcal{K}$ and $\phi \in L$, we let $K + \phi$ denote the *expansion* of K by ϕ, i.e., $K + \phi = Cn(K \cup \{\phi\})$.

We let $\mathcal{W}$ denote the set of propositional worlds associated to L, i.e., the set of truth-assignments to the propositional variables in L. For any set $X \subseteq L$ of sentences we denote by $[X]$ the set of worlds in $\mathcal{W}$ which satisfy all the sentences in X (writing $[\phi]$ rather than $[\{\phi\}]$ for the case of singletons). Given a set $S \subseteq \mathcal{W}$ of worlds we write $Th(S)$ to denote the set of sentences in L which are satisfied by all the worlds in S. A *total pre-order* on $\mathcal{W}$ is any binary relation $\leq$ on $\mathcal{W}$ which is reflexive, transitive and connected (for all $w_1, w_2 \in \mathcal{W}$ either $w_1 \leq w_2$ or $w_2 \leq w_1$). For each such order $\leq$ we let $<$ denote its strict part and $\sim$ denote its symmetric part, i.e., we have $w_1 < w_2$ iff both $w_1 \leq w_2$ and $w_2 \not\leq w_1$, and $w_1 \sim w_2$ iff both $w_1 \leq w_2$ and $w_2 \leq w_1$. Given a total pre-order $\leq$ on $\mathcal{W}$ and given $S \subseteq \mathcal{W}$ we will use $\min(S, \leq)$ to denote the set of worlds which are minimal in S under $\leq$, i.e., $\min(S, \leq) = \{w \in S \mid w \leq w' \text{ for all } w' \in S\}$. We will say that a total pre-order $\leq$ on $\mathcal{W}$ is *anchored on S* if S contains precisely the minimal elements of $\mathcal{W}$ under $\leq$, i.e., if $S = \min(\mathcal{W}, \leq)$.

2   AGM and Epistemic Frames

The original AGM theory of revision is a theory about how to revise a fixed generic belief set K by any given sentence. In this paper we simplify by assuming all revision input sentences are consistent. (For this reason the usual, but for us vacuous, pre-condition "if ϕ is consistent" is absent from our formulation of AGM postulate **(K*5)** below.) At the centre of this theory is the list of *AGM revision postulates (relative to K)* which seek to rationally constrain the outcome of such a revision. Using $K * \phi$ as usual to denote the result of revising K by $\phi \in L^+$, the full list of these postulates is:

(K*1) $K * \phi = Cn(K * \phi)$
(K*2) If $\phi_1 \leftrightarrow \phi_2 \in Cn(\emptyset)$ then $K * \phi_1 = K * \phi_2$
(K*3) $\phi \in K * \phi$
(K*4) If $\neg\phi \notin K$ then $K * \phi = K + \phi$
(K*5) $K * \phi$ is consistent
(K*6) If $\neg\phi \notin K * \theta$ then $K * (\theta \wedge \phi) = (K * \theta) + \phi$

Note the presence of **(K*3)** – the "success" postulate – which says that the input sentence is always accepted. For a given belief set K, we shall call any function $*$ which satisfies the above postulates a *simple AGM revision function for K*. It is well-known that requiring these postulates to hold is equivalent to requiring that, when performing an operation of revision on his belief set K, an agent acts as though he has a total pre-order $\leq$ on the set of worlds $\mathcal{W}$ representing some subjective assessment of their relative *plausibility*, with the worlds in $[K]$ being the most plausible, i.e., $\leq$-minimal. Given the input sentence ϕ, the agent then takes as his new belief set the set of sentences true in all the most plausible worlds satisfying ϕ. Precisely we have:

Theorem 1 ([11, 15]). *Let $K \in \mathcal{K}$ and $*$ be an operator which, for each $\phi \in L^+$, returns a new set of sentences $K * \phi$. Then $*$ is a simple AGM revision function for K iff there exists some total pre-order $\leq$ on $\mathcal{W}$, anchored on $[K]$, such that, for all $\phi \in L^+$, $K * \phi = Th(\min([\phi], \leq))$.*

In this paper we will make extensive use of the above equivalence.

2.1   Epistemic Frames

One of the morals of the work already done on attempting to extend the AGM framework to cover iterated revision (see, e.g. [8, 14, 18, 21]) is that, in order to be able to formally say anything interesting about iterated revision, it is necessary to move away from the AGM representation of an agent's epistemic state as a simple belief set, and instead assume that revision is carried out on some more comprehensive object of which the belief set is but one ingredient. We will initially follow [8] in taking an abstract view of epistemic states. As in that paper, we assume a set Ep of epistemic states as primitive and assume that from each such state $\mathbb{E} \in Ep$ we can extract a belief set $\triangle(\mathbb{E})$ representing the agent's regular beliefs in $\mathbb{E}$. Unlike in [8] however, we also explicitly assume that we can extract a second belief set $\blacktriangle(\mathbb{E}) \subseteq \triangle(\mathbb{E})$ representing the agent's *core* beliefs in $\mathbb{E}$. This is all captured by the definition of an *epistemic frame*:

Definition 1. *An* epistemic frame *is a triple $\langle Ep, \triangle, \blacktriangle \rangle$, where Ep is a set, whose elements will be called* epistemic states, *and $\triangle : Ep \to \mathcal{K}$ and $\blacktriangle : Ep \to \mathcal{K}$ are functions such that, for all $\mathbb{E} \in Ep$, $\blacktriangle(\mathbb{E}) \subseteq \triangle(\mathbb{E})$.*

For most of this paper we will assume that we are working with some arbitrary, but fixed, epistemic frame $\langle Ep, \triangle, \blacktriangle \rangle$ in the background. Not until Sect. 6 will we get more specific and employ a more concrete representation of an epistemic frame. An obvious fact which is worth keeping in mind is that, since $\blacktriangle(\mathbb{E}) \subseteq \triangle(\mathbb{E})$, we always have $[\triangle(\mathbb{E})] \subseteq [\blacktriangle(\mathbb{E})]$. The set $\blacktriangle(\mathbb{E})$ can in general be *any* sub(belief)set of $\triangle(\mathbb{E})$. As two special cases, at opposite extremes, we have $\blacktriangle(\mathbb{E}) = Cn(\emptyset)$, i.e., the only core beliefs are the tautologies, and $\blacktriangle(\mathbb{E}) = \triangle(\mathbb{E})$, i.e., *all* regular beliefs are also core beliefs. One of our main aims in this paper will be to try and formulate rational constraints on the behaviour of both the regular beliefs $\triangle(\mathbb{E})$ **and** the core beliefs $\blacktriangle(\mathbb{E})$ under operations of change **to the underlying epistemic state** $\mathbb{E}$. We begin with the case when the operation of change is triggered by a regular belief input.

3   Regular Revision Inputs

In this section we consider the usual case where the revision input is a (consistent) sentence to be included in the regular belief set $\triangle(\mathbb{E})$. Given an epistemic state $\mathbb{E} \in Ep$ and a regular input $\phi \in L^+$, we shall let $\mathbb{E} \circ \phi$ denote the resulting epistemic state. We consider the single-step case and the iterated case in turn.

3.1   Single-Step Regular Revision

As indicated in the introduction, we follow the spirit of screened revision and assume that the new regular belief set $\triangle(\mathbb{E} \circ \phi)$ is given by

$$\triangle(\mathbb{E} \circ \phi) = \begin{cases} \triangle(\mathbb{E}) *_\triangle^\mathbb{E} \phi & \text{if } \neg\phi \notin \blacktriangle(\mathbb{E}) \\ \triangle(\mathbb{E}) \text{ otherwise.} \end{cases}$$

where, for each epistemic state $\mathbb{E}$, $*_\triangle^\mathbb{E}$ is a simple AGM revision function for $\triangle(\mathbb{E})$. This is also very similar to the definition of *endorsed core beliefs revision* in [13]. The difference is that in that paper the function $*_\triangle^\mathbb{E}$ is not assumed to satisfy the postulate **(K*6)** from Sect. 2. By Theorem 1, the above method is equivalent to assuming that for each $\mathbb{E}$ there exists some total pre-order $\leq_\triangle^\mathbb{E}$ on $\mathcal{W}$, anchored on $[\triangle(\mathbb{E})]$, such that

$$\triangle(\mathbb{E} \circ \phi) = \begin{cases} Th(\min([\phi], \leq_\triangle^\mathbb{E})) \text{ if } \neg\phi \notin \blacktriangle(\mathbb{E}) \\ \triangle(\mathbb{E}) \text{ otherwise.} \end{cases} \tag{1}$$

We remark that the subscript on $\leq_\triangle^\mathbb{E}$ does not actually denote the $\triangle$-function itself, but is merely a decoration to remind us that this order is being used to revise the *regular* beliefs in $\mathbb{E}$. We now make the following definition:

Definition 2. *Let $\circ : Ep \times L^+ \to Ep$ be a function. Then $\circ$ is a* regular revision operator *(on the epistemic frame $\langle Ep, \triangle, \blacktriangle \rangle$) if, for each $\mathbb{E} \in Ep$, there exists a total pre-order $\leq_\triangle^\mathbb{E}$ on $\mathcal{W}$, anchored on $[\triangle(\mathbb{E})]$, such that $\triangle(\mathbb{E} \circ \phi)$ may be determined as in (1) above. We call $\leq_\triangle^\mathbb{E}$ the* regular pre-order associated to $\mathbb{E}$ *(according to $\circ$).*

For some properties satisfied by this general type of construction the reader is referred to [13, 19]. One intuitive property which is not guaranteed to hold under the above definition as it stands[2] is the following, which essentially corresponds to the rule (Strong Regularity) from [13]:

(SR) $\blacktriangle(\mathbb{E}) \subseteq \triangle(\mathbb{E} \circ \phi)$

---

[2] For a counter-example suppose $\mathbb{E}$ is such that $\triangle(\mathbb{E}) = Cn(p)$ and $\blacktriangle(\mathbb{E}) = Cn(p \vee q)$ where p, q are distinct propositional variables, and suppose $*_\triangle^\mathbb{E}$ is the "trivial" simple AGM revision function for $\triangle(\mathbb{E})$ given by $\triangle(\mathbb{E}) *_\triangle^\mathbb{E} \phi = \triangle(\mathbb{E}) + \phi$ if $\neg\phi \notin \triangle(\mathbb{E})$, $\triangle(\mathbb{E}) *_\triangle^\mathbb{E} \phi = Cn(\phi)$ otherwise. Then $\blacktriangle(\mathbb{E}) \not\subseteq \triangle(\mathbb{E} \circ \neg p) = Cn(\neg p)$.

This postulate states that the set of core beliefs are retained as regular beliefs after revision, while leaving open the question of whether they are again retained as *core* beliefs. For this property to hold of a regular revision operator ○ we require that the pre-orders $\leq_\triangle^\mathbb{E}$ associated with each $\mathbb{E}$ satisfy an extra condition, namely that $\leq_\triangle^\mathbb{E}$ considers all the worlds in $[\blacktriangle(\mathbb{E})]$ as strictly more plausible than all the worlds *not* in $[\blacktriangle(\mathbb{E})]$.[3]

Proposition 1. *Let* ○ *be a regular revision operator. Then* ○ *satisfies* **(SR)** *iff, for each* $\mathbb{E} \in Ep$, $\leq_\triangle^\mathbb{E}$ *satisfies* $w_1 <_\triangle^\mathbb{E} w_2$ *whenever* $w_1 \in [\blacktriangle(\mathbb{E})]$ *and* $w_2 \notin [\blacktriangle(\mathbb{E})]$.

The reader may have noticed that, since inputs which contradict $\blacktriangle(\mathbb{E})$ are simply rejected, the $\leq_\triangle^\mathbb{E}$-ordering of the worlds outside of $[\blacktriangle(\mathbb{E})]$ never plays any role in determining the new regular belief set. It *will*, however, play a role later on when we come to look at core revision.

What effect should performing a regular revision ○ have on the core belief set? In this paper we take the position that ○ is concerned exclusively with changes to $\triangle(\mathbb{E})$, and so the core belief set does not change at all.

(X1) $\blacktriangle(\mathbb{E} \circ \phi) \subseteq \blacktriangle(\mathbb{E})$ (Core Non-expansion)
(X2) $\blacktriangle(\mathbb{E}) \subseteq \blacktriangle(\mathbb{E} \circ \phi)$ (Core Preservation)

Thus **(X1)**, respectively **(X2)**, says that no core beliefs are added, respectively lost, during an operation of regular revision.

Definition 3. *Let* ○ *be a regular revision operator. Then* ○ *is* core-invariant *iff* ○ *satisfies both* **(X1)** *and* **(X2)**.

Since clearly **(X2)** implies **(SR)**, we have that every core-invariant regular revision operator satisfies **(SR)**. The reasonableness of core-invariance may be questioned. For example a consequence of **(X2)** is that we automatically get that if $\neg\phi \in \blacktriangle(\mathbb{E})$ then $\phi \notin \triangle(\mathbb{E} \circ \phi \circ \phi \circ \cdots \circ \phi)$, and this holds regardless of how many times we revise by ϕ, be it one or one billion. It might be expected here that repeatedly receiving ϕ might have the effect of gradually "loosening" $\neg\phi$ from the core beliefs until eventually at some point it "falls out", leading ϕ to become acceptable. Similarly, rule **(X1)** precludes the situation in which repeated input of a non-core belief eventually leads to the admittance of that belief into the core. On the other hand there exist situations in which core-invariance does seem reasonable in these cases. An example is when the regular belief inputs are assumed to be coming from a single source throughout, i.e., the source is just repeating itself. Weaker alternatives to the rules **(X1)** and **(X2)** which come to mind are:

(wX1) $\blacktriangle(\mathbb{E} \circ \phi) \subseteq \blacktriangle(\mathbb{E}) + \phi$ (Weak Core Non-expansion)
(wX2) $\blacktriangle(\mathbb{E}) \subseteq \blacktriangle(\mathbb{E} \circ \phi) + \neg\phi$ (Weak Core Preservation)

In terms of propositional worlds, **(wX1)** is equivalent to requiring $[\blacktriangle(\mathbb{E})] \cap [\phi] \subseteq [\blacktriangle(\mathbb{E} \circ \phi)]$, while **(wX2)**, which is reminiscent of the "recovery" postulate

---

[3] Due to space limitations, proofs are omitted from this version of the paper.

from belief contraction [10], is equivalent to requiring $[\blacktriangle(\mathbb{E} \circ \phi)] \subseteq [\blacktriangle(\mathbb{E})] \cup [\phi]$. Thus **(wX1)** says that, in the transformation of $[\blacktriangle(\mathbb{E})]$ into $[\blacktriangle(\mathbb{E} \circ \phi)]$, the only worlds which can possibly be *removed* from $[\blacktriangle(\mathbb{E})]$ are those in $[\neg\phi]$, while **(wX2)** says that the only worlds which can possibly be *added* are those in $[\phi]$. For this paper, however, we will assume that both **(X1)** and **(X2)** hold throughout, and so we will make no further reference to the above weaker versions.

3.2   Iterating Regular Revision

Now we consider iteration of $\circ$. How should $\blacktriangle(\mathbb{E})$ and $\triangle(\mathbb{E})$ behave under sequences of regular inputs? Clearly since we are accepting both **(X1)** and **(X2)** this question is already answered in the case of $\blacktriangle(\mathbb{E})$ — the core beliefs remain constant throughout. What about the regular beliefs $\triangle(\mathbb{E})$? Here we take our lead from the work on iterated AGM ("prioritised") revision by Darwiche and Pearl [8]. They suggest a list of four postulates to rationally constrain the beliefs under iterated AGM revision (we will write "$\mathbb{E} \circ \theta \circ \phi$" rather than "$(\mathbb{E} \circ \theta) \circ \phi$" etc.):

(C1)$_\triangle$ If $\phi \rightarrow \theta \in Cn(\emptyset)$ then $\triangle(\mathbb{E} \circ \theta \circ \phi) = \triangle(\mathbb{E} \circ \phi)$
(C2)$_\triangle$ If $\phi \rightarrow \neg\theta \in Cn(\emptyset)$ then $\triangle(\mathbb{E} \circ \theta \circ \phi) = \triangle(\mathbb{E} \circ \phi)$
(C3)$_\triangle$ If $\theta \in \triangle(\mathbb{E} \circ \phi)$ then $\theta \in \triangle(\mathbb{E} \circ \theta \circ \phi)$
(C4)$_\triangle$ If $\neg\theta \notin \triangle(\mathbb{E} \circ \phi)$ then $\neg\theta \notin \triangle(\mathbb{E} \circ \theta \circ \phi)$

Briefly, these postulates can be explained as follows: The rule **(C1)**$_\triangle$ says that if two inputs are received, the second being more specific than the first, then the first is rendered redundant (at least regarding its effects on the regular belief set). Rule **(C2)**$_\triangle$ says that if two contradictory inputs are received, then the most recent one prevails. Rule **(C3)**$_\triangle$ says that an input θ should be in the regular beliefs after receiving the subsequent input ϕ if θ would have been believed given input ϕ to begin with. Finally **(C4)**$_\triangle$ says that if θ is not contradicted after receipt of input ϕ, then it should still be uncontradicted if input ϕ is preceded by input θ itself.[4] Which of these postulates are suitable for core-invariant regular revision? While **(C3)**$_\triangle$ and **(C4)**$_\triangle$ seem to retain their validity in our setting, there is a slight problem with **(C1)**$_\triangle$ and **(C2)**$_\triangle$ concerning the case when ϕ is taken to be a core-contravening sentence, i.e., when $\neg\phi \in \blacktriangle(\mathbb{E}) = \blacktriangle(\mathbb{E} \circ \theta)$. Consider momentarily the following two properties:

(wC1)$_\triangle$ If $\neg\phi \in \blacktriangle(\mathbb{E})$ and $\phi \rightarrow \theta \in Cn(\emptyset)$ then $\triangle(\mathbb{E} \circ \theta) = \triangle(\mathbb{E})$
(wC2)$_\triangle$ If $\neg\phi \in \blacktriangle(\mathbb{E})$ and $\phi \rightarrow \neg\theta \in Cn(\emptyset)$ then $\triangle(\mathbb{E} \circ \theta) = \triangle(\mathbb{E})$

Then it can easily be shown that any core-invariant regular revision operator $\circ$ which satisfies **(C1)**$_\triangle$, respectively **(C2)**$_\triangle$, also satisfies **(wC1)**$_\triangle$, respectively **(wC2)**$_\triangle$. However one can easily find examples of core-invariant regular revision operators which fail to satisfy the latter two properties. For example let p and q be propositional variables and suppose $\mathbb{E}$ is such that $\blacktriangle(\mathbb{E}) = \triangle(\mathbb{E}) = Cn(\neg p)$.

---

[4] We remark that these postulates have not been *totally* immune to criticism in the literature. In particular **(C2)**$_\triangle$ is viewed by some as problematic (see [5, 7, 17]).

Then clearly we have $\neg p \in \blacktriangle(\mathbb{E})$ and $p \rightarrow (p \vee q), p \rightarrow \neg(\neg p \wedge q) \in Cn(\emptyset)$. But $\triangle(\mathbb{E} \circ (p \vee q)) = \triangle(\mathbb{E}) + (p \vee q) = Cn(\neg p \wedge q)$ (contradicting $(\mathbf{wC1})_\triangle$), while $\triangle(\mathbb{E} \circ (\neg p \wedge q)) = \triangle(\mathbb{E}) + (\neg p \wedge q) = Cn(\neg p \wedge q)$ (contradicting $(\mathbf{wC2})_\triangle$). Thus we conclude that $(\mathbf{C1})_\triangle$ and $(\mathbf{C2})_\triangle$ are not suitable as they stand. Instead we propose to modify them so that they apply only when $\neg \phi \notin \blacktriangle(\mathbb{E})$.

$(\mathbf{C1'})_\triangle$ If $\neg \phi \notin \blacktriangle(\mathbb{E})$ and $\phi \rightarrow \theta \in Cn(\emptyset)$ then $\triangle(\mathbb{E} \circ \theta \circ \phi) = \triangle(\mathbb{E} \circ \phi)$
$(\mathbf{C2'})_\triangle$ If $\neg \phi \notin \blacktriangle(\mathbb{E})$ and $\phi \rightarrow \neg \theta \in Cn(\emptyset)$ then $\triangle(\mathbb{E} \circ \theta \circ \phi) = \triangle(\mathbb{E} \circ \phi)$

We tend to view $(\mathbf{C1'})_\triangle$, $(\mathbf{C2'})_\triangle$, $(\mathbf{C3})_\triangle$ and $(\mathbf{C4})_\triangle$ as being minimal conditions on iterated regular revision. An interesting consequence of $(\mathbf{C2'})_\triangle$ is revealed by the following proposition.

Proposition 2. *Let $\circ$ be a core-invariant regular revision operator which satisfies $(\mathbf{C2'})_\triangle$. Then, for all $\mathbb{E} \in Ep$ and $\theta, \phi \in L^+$, we have that $\neg \theta \in \blacktriangle(\mathbb{E})$ implies $\triangle(\mathbb{E} \circ \theta \circ \phi) = \triangle(\mathbb{E} \circ \phi)$.*

The above proposition says that not only does revising by a core-contravening sentence θ have no effect on the regular belief set, it also has no impact on the result of revising by any subsequent *regular* inputs. (As we will see later, this does not necessarily mean that core-contravening regular inputs are *totally* devoid of impact.)

As in our current set-up, Darwiche and Pearl assume that the new belief set $\triangle(\mathbb{E} \circ \theta)$ resulting from the single revision by θ is determined AGM-style by a total pre-order $\leq_\triangle^\mathbb{E}$ anchored on $[\triangle(\mathbb{E})]$. Likewise the new belief set $\triangle(\mathbb{E} \circ \theta \circ \phi)$ following a *subsequent* revision by ϕ is then determined by the total pre-order $\leq_\triangle^{\mathbb{E} \circ \theta}$ anchored on $[\triangle(\mathbb{E} \circ \theta)]$. Thus the question of which properties of iterated revision are satisfied is essentially the same as asking what the new pre-order $\leq_\triangle^{\mathbb{E} \circ \theta}$ looks like. In a result in [8], Darwiche and Pearl show how each of their postulates $(\mathbf{C1})_\triangle$–$(\mathbf{C4})_\triangle$ regulates a different aspect of the relationship between $\leq_\triangle^\mathbb{E}$ and the new regular pre-order $\leq_\triangle^{\mathbb{E} \circ \theta}$. The following proposition, which may be viewed as a *generalisation* of Darwiche and Pearl's result (roughly speaking, Darwiche and Pearl are looking at the special case when $\blacktriangle(\mathbb{E}) = Cn(\emptyset)$), does the same for $(\mathbf{C1'})_\triangle$, $(\mathbf{C2'})_\triangle$, $(\mathbf{C3})_\triangle$ and $(\mathbf{C4})_\triangle$ in our non-prioritised setting.

Proposition 3. *Let $\circ$ be a core-invariant regular revision operator. Then $\circ$ satisfies $(\mathbf{C1'})_\triangle$, $(\mathbf{C2'})_\triangle$, $(\mathbf{C3})_\triangle$ and $(\mathbf{C4})_\triangle$ iff each of the following conditions holds for all $\mathbb{E} \in Ep$ and $\theta \in L^+$:*

(1) For all $w_1, w_2 \in [\blacktriangle(\mathbb{E})] \cap [\theta]$, $w_1 \leq_\triangle^{\mathbb{E} \circ \theta} w_2$ iff $w_1 \leq_\triangle^\mathbb{E} w_2$
(2) For all $w_1, w_2 \in [\blacktriangle(\mathbb{E})] \cap [\neg\theta]$, $w_1 \leq_\triangle^{\mathbb{E} \circ \theta} w_2$ iff $w_1 \leq_\triangle^\mathbb{E} w_2$
(3) For all $w_1, w_2 \in [\blacktriangle(\mathbb{E})]$, if $w_1 \in [\theta], w_2 \in [\neg\theta]$ and $w_1 <_\triangle^\mathbb{E} w_2$, then
$$w_1 <_\triangle^{\mathbb{E} \circ \theta} w_2$$
(4) For all $w_1, w_2 \in [\blacktriangle(\mathbb{E})]$, if $w_1 \in [\theta], w_2 \in [\neg\theta]$ and $w_1 \leq_\triangle^\mathbb{E} w_2$, then
$$w_1 \leq_\triangle^{\mathbb{E} \circ \theta} w_2$$

Thus, according to the above proposition, $(\mathbf{C1'})_\triangle$ corresponds to the requirement that, in the transformation from $\leq_\triangle^\mathbb{E}$ to $\leq_\triangle^{\mathbb{E} \circ \theta}$, the relative ordering between the $[\theta]$-worlds in $[\blacktriangle(\mathbb{E})]$ remains unchanged. $(\mathbf{C2'})_\triangle$ corresponds to the same requirement but with regard to the $[\neg\theta]$-worlds in $[\blacktriangle(\mathbb{E})]$. $(\mathbf{C3})_\triangle$ corresponds to the requirement that if a given $[\theta]$-world in $[\blacktriangle(\mathbb{E})]$ was regarded as strictly more plausible than a given $[\neg\theta]$-world in $[\blacktriangle(\mathbb{E})]$ *before* receipt of the input θ, then this relation should be preserved *after* receipt of θ. Finally $(\mathbf{C4})_\triangle$ matches the same requirement as $(\mathbf{C3})_\triangle$, but with "at least as plausible as" substituted for "strictly more plausible than". Note how each property only constrains the transformation from $\leq_\triangle^{\mathbb{E}}$ to $\leq_\triangle^{\mathbb{E}\circ\theta}$ **within** $[\blacktriangle(\mathbb{E})]$. We will later see some conditions which constrain the movement of the other worlds.

The Darwiche and Pearl postulates form our starting point in the study of iterated revision. However, other postulates have been suggested. In particular, another postulate of interest which may be found in the literature on iterated AGM revision (cf. the rule (Recalcitrance) in [21]) is:

$(\mathbf{C5})_\triangle$ If $\phi \to \neg\theta \notin Cn(\emptyset)$ then $\theta \in \triangle(\mathbb{E} \circ \theta \circ \phi)$

Note that this postulate is in fact a strengthening of $(\mathbf{C3})_\triangle$ and $(\mathbf{C4})_\triangle$. (This will also soon follow from Proposition 5.) In fact $(\mathbf{C5})_\triangle$ might just have well have been called "strong success" since it also implies that $\theta \in \triangle(\mathbb{E} \circ \theta)$ for all $\theta \in L^+$. (Hint: substitute $\top$ for ϕ.) For this reason the postulate, as it stands, is obviously *not* suitable in our non-prioritised setting. However the following weaker version will be of interest to us:

$(\mathbf{C5'})_\triangle$ If $\phi \to \neg\theta \notin \blacktriangle(\mathbb{E})$ then $\theta \in \triangle(\mathbb{E} \circ \theta \circ \phi)$.

$(\mathbf{C5'})_\triangle$ entails that if, having received a regular input θ, we *do* decide to accept it, then we do so wholeheartedly (or as wholeheartedly as we can without actually elevating it to the status of a core belief!) in that the only way it can be dislodged from the belief set by a succeeding regular input is if that input contradicts it given the core beliefs $\blacktriangle(\mathbb{E})$. This postulate too can be translated into a somewhat plausible constraint on the new regular pre-order $\leq_\triangle^{\mathbb{E}\circ\theta}$.

Proposition 4. *Let $\circ$ be a core-invariant regular revision operator. Then $\circ$ satisfies* $(\mathbf{C5'})_\triangle$ *iff, for each $\mathbb{E} \in Ep$ and $\theta \in L^+$, and for all $w_1, w_2 \in [\blacktriangle(\mathbb{E})]$, if $w_1 \in [\theta]$ and $w_2 \in [\neg\theta]$ then $w_1 <_\triangle^{\mathbb{E}\circ\theta} w_2$.*

Thus $(\mathbf{C5'})_\triangle$ corresponds to the property that all the $[\theta]$-worlds in $[\blacktriangle(\mathbb{E})]$ are deemed strictly more plausible by $\leq_\triangle^{\mathbb{E}\circ\theta}$ than all the $[\neg\theta]$-worlds in $[\blacktriangle(\mathbb{E})]$. $(\mathbf{C5'})_\triangle$ is related to our previous postulates in the following way:

Proposition 5. *Let $\circ$ be a core-invariant regular revision operator which satisfies* $(\mathbf{C5'})_\triangle$*. Then $\circ$ satisfies* $(\mathbf{C3})_\triangle$ *and* $(\mathbf{C4})_\triangle$*.*

4   Core Belief Inputs

So far we have assumed that the set of core beliefs in an epistemic state $\mathbb{E}$ remains constant under regular revision inputs. In this section we want to consider the case when the core beliefs are themselves subject to revision by external inputs. To do this we shall now assume that we are given a second type of revision operator on epistemic states which we denote by $\bullet$. Given $\mathbb{E} \in Ep$ and $\phi \in L^+$, $\mathbb{E} \bullet \phi$ will denote the result of revising $\mathbb{E}$ so that ϕ is included as a core belief.[5] The operator $\bullet$ is distinct from $\circ$, though intuitively we should expect some interaction between the two. Once again we consider single-step revision and iterated revision in turn.

4.1   Single-Step Core Revision

What constraints should we put on $\blacktriangle(\mathbb{E} \bullet \phi)$? Well first of all, in order to simplify matters and unlike for $\circ$, we shall assume that **every revision using $\bullet$ is successful**, i.e., $\phi \in \blacktriangle(\mathbb{E} \bullet \phi)$.[6] For example core belief inputs might correspond to information from a source which the agent deems to be highly reliable or trustworthy, such as first-hand observations. A reasonable possibility is then to treat the core beliefs as we would any other belief set in this case and assume that the new core can be obtained by applying some simple AGM revision function for $\blacktriangle(\mathbb{E})$. Equivalently, by Theorem 1, we assume, for each $\mathbb{E} \in Ep$, the existence of a total pre-order $\leq_\blacktriangle^\mathbb{E}$ on $\mathcal{W}$, anchored on $[\blacktriangle(\mathbb{E})]$, such that, for all $\phi \in L^+$,

$$\blacktriangle(\mathbb{E} \bullet \phi) = Th(\min([\phi], \leq_\blacktriangle^\mathbb{E})). \tag{2}$$

Definition 4. *Let* $\bullet : Ep \times L^+ \to Ep$ *be a function. Then* $\bullet$ *is a* core revision operator *(on the epistemic frame* $\langle Ep, \triangle, \blacktriangle \rangle$*) if, for each* $\mathbb{E} \in Ep$*, there exists a total pre-order* $\leq_\blacktriangle^\mathbb{E}$ *on* $\mathcal{W}$*, anchored on* $[\blacktriangle(\mathbb{E})]$*, such that* $\blacktriangle(\mathbb{E} \bullet \phi)$ *may be determined as in (2) above. We call* $\leq_\blacktriangle^\mathbb{E}$ *the* core pre-order *associated to* $\mathbb{E}$ *(according to* $\bullet$*).*

So, to have both a regular revision operator *and* a core revision operator on an epistemic frame $\langle Ep, \triangle, \blacktriangle \rangle$ means to assume that each epistemic state $\mathbb{E} \in Ep$ comes equipped with **two** total pre-orders $\leq_\triangle^\mathbb{E}$ and $\leq_\blacktriangle^\mathbb{E}$, anchored on $[\triangle(\mathbb{E})]$ and $[\blacktriangle(\mathbb{E})]$ respectively. The interplay between these two orders will be of concern throughout the rest of the paper.

What constraints should we be putting on $\triangle(\mathbb{E} \bullet \phi)$? This question isn't so easy to answer. Here we need to keep in mind that we must have $\blacktriangle(\mathbb{E} \bullet \phi) \subseteq \triangle(\mathbb{E} \bullet \phi)$ and so, since we are assuming we always have $\phi \in \blacktriangle(\mathbb{E} \bullet \phi)$, we necessarily require $\phi \in \triangle(\mathbb{E} \bullet \phi)$. Hence if $\phi \notin \triangle(\mathbb{E})$ then some changes to the regular beliefs will certainly be necessary. In the case when $\neg\phi \notin \blacktriangle(\mathbb{E})$ it seems reasonable to expect that $\triangle(\mathbb{E})$ should be revised just as if ϕ was a regular belief input, i.e,:

(Y1) If $\neg\phi \notin \blacktriangle(\mathbb{E})$ then $\triangle(\mathbb{E} \bullet \phi) = \triangle(\mathbb{E} \circ \phi)$ (Cross-Vacuity)

---

[5] To put it another way in terms of revising epistemic states by *conditional* beliefs [4, 16]: whereas a regular revision by ϕ may be equated with a revision by the conditional $\top \Rightarrow \phi$, a *core* revision by ϕ may be equated with a revision by the conditional $\neg\phi \Rightarrow \bot$.

[6] An interesting alternative could be to reject ϕ from the core belief set, but include it instead merely as a regular belief.

This postulate gives us our basic point of contact between core revision and regular revision.

What should we do if $\neg\phi \in \blacktriangle(\mathbb{E})$? In this case we can't set $\triangle(\mathbb{E}\bullet\phi) = \triangle(\mathbb{E}\circ\phi)$ since ϕ is not contained in the right-hand side. One possibility could be to just throw away the distinctions between core belief and regular belief in this case by setting

(sY2) If $\neg\phi \in \blacktriangle(\mathbb{E})$ then $\triangle(\mathbb{E} \bullet \phi) = \blacktriangle(\mathbb{E} \bullet \phi)$.   (Regular Collapse)

However this seems a bit drastic. A more interesting possibility which we intend to explore in future work could be to adopt a Levi-style approach (cf. the Levi Identity [10]) and decompose the operation into two steps: first remove $\neg\phi$ from $\blacktriangle(\mathbb{E})$ using some sort of "core contraction" operation, and *then* revise by ϕ using $\circ$. For now, though, we take a different approach. Note that **(Y1)** says that, in the case when $\neg\phi \notin \blacktriangle(\mathbb{E})$, we should just use the pre-order $\leq_\triangle^\mathbb{E}$ to determine the new regular belief set. Why not just use $\leq_\triangle^\mathbb{E}$ also in the case when $\neg\phi \in \blacktriangle(\mathbb{E})$? That is we just set, in *all* cases

$$\triangle(\mathbb{E} \bullet \phi) = Th(\min([\phi], \leq_\triangle^\mathbb{E})). \tag{3}$$

However we need to be careful here, for remember we must have $\blacktriangle(\mathbb{E} \bullet \phi) \subseteq \triangle(\mathbb{E} \bullet \phi)$. This will be ensured if we require the two pre-orders $\leq_\triangle^\mathbb{E}$ and $\leq_\blacktriangle^\mathbb{E}$ to *cohere* with one another in a certain respect. Namely if we require

$$\leq_\triangle^\mathbb{E} \subseteq \leq_\blacktriangle^\mathbb{E},$$

i.e., that $\leq_\triangle^\mathbb{E}$ is a *refinement* of $\leq_\blacktriangle^\mathbb{E}$. This is confirmed by the following result.

Proposition 6. *Let* $\mathbb{E} \in Ep$ *and let* $\leq_\triangle^\mathbb{E}, \leq_\blacktriangle^\mathbb{E}$ *be two total pre-orders on* $\mathcal{W}$ *anchored on* $[\triangle(\mathbb{E})]$ *and* $[\blacktriangle(\mathbb{E})]$ *respectively. If* $\leq_\triangle^\mathbb{E} \subseteq \leq_\blacktriangle^\mathbb{E}$ *then, for all* $\phi \in L^+$, *we have* $Th(\min([\phi], \leq_\blacktriangle^\mathbb{E})) \subseteq Th(\min([\phi], \leq_\triangle^\mathbb{E}))$.

As we will shortly see in Theorem 2, defining $\triangle(\mathbb{E} \bullet \phi)$ as in (3) above has the consequence that, in addition to **(Y1)**, the following two properties are satisfied.

(Y2) If $\neg\phi \notin \triangle(\mathbb{E} \bullet \theta)$ then $\triangle(\mathbb{E} \bullet (\theta \wedge \phi)) = \triangle(\mathbb{E} \bullet \theta) + \phi$
                                    (Cross-Conjunction 1)
(Y3) If $\phi_1 \leftrightarrow \phi_2 \in Cn(\emptyset)$ then $\triangle(\mathbb{E} \bullet \phi_1) = \triangle(\mathbb{E} \bullet \phi_2)$ (Cross-Extensionality)

The first property above is similar to the AGM postulate **(K*6)** from Sect. 2. It says that the new regular belief set after core-revising by $\theta \wedge \phi$ should be obtainable by first core-revising by θ and then simply expanding the resultant regular belief set $\triangle(\mathbb{E} \bullet \theta)$ by ϕ, *provided* ϕ is consistent with $\triangle(\mathbb{E} \bullet \theta)$. It is easy to see that, for core-revision operators, **(sY2)** implies **(Y2)**. The second property above expresses the reasonable requirement that core-revising by logically equivalent sentences should yield the same regular belief set. We make the following definition:

Definition 5. *Let* $\circ$ *and* $\bullet$ *be a core-invariant regular revision operator and a core revision operator on the epistemic frame* $\langle Ep, \triangle, \blacktriangle \rangle$ *respectively. If* $\circ$ *and* $\bullet$ *together satisfy* **(Y1)** *and* $\bullet$ *satisfies* **(Y2)** *and* **(Y3)** *then we call the pair* $\langle \circ, \bullet \rangle$ *a revision system (on* $\langle Ep, \triangle, \blacktriangle \rangle$*).*

The next theorem is one of the main results of this paper. It gives a characterisation for revision systems.

Theorem 2. *Let* $\langle Ep, \triangle, \blacktriangle \rangle$ *be an epistemic frame and let* $\circ, \bullet : Ep \times L^+ \to Ep$ *be two functions. Then the following are equivalent:*
(i). $\langle \circ, \bullet \rangle$ *is a revision system on* $\langle Ep, \triangle, \blacktriangle \rangle$*.*
(ii). For each $\mathbb{E} \in Ep$ *there exist a total pre-order* $\leq_\triangle^\mathbb{E}$ *on* $\mathcal{W}$ *anchored on* $[\triangle(\mathbb{E})]$*, and a total pre-order* $\leq_\blacktriangle^\mathbb{E}$ *on* $\mathcal{W}$ *anchored on* $[\blacktriangle(\mathbb{E})]$ *such that* $\leq_\triangle^\mathbb{E} \subseteq \leq_\blacktriangle^\mathbb{E}$ *and, for all* $\phi \in L^+$*,*

$$\triangle(\mathbb{E} \bullet \phi) = Th(\min([\phi], \leq_\triangle^\mathbb{E})) \qquad\qquad \blacktriangle(\mathbb{E} \bullet \phi) = Th(\min([\phi], \leq_\blacktriangle^\mathbb{E}))$$

$$\triangle(\mathbb{E} \circ \phi) = \begin{cases} Th(\min([\phi], \leq_\triangle^\mathbb{E})) \ if \ \neg\phi \notin \blacktriangle(\mathbb{E}) \\ \triangle(\mathbb{E}) \qquad\qquad\quad otherwise \end{cases} \quad \blacktriangle(\mathbb{E} \circ \phi) = \blacktriangle(\mathbb{E})$$

Proof (Sketch). To show that (i) implies (ii), let $\langle \circ, \bullet \rangle$ be a revision system on $\langle Ep, \triangle, \blacktriangle \rangle$. Then, by definition, $\circ$ is a core-invariant regular revision operator. Hence there exists, for each $\mathbb{E} \in Ep$, a total pre-order $\leq_r^\mathbb{E}$ on $\mathcal{W}$ anchored on $[\triangle(\mathbb{E})]$ such that, for all $\phi \in L^+$,

$$\triangle(\mathbb{E} \circ \phi) = \begin{cases} Th(\min([\phi], \leq_r^E)) \ if \ \neg\phi \notin \blacktriangle(\mathbb{E}) \\ \triangle(\mathbb{E}) \qquad\qquad\quad otherwise \end{cases}$$

and $\blacktriangle(\mathbb{E} \circ \phi) = \blacktriangle(\mathbb{E})$. We also know that $\bullet$ is a core revision operator. Hence, for each $\mathbb{E} \in Ep$ there also exists a total pre-order $\leq_\blacktriangle^\mathbb{E}$ on $\mathcal{W}$ anchored on $[\blacktriangle(\mathbb{E})]$ such that, for all $\phi \in L^+$ we have $\blacktriangle(\mathbb{E} \bullet \phi) = Th(\min([\phi], \leq_\blacktriangle^\mathbb{E}))$. It might be hoped now that $\leq_r^\mathbb{E}$ and $\leq_\blacktriangle^\mathbb{E}$ then give us our required pair of pre-orders, however we first need to make some modification to $\leq_r^\mathbb{E}$. We define a new ordering $\leq_\triangle^\mathbb{E}$ which agrees with $\leq_r^\mathbb{E}$ within $[\blacktriangle(\mathbb{E})]$ and likewise makes all $[\blacktriangle(\mathbb{E})]$-worlds more plausible than all non-$[\blacktriangle(\mathbb{E})]$-worlds. However, $\leq_\triangle^\mathbb{E}$ orders the non-$[\blacktriangle(\mathbb{E})]$-worlds differently from $\leq_r^\mathbb{E}$. Precisely we set, for $w_1, w_2 \in \mathcal{W}$,

$$w_1 \leq_\triangle^\mathbb{E} w_2 \ iff \ w_1, w_2 \in [\blacktriangle(\mathbb{E})] \ and \ w_1 \leq_r^\mathbb{E} w_2$$
$$or \ w_1 \in [\blacktriangle(\mathbb{E})] \ and \ w_2 \notin [\blacktriangle(\mathbb{E})]$$
$$or \ w_1, w_2 \notin [\blacktriangle(\mathbb{E})] \ and \ \neg\alpha_1 \notin \triangle(\mathbb{E} \bullet (\alpha_1 \vee \alpha_2))$$

In the last line here, α_i is any sentence such that $[\alpha_i] = \{w_i\}$ $(i = 1, 2)$. (By **(Y3)** the precise choice of α_i is irrelevant.) It can then be shown that $\leq_\triangle^\mathbb{E}$ is a total pre-order anchored on $[\triangle(\mathbb{E})]$ and that $\leq_\triangle^\mathbb{E}$ and $\leq_\blacktriangle^\mathbb{E}$ then give the required pair of pre-orders.

The proof that (ii) implies (i) is straightforward. $\qquad\qquad\qquad\qquad\qquad \square$

For the rest of this section we assume $\langle \circ, \bullet \rangle$ to be an arbitrary but fixed revision system.

4.2   Iterating Core Beliefs Revision

What should be the effect on $\blacktriangle(\mathbb{E})$ and $\triangle(\mathbb{E})$ of iterated applications of $\bullet$? For the former, since we are assuming $\bullet$ behaves just like an AGM revision operator with regard to $\blacktriangle(\mathbb{E})$, the iterated AGM revision postulates mentioned in Sect. 3.2 are again relevant. Rephrased in terms of core revision, they are:

(C1)$_\blacktriangle$ If $\phi \to \theta \in Cn(\emptyset)$ then $\blacktriangle(\mathbb{E} \bullet \theta \bullet \phi) = \blacktriangle(\mathbb{E} \bullet \phi)$
(C2)$_\blacktriangle$ If $\phi \to \neg\theta \in Cn(\emptyset)$ then $\blacktriangle(\mathbb{E} \bullet \theta \bullet \phi) = \blacktriangle(\mathbb{E} \bullet \phi)$
(C3)$_\blacktriangle$ If $\theta \in \blacktriangle(\mathbb{E} \bullet \phi)$ then $\theta \in \blacktriangle(\mathbb{E} \bullet \theta \bullet \phi)$
(C4)$_\blacktriangle$ If $\neg\theta \notin \blacktriangle(\mathbb{E} \bullet \phi)$ then $\neg\theta \notin \blacktriangle(\mathbb{E} \bullet \theta \bullet \phi)$
(C5)$_\blacktriangle$ If $\phi \to \theta \notin Cn(\emptyset)$ then $\theta \in \blacktriangle(\mathbb{E} \bullet \theta \bullet \phi)$

We take **(C1)$_\blacktriangle$**–**(C4)$_\blacktriangle$** to be minimal requirements. We remind the reader that **(C5)$_\blacktriangle$** implies both **(C3)$_\blacktriangle$** and **(C4)$_\blacktriangle$**. The characterisation result of Darwiche and Pearl already tells us how each of **(C1)$_\blacktriangle$**–**(C4)$_\blacktriangle$** regulates a certain aspect of the relationship between $\leq_\blacktriangle^\mathbb{E}$ and the new *core* pre-order $\leq_\blacktriangle^{\mathbb{E}\bullet\theta}$. **(C5)$_\blacktriangle$** also corresponds to a constraint on $\leq_\blacktriangle^{\mathbb{E}\bullet\theta}$. The proof of this correspondence is implicit in [21].

Proposition 7 ([8, 21]). *Let $\bullet$ be a core revision operator. Then $\bullet$ satisfies* **(C1)$_\blacktriangle$**–**(C5)$_\blacktriangle$** *iff each of the following conditions hold for all $\mathbb{E} \in Ep$ and $\theta \in L^+$:*

(1) For all $w_1, w_2 \in [\theta]$, $w_1 \leq_\blacktriangle^{\mathbb{E}\bullet\theta} w_2$ iff $w_1 \leq_\blacktriangle^\mathbb{E} w_2$
(2) For all $w_1, w_2 \in [\neg\theta]$, $w_1 \leq_\blacktriangle^{\mathbb{E}\bullet\theta} w_2$ iff $w_1 \leq_\blacktriangle^\mathbb{E} w_2$
(3) For all $w_1, w_2 \in W$, if $w_1 \in [\theta], w_2 \in [\neg\theta]$ and $w_1 <_\blacktriangle^\mathbb{E} w_2$, then $w_1 <_\blacktriangle^{\mathbb{E}\bullet\theta} w_2$
(4) For all $w_1, w_2 \in W$, if $w_1 \in [\theta], w_2 \in [\neg\theta]$ and $w_1 \leq_\blacktriangle^\mathbb{E} w_2$, then $w_1 \leq_\blacktriangle^{\mathbb{E}\bullet\theta} w_2$
(5) For all $w_1, w_2 \in W$, if $w_1 \in [\theta]$ and $w_2 \in [\neg\theta]$ then $w_1 <_\blacktriangle^{\mathbb{E}\bullet\theta} w_2$

For the case of $\triangle(\mathbb{E})$ we expect that the behaviour of the regular belief set under a sequence of core inputs should be connected in some way with the behaviour of the core itself. But how? Here we present one idea, which is perhaps best motivated directly in terms of the two pre-orders $\leq_\triangle^\mathbb{E}$ and $\leq_\blacktriangle^\mathbb{E}$ which we take to underlie a given epistemic state $\mathbb{E}$. First note that the question of how $\triangle(\mathbb{E})$ should behave under sequences of core inputs essentially reduces to the question of what the new regular pre-order $\leq_\triangle^{\mathbb{E}\bullet\theta}$ following the core input θ should look like. One constraint on $\leq_\triangle^{\mathbb{E}\bullet\theta}$ is already in place, namely that $\leq_\blacktriangle^{\mathbb{E}\bullet\theta} \subseteq \leq_\triangle^{\mathbb{E}\bullet\theta}$. Our idea is to carry over as much of the structure of $\leq_\triangle^\mathbb{E}$ to $\leq_\triangle^{\mathbb{E}\bullet\theta}$ as possible, while obeying this constraint. This can be achieved by defining $\leq_\triangle^{\mathbb{E}\bullet\theta}$ simply to be the lexicographic *refinement* of $\leq_\blacktriangle^{\mathbb{E}\bullet\theta}$ by $\leq_\triangle^\mathbb{E}$, i.e., for all $w_1, w_2 \in W$,

$$w_1 \leq_\triangle^{\mathbb{E}\bullet\theta} w_2 \text{ iff either } w_1 <_\blacktriangle^{\mathbb{E}\bullet\theta} w_2$$
$$\text{or } w_1 \sim_\blacktriangle^{\mathbb{E}\bullet\theta} w_2 \text{ and } w_1 \leq_\triangle^\mathbb{E} w_2.$$

We remark that the idea of combining pre-orders using lexicographic refinement crops up several times in the literature on belief revision, e.g. [20]. It turns out that this behaviour may be characterised here by the following property:

(Z1) If $\blacktriangle(\mathbb{E} \bullet \theta \bullet \phi) \subseteq \blacktriangle(\mathbb{E} \bullet \phi)$ then $\triangle(\mathbb{E} \bullet \theta \bullet \phi) = \triangle(\mathbb{E} \bullet \phi)$ (Coupling)

This property says that if a core input ϕ can be preceded by another core input θ *without increasing* the set of core beliefs, then preceding ϕ by θ should lead to precisely the same set of regular beliefs.

Proposition 8. $\bullet$ *satisfies* **(Z1)** *iff, for each* $\mathbb{E} \in Ep$ *and* $\theta \in L^+$, $\leq_{\triangle}^{\mathbb{E} \bullet \theta}$ *is equal to the lexicographic refinement of* $\leq_{\blacktriangle}^{\mathbb{E} \bullet \theta}$ *by* $\leq_{\triangle}^{\mathbb{E}}$.

The appearance of "$\subseteq$" rather than "$=$" in the antecedent of **(Z1)** may seem surprising. However, as the next proposition shows, if $\bullet$ satisfies certain other properties then there is no difference.

Proposition 9. *Let* $\bullet$ *be a core revision operator which satisfies* **(C1)**$_{\blacktriangle}$, **(C2)**$_{\blacktriangle}$ *and* **(C5)**$_{\blacktriangle}$. *Then, for all* $\mathbb{E} \in Ep$ *and* $\theta, \phi \in L^+$, *we have* $\blacktriangle(\mathbb{E} \bullet \theta \bullet \phi) \subseteq \blacktriangle(\mathbb{E} \bullet \phi)$ *iff* $\blacktriangle(\mathbb{E} \bullet \theta \bullet \phi) = \blacktriangle(\mathbb{E} \bullet \phi)$.

Finally in this section we have the following result, which attests to the strength of **(Z1)**.

Proposition 10. *If* $\bullet$ *satisfies* **(Z1)** *and* **(C1)**$_{\blacktriangle}$, **(C2)**$_{\blacktriangle}$ *and* **(C5)**$_{\blacktriangle}$, *then* $\bullet$ *also satisfies* **(C1)**$_{\triangle}$, **(C2)**$_{\triangle}$ *and* **(C5)**$_{\triangle}$ *(with* $\bullet$ *in place of* $\circ$*).*

5   Mixing Regular and Core Revision

In this section we explore the possibility of performing mixed sequences of revisions, containing both regular and core inputs. We give a number of possible properties, relating each one to a condition in terms of the underlying regular and core pre-orders, and also showing how they relate to the postulates of the previous sections. In the following section we will give a concrete pair of revision operators which in fact satisfies all the postulates from this section. The intuitive simplicity of these constructed operators will lend some support to the reasonableness of these postulates. Throughout this section we again assume $\langle \circ, \bullet \rangle$ is an arbitrary but fixed revision system.

Recall that we have assumed regular revision to be core-invariant, i.e., a regular belief input leaves the core belief set unchanged. Our first property seeks to lessen the impact of regular inputs on the core even further. It says that a regular input also has no influence on what the core beliefs should look like after the *next core input*.

(M0) $\blacktriangle(\mathbb{E} \circ \theta \bullet \phi) = \blacktriangle(\mathbb{E} \bullet \phi)$ (Core-Conditional-Invariance)

Since $\circ$ is core-invariant, the right-hand-side here is equal to $\blacktriangle(\mathbb{E} \bullet \phi \circ \theta)$. Hence, for a revision system, **(M0)** is equivalent to the following property which says that when receiving two consecutive inputs, one of which is a regular input and the other a core input, the resulting core belief set is actually independent of the order in which these two inputs are received.

(Comm)$_\blacktriangle$ $\blacktriangle(\mathbb{E} \circ \theta \bullet \phi) = \blacktriangle(\mathbb{E} \bullet \phi \circ \theta)$

In terms of the underlying pre-orders, the property **(M0)** corresponds to the simple requirement that the entire core pre-order remains unchanged under regular revision. Thus **(M0)** can be seen as a kind of "scaling-up" of the property of core-invariance to apply to the iterated case.

Proposition 11. $\langle \circ, \bullet \rangle$ *satisfies* **(M0)** *iff, for all* $\mathbb{E} \in Ep$ *and* $\theta \in L^+$, *we have* $\leq_\blacktriangle^{\mathbb{E}\circ\theta} = \leq_\blacktriangle^{\mathbb{E}}$.

For the next batch of postulates, we consider again Darwiche and Pearl's original postulates **(C1)**$_\triangle$–**(C4)**$_\triangle$ for iterated regular revision listed at the beginning of Sect. 3.2. The postulates **(C1)**$_\triangle$ and **(C2)**$_\triangle$ provide conditions under which the effects of a regular input on the regular belief set should be "overruled" by a succeeding regular input. Namely when the second is more specific than the first (**(C1)**$_\triangle$), or when it contradicts it (**(C2)**$_\triangle$). Since core inputs can be viewed as carrying more "weight" than regular inputs, it seems reasonable to suggest that in both cases this overruling should also occur when the second input is upgraded from being a regular input to being a *core* input:

(M1) If $\phi \rightarrow \theta \in Cn(\emptyset)$ then $\triangle(\mathbb{E} \circ \theta \bullet \phi) = \triangle(\mathbb{E} \bullet \phi)$
(M2) If $\phi \rightarrow \neg\theta \in Cn(\emptyset)$ then $\triangle(\mathbb{E} \circ \theta \bullet \phi) = \triangle(\mathbb{E} \bullet \phi)$

We propose similar modifications to **(C3)**$_\triangle$ and **(C4)**$_\triangle$. If θ would be a regular belief after receiving the core input ϕ then preceding this core input with the regular input θ should not change this fact. Similarly if θ is not discounted as a regular belief after receiving the core input ϕ then preceding this core input with the regular input θ should not change this fact.

(M3) If $\theta \in \triangle(\mathbb{E} \bullet \phi)$ then $\theta \in \triangle(\mathbb{E} \circ \theta \bullet \phi)$
(M4) If $\neg\theta \notin \triangle(\mathbb{E} \bullet \phi)$ then $\neg\theta \notin \triangle(\mathbb{E} \circ \theta \bullet \phi)$

Recall Proposition 3 which showed how each of the postulates **(C1′)**$_\triangle$, **(C2′)**$_\triangle$, **(C3)**$_\triangle$ and **(C4)**$_\triangle$ corresponded to a constraint on the new regular pre-order $\leq_\triangle^{\mathbb{E}\circ\theta}$ within $[\blacktriangle(\mathbb{E})]$. The following result shows how each of **(M1)**–**(M4)** corresponds to the same constraints on $\leq_\triangle^{\mathbb{E}\circ\theta}$, but extended to apply to the *whole* of $\mathcal{W}$.

Proposition 12. $\langle \circ, \bullet \rangle$ *satisfies* **(M1)**–**(M4)** *iff the following hold, for each* $\mathbb{E} \in Ep$ *and* $\theta \in L^+$:

(1) For all $w_1, w_2 \in [\theta]$, $w_1 \leq_\triangle^{\mathbb{E}\circ\theta} w_2$ iff $w_1 \leq_\triangle^{\mathbb{E}} w_2$
(2) For all $w_1, w_2 \in [\neg\theta]$, $w_1 \leq_\triangle^{\mathbb{E}\circ\theta} w_2$ iff $w_1 \leq_\triangle^{\mathbb{E}} w_2$
(3) For all $w_1, w_2 \in \mathcal{W}$, if $w_1 \in [\theta], w_2 \in [\neg\theta]$ and $w_1 <_\triangle^{\mathbb{E}} w_2$, then $w_1 <_\triangle^{\mathbb{E}\circ\theta} w_2$
(4) For all $w_1, w_2 \in \mathcal{W}$, if $w_1 \in [\theta], w_2 \in [\neg\theta]$ and $w_1 \leq_\triangle^{\mathbb{E}} w_2$, then $w_1 \leq_\triangle^{\mathbb{E}\circ\theta} w_2$

It is then easy to see:

Proposition 13. *The postulates* **(M1)**–**(M4)** *imply* **(C1′)**$_\triangle$, **(C2′)**$_\triangle$, **(C3)**$_\triangle$ *and* **(C4)**$_\triangle$ *respectively.*

Now suppose the agent receives a regular input θ followed by a core input ϕ. Then our next property says θ should be in the regular beliefs as long as it is consistent with the resultant core beliefs.

(M5) If $\neg\theta \notin \blacktriangle(\mathbb{E} \circ \theta \bullet \phi)$ then $\theta \in \triangle(\mathbb{E} \circ \theta \bullet \phi)$ (Retro-success)

Moreover this should hold even if θ is *not* believed after the first revision. Thus this property allows a regular input which may initially have been rejected to be accepted *retrospectively* as long as it does not contradict the new core beliefs. To roughly illustrate, let us return to Example 1 from the introduction. Suppose your initial epistemic state is $\mathbb{E}$ and let θ and ϕ stand for the information "your son ate lunch at school with King Gustav" and "King Gustav was at a local school" respectively. Assume that your expectations of the King's behaviour are such that you very strongly believe that, of all the places where the King might be, a local school is not one of them, i.e., $\neg\phi \in \blacktriangle(\mathbb{E})$. You also strongly believe that no-one can be in two places at the same time, i.e., $\neg\phi \to \neg\theta \in \blacktriangle(\mathbb{E})$, and so you deduce $\neg\theta \in \blacktriangle(\mathbb{E})$. Assuming you always treat information provided to you by your son as a regular input, your epistemic state after your son provides you with θ is $\mathbb{E} \circ \theta$. However, since $\neg\theta \in \blacktriangle(\mathbb{E})$, we have $\theta \notin \triangle(\mathbb{E} \circ \theta) = \triangle(\mathbb{E})$. Now suppose you receive the information ϕ from the TV news. Since you take the TV news to be highly reliable, you treat this as a core input, and so your epistemic state after this input is $\mathbb{E} \circ \theta \bullet \phi$. Now, because $\neg\phi$ is necessarily removed from the core at this point, your grounds for deducing that $\neg\theta$ is a core belief have been taken away. If this was really your *only* argument for deducing $\neg\theta \in \blacktriangle(\mathbb{E})$ then you should now have $\neg\theta \notin \blacktriangle(\mathbb{E} \circ \theta \bullet \phi)$. **(M5)** says that, in this case, $\theta \in \triangle(\mathbb{E} \circ \theta \bullet \phi)$, i.e., you should now believe your son's information.

In terms of pre-orders **(M5)** corresponds to the following property:

Proposition 14. $\langle \circ, \bullet \rangle$ *satisfies* **(M5)** *iff, for all* $\mathbb{E} \in Ep$ *and* $\theta \in L^+$ *and for all* $w_1, w_2 \in \mathcal{W}$ *such that* $w_1 \in [\neg\theta]$ *and* $w_2 \in [\theta]$, *we have* $w_1 \leq_{\triangle}^{\mathbb{E}\circ\theta} w_2$ *implies* $w_1 <_{\blacktriangle}^{\mathbb{E}\circ\theta} w_2$.

The above property says that after a regular revision by θ, and for each $[\theta]$-world w_2, the only $[\neg\theta]$-worlds considered at least as plausible as w_2 by the new regular pre-order are those considered *strictly* more plausible than w_2 by the new core pre-order. (Recall that we always have $\leq_{\triangle}^{\mathbb{E}\circ\theta} \subseteq \leq_{\blacktriangle}^{\mathbb{E}\circ\theta}$ and so we necessarily have that $w_1 <_{\blacktriangle}^{\mathbb{E}\circ\theta} w_2$ implies $w_1 \leq_{\triangle}^{\mathbb{E}\circ\theta} w_2$.) The following proposition reveals **(M5)** to be quite a strong property.

Proposition 15. *Let* $\langle \circ, \bullet \rangle$ *be a revision system which satisfies* **(M5)**. *Then* $\circ$ *satisfies* **(C5′)**$_{\triangle}$. *If, in addition,* $\langle \circ, \bullet \rangle$ *satisfies* **(M0)** *then* $\langle \circ, \bullet \rangle$ *also satisfies* **(M3)** *and* **(M4)**.

Our next postulate is inspired by similar considerations to those behind the AGM revision postulate **(K*6)** (see [10]). Suppose an agent receives a core input θ followed by a regular input ϕ. Then if ϕ is consistent with the core beliefs *after* the first revision, then the regular belief set after the second revision should be just the same as if the agent had received θ and ϕ *together* as a core input.

(**M6**) If $\neg\phi \notin \blacktriangle(\mathbb{E} \bullet \theta)$ then $\triangle(\mathbb{E} \bullet \theta \circ \phi) = \triangle(\mathbb{E} \bullet (\theta \wedge \phi))$

(Cross-Conjunction 2)

It is easy to see that, for a revision system, this postulate implies (**Y2**) (i.e., (Cross-Conjunction 1)). In terms of pre-orders (**M6**) can be understood as follows:

Proposition 16. $\langle \circ, \bullet \rangle$ *satisfies* (**M6**) *iff, for all* $\mathbb{E} \in Ep$ *and* $\theta \in L^+$ *and for all* $w_1, w_2 \in [\blacktriangle(\mathbb{E} \bullet \theta)]$, *we have* $w_1 \leq_\triangle^{\mathbb{E}\bullet\theta} w_2$ *iff* $w_1 \leq_\triangle^{\mathbb{E}} w_2$.

The above condition places a constraint on the new regular pre-order $\leq_\triangle^{\mathbb{E}\bullet\theta}$ which follows a core input θ. It says that within $[\blacktriangle(\mathbb{E} \bullet \theta)]$ the old regular ordering $\leq_\triangle^{\mathbb{E}}$ is preserved. It is quite easy to see that if $\leq_\triangle^{\mathbb{E}\bullet\theta}$ is taken to be the lexicographic refinement of $\leq_\blacktriangle^{\mathbb{E}\bullet\theta}$ by $\leq_\triangle^{\mathbb{E}}$ as in Proposition 8 then this constraint is satisfied. Thus, using propositions 8 and 16, we see that

Proposition 17. *If* $\bullet$ *satisfies* (**Z1**) *then* $\langle \circ, \bullet \rangle$ *satisfies* (**M6**).

Our last property is the regular-belief analogue of (**Comm**)$_\blacktriangle$:

(**Comm**)$_\triangle$ $\triangle(\mathbb{E} \circ \theta \bullet \phi) = \triangle(\mathbb{E} \bullet \phi \circ \theta)$

Rather surprisingly it turns out that, in the presence of (**M0**), this rule is equivalent to the conjunction of some of our previous postulates:

Proposition 18. *Let* $\langle \circ, \bullet \rangle$ *be a revision system which satisfies* (**M0**). *Then* $\langle \circ, \bullet \rangle$ *satisfies* (**Comm**)$_\triangle$ *iff* $\langle \circ, \bullet \rangle$ *satisfies* (**M1**), (**M2**), (**M5**) *and* (**M6**).

We remark that the derivations of (**M1**) and (**M2**) from (**Comm**)$_\triangle$ do not require (**M0**). Figure 1 summarises the postulates from this section, together with their inferential interrelations and their relations with some of our previous postulates. A dashed line indicates a derivation which requires the presence of the postulate (**M0**).

6   A Construction

We now give an explicit construction of a pair of operators $\circ$ and $\bullet$ which display much of the behaviour described in the previous sections. For this construction we will use a specific representation of an epistemic frame. For the set of epistemic states the basic idea is that the agent keeps two separate lists of sentences, one which records all the regular inputs he receives, and one which records all the core inputs he receives. We take an agent's epistemic state simply *to be* this record. More precisely we take for the set of epistemic states the set $seq(L^+)^2$ of all pairs of finite sequences of consistent sentences $\mathcal{E} = \langle (\beta_1, \ldots, \beta_m), (\alpha_1, \ldots, \alpha_n) \rangle$, where the β_i are the regular inputs which the agent has so far received, and the α_i are the core inputs.[7] Revision of such an epistemic state is then a trivial affair

---

[7] In the literature on iterated AGM revision, the idea of taking an epistemic state to be a *single* sequence of sentences reflecting the revision history has already been suggested in, e.g., [6, 17, 18].

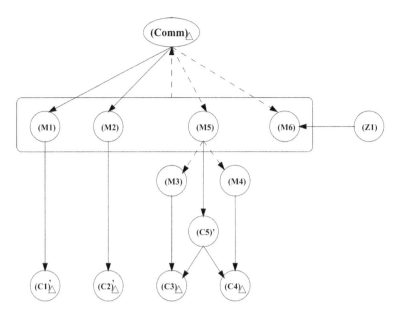

Fig. 1. The inferential interrelations of the postulates for a revision system

– we obtain our new epistemic state by simply adding the sentence received as input to the end of the appropriate list, depending on whether the sentence is received as a regular input or a core input. More precisely, given $\mathcal{E} = \langle \boldsymbol{b}, \boldsymbol{a} \rangle$ where $\boldsymbol{b} = (\beta_1, \ldots, \beta_m)$ and $\boldsymbol{a} = (\alpha_1, \ldots, \alpha_n)$, we define

$$\mathcal{E} \circ \phi = \langle \boldsymbol{b} \cdot \phi,\ \boldsymbol{a} \rangle \quad \text{and} \quad \mathcal{E} \bullet \phi = \langle \boldsymbol{b},\ \boldsymbol{a} \cdot \phi \rangle$$

where $\cdot$ denotes sequence concatenation. The properties of the revision operators are then determined entirely by the particular ways we choose to extract the belief set $\triangle(\mathcal{E})$ and the core belief set $\blacktriangle(\mathcal{E})$ from any given $\mathcal{E} \in seq(L^+)^2$.[8] However there are some things we can say immediately about $\bullet$ and $\circ$ without referring to $\blacktriangle(\mathcal{E})$ or $\triangle(\mathcal{E})$, for clearly we have $\mathcal{E} \circ \theta \bullet \phi = \langle \boldsymbol{b} \cdot \theta,\ \boldsymbol{a} \cdot \phi \rangle = \mathcal{E} \bullet \phi \circ \theta$. Hence we already have:

Proposition 19. $\bullet$ *and* $\circ$ *defined above together satisfy the rules* **(Comm)**$_{\blacktriangle}$ *and* **(Comm)**$_{\triangle}$.

Hence, once we have shown that $\langle \circ, \bullet \rangle$ is a revision system on the epistemic frame $\langle seq(L^+)^2, \triangle, \blacktriangle \rangle$, it will immediately follow that $\langle \circ, \bullet \rangle$ satisfies many of the postulates we have considered, including *all* those of the previous section (cf. Fig. 1 and recall that, for a revision system, **(Comm)**$_{\blacktriangle}$ is equivalent to **(M0)**).

---

[8] The general approach of leaving all the work in performing revision to some operation of *retrieval* on epistemic states is known as the *vertical* approach to belief revision [22].

Now let's determine $\triangle(\mathcal{E})$ and $\blacktriangle(\mathcal{E})$. Here we again assume $\mathcal{E}$ is of the form $\langle(\beta_1,\ldots,\beta_m),(\alpha_1,\ldots,\alpha_n)\rangle$. Turning first to $\blacktriangle(\mathcal{E})$, we first form an increasing sequence of sets of sentences Γ_i starting with $\Gamma_0 = \emptyset$ by setting, for each $i = 1,\ldots,n$,

$$\Gamma_i = \begin{cases} \Gamma_{i-1} \cup \{\alpha_{n+1-i}\} & \text{if this is consistent} \\ \Gamma_{i-1} & \text{otherwise} \end{cases}$$

In other words we form the set of sentences Γ_n by taking the sequence $(\alpha_1,\ldots,\alpha_n)$ and, starting at the end with α_n, working our way backwards through the sequence and adding each sentence as we go, provided it is consistent with the sentences already collected up to that point. We then define

$$\blacktriangle(\mathcal{E}) = Cn(\Gamma_n).$$

Note that $\blacktriangle(\mathcal{E})$ so defined is bound to be consistent. Also note that in the construction of $\blacktriangle(\mathcal{E})$ no mention is made of the first sequence $(\beta_1,\ldots,\beta_m)$ of $\mathcal{E}$, i.e., $\blacktriangle(\mathcal{E})$ depends only the second sequence $(\alpha_1,\ldots,\alpha_n)$. This is all that is needed to show:

Proposition 20. $\circ$ *satisfies* **(X1)** *and* **(X2)**.

We also have

Proposition 21. $\bullet$ *is a core revision operator (on the epistemic frame* $\langle seq(L^+)^2, \triangle, \blacktriangle\rangle)$ *according to Definition 4. Furthermore* $\bullet$ *satisfies* **(C1)**$_\blacktriangle$, **(C2)**$_\blacktriangle$ *and* **(C5)**$_\blacktriangle$.

We now turn to our definition of $\triangle(\mathcal{E})$. This is a matter of just picking up from where we left off in the construction of $\blacktriangle(\mathcal{E})$. We define an increasing sequence of sets of sentences Θ_i, starting with $\Theta_0 = \Gamma_n$, by setting, for $i = 1,\ldots,m$,

$$\Theta_i = \begin{cases} \Theta_{i-1} \cup \{\beta_{m+1-i}\} & \text{if this is consistent} \\ \Theta_{i-1} & \text{otherwise} \end{cases}$$

We then set

$$\triangle(\mathcal{E}) = Cn(\Theta_m).$$

Clearly, since $\Gamma_n \subseteq \Theta_m$, we get $\blacktriangle(\mathcal{E}) \subseteq \triangle(\mathcal{E})$. Also since Γ_n is consistent we also have $\triangle(\mathcal{E})$ is consistent. We would now like to show that $\circ$ forms a core-invariant regular revision operator on the epistemic frame $\langle seq(L^+)^2, \triangle, \blacktriangle\rangle$. As a first step we show the following:

Proposition 22. *For all* $\mathcal{E} \in seq(L^+)^2$ *and all* $\phi \in L^+$ *we have*

$$\triangle(\mathcal{E} \circ \phi) = \begin{cases} \triangle(\mathcal{E} \bullet \phi) & \text{if } \neg\phi \notin \blacktriangle(\mathcal{E}) \\ \triangle(\mathcal{E}) & \text{otherwise} \end{cases}$$

(In particular $\circ$ *and* $\bullet$ *together satisfy* **(Y1)**.*)*

Next we have the following result:

Proposition 23. *For each $\mathcal{E} \in seq(L^+)^2$ there exists some simple AGM revision function $*$ for $\triangle(\mathcal{E})$ such that $\triangle(\mathcal{E} \bullet \phi) = \triangle(\mathcal{E}) * \phi$ for all $\phi \in L^+$.*

Combining propositions 20, 22 and 23 then allows us to prove:

Proposition 24. $\circ$ *is a core-invariant regular revision operator (on the epistemic frame $\langle seq(L^+)^2, \triangle, \blacktriangle \rangle$) according to Definitions 2 and 3.*

Meanwhile, as a corollary to Proposition 23 we have:

Corollary 1. $\bullet$ *satisfies* **(Y2)** *and* **(Y3)**.

We remark, however, that $\bullet$ does *not* satisfy **(sY3)**, i.e., core-revising $\mathcal{E}$ by a core-contravening sentence will not necessarily erase the distinctions between core belief set and regular belief set. To see this let $\mathcal{E} = \langle (p \wedge q), (\neg q) \rangle$. Then $\blacktriangle(\mathcal{E}) = \triangle(\mathcal{E}) = Cn(\neg q)$. Consider core-revising by q. Then obviously $\neg q \in \blacktriangle(\mathcal{E})$ but $\mathcal{E} \bullet q = \langle (p \wedge q), (\neg q, q) \rangle$, leading to $\blacktriangle(\mathcal{E} \bullet q) = Cn(q) \neq Cn(p \wedge q) = \triangle(\mathcal{E} \bullet q)$.
    We are now finally in a position to show:

Theorem 3. *The pair $\langle \circ, \bullet \rangle$ forms a revision system (on the epistemic frame $\langle seq(L^+)^2, \triangle, \blacktriangle \rangle$).*

    As a final piece in our jigsaw we have the following result regarding the behaviour of $\triangle(\mathcal{E})$ under iterated core revision.

Proposition 25. $\bullet$ *satisfies* **(Z1)**. *Hence* $\bullet$ *satisfies* **(C1')**$_\triangle$, **(C2')**$_\triangle$ *and* **(C5')**$_\triangle$ *(with* $\bullet$ *in place of* $\circ$*).*

The second part of this proposition follows from propositions 10 and 21.

7   Conclusion

We have taken a close look at iterated non-prioritised revision, using as a starting point a basic model of non-prioritised revision which makes use of a set of core beliefs amongst the set of regular beliefs. We considered two types of revision operator on epistemic states: a normal, regular revision corresponding to a direction to include a given sentence in the regular beliefs, and a core revision operator. We presented some postulates for the iteration of both these operators, including some for the particularly interesting case of mixed iterated sequences of revisions consisting of both types of revision operations. In many cases we have shown how these postulates correspond to conditions on the dynamics of the plausibility orderings on worlds which underlie an agent's epistemic state. Finally we provided a construction which illustrated some of the ideas.
    As further work we would like to examine also operations of contraction in this context. Of particular interest would be an operation of "core contraction" in which a sentence is removed from the core beliefs. What effect should such an operation have on the *regular* belief set? For instance, should the sentence removed from the core be retained as a regular belief? Also, in this paper we considered revision systems consisting of just *two* revision operators $\langle \circ, \bullet \rangle$ with $\circ$

being, in a sense, *dominated* by •. A possible extension would be to consider an extended *family* of revision operators $\langle \circ_1, \circ_2, \ldots, \circ_n \rangle$ of increasing "strength", perhaps with each $\circ_i$ corresponding to a different source of information. Finally we would like to investigate other natural ways of constructing a revision system $\langle \circ, \bullet \rangle$ which perhaps do not satisfy some of the stronger postulates considered here (such as **(Comm)**$_\triangle$).

Acknowledgements

This work is supported by the DFG research project "Computationale Dialektik". The motivation to begin the work came from some interesting discussions with Emil Weydert. Thanks are also due to Gerhard Brewka for his general support and to the anonymous referees for their detailed comments.

References

1. C. Alchourrón, P. Gärdenfors and D. Makinson, On the logic of theory change: Partial meet contraction and revision functions, *Journal of Symbolic Logic* **50** (1985) 510–530.
2. R. Booth, A negotiation-style framework for non-prioritised revision, in: *Proceedings of the Eighth Conference on Theoretical Aspects of Rationality and Knowledge (TARK 2001)* (2001) 137–150.
3. C. Boutilier, Iterated revision and minimal change of conditional beliefs, *Journal of Philosophical Logic* **25** (1996) 262–305.
4. C. Boutilier and M. Goldszmidt, Revision by conditional beliefs, in: *Proceedings of the Eleventh National Conference on Artificial Intelligence (AAAI'93)* (1993) 649–654.
5. J. Cantwell, Some logics of iterated belief change, *Studia Logica* **63** (1999) 49–84.
6. S. Chopra, K. Georgatos and R. Parikh, Relevance sensitive non-monotonic inference on belief sequences, *Journal of Applied Non-Classical Logics* **11**(1-2) (2001) 131–150.
7. S. Chopra, A. K. Ghose and T. A. Meyer, Iterated revision and recovery: a unified treatment via epistemic states, in: *Proceedings of the Fifteenth European Conference on Artificial Intelligence (ECAI 2002)* (2002) 541–545.
8. A. Darwiche and J. Pearl, On the logic of iterated belief revision, *Artificial Intelligence* **89** (1997) 1–29.
9. E. Fermé, *Revising the AGM Postulates*, PhD Thesis, Universidad de Buenos Aires (1999).
10. P. Gärdenfors, *Knowledge in Flux*, MIT Press (1988).
11. A. Grove, Two modellings for theory change, *Journal of Philosophical Logic*, **17** (1988) 157–170.
12. S. O. Hansson, A survey of non-prioritized belief revision, *Erkenntnis* **50** (1999) 413–427.
13. S. O. Hansson, E. Fermé, J. Cantwell and M. Falappa, Credibility-limited revision, *Journal of Symbolic Logic* **66**(4) (2001) 1581–1596.
14. A. Herzig, S. Konieczny and L. Perrussel, On iterated revision in the AGM framework, in: *Proceedings of the Seventh European Conference on Symbolic and Quantitative Approaches to Reasoning with Uncertainty (ECSQARU'03)* (2003).

15. H. Katsuno and A. O. Mendelzon, Propositional knowledge base revision and minimal change, *Artificial Intelligence* **52** (1991) 263–294.

16. G. Kern-Isberner, Postulates for conditional belief revision, in: *Proceedings of the Sixteenth International Conference on Artificial Intelligence (IJCAI'99)* (1999) 186–191.

17. S. Konieczny and R. Pino Pérez, A framework for iterated revision, *Journal of Applied Non-Classical Logics* **10**(3-4) (2000) 339–367.

18. D. Lehmann, Belief revision, revised, in: *Proceedings of the Fourteenth International Joint Conference on Artificial Intelligence (IJCAI'95)* (1995) 1534–1540.

19. D. Makinson, Screened revision, *Theoria* **63** (1997) 14–23.

20. A. C. Nayak, Iterated belief change based on epistemic entrenchment, *Erkenntnis* **41** (1994) 353–390.

21. A. C. Nayak, M. Pagnucco and P. Peppas, Dynamic belief revision operators, *Artificial Intelligence* **146** (2003) 193–228.

22. H. Rott, *Change, Choice and Inference: A Study of Belief Revision and Nonmonotonic Reasoning*, Oxford University Press (2001).

Assertions, Conditionals, and Defaults

Rainer Osswald

Praktische Informatik VII, FernUniversität in Hagen,
Universitätsstraße 1, 58084 Hagen, Germany
rainer.osswald@fernuni-hagen.de

Abstract. A logical framework that emphasizes the impact of *affirmative* assertions is investigated with respect to providing an interpretation for conditionals and defaults. We consider theories consisting essentially of monadic subsumption statements. For every such theory there is an associated *domain of information states* ordered by a *specialization relation*. We show how to naturally interpret *intuitionistic conditionals* in this information domain. Moreover, we study several ways to cope with *defaults*. In particular, we adapt the approaches of Poole and Reiter to our framework.

1 Introduction

The logic of *positive* or *affirmative* assertions has been proposed as an adequate programming logic to formally describe or specify the behavior of computational devices [21, 23]. The underlying idea is that only the presence and not the absence of a property is "finitely observable". In this paper, we adopt the logic of affirmative assertions as a suitable paradigm for information processing. According to this view, properties are always affirmed to hold of an object or situation, and not denied to hold. As a consequence, the information about an entity accumulated during processing is *persistent* in the sense that once affirmed, a property will stay affirmed. Information processing is thus considered as *incremental* and the information accumulated at a certain state as *partial*.

Under such a perspective, the absence of a property is not informative because it just indicates lack of information. A different thing is to deny a property persistently, which means to affirm that the entity under consideration surely does not have this property. But notice that persistent denial (in general) cannot be based on observations. A second option is to affirm the negated property by considering the absence of a property as an observable property itself. Both options and the corresponding versions for conditionals will be discussed in detail.

In general, information processing is based on *background knowledge*. In this paper, we keep to a simple knowledge representation language, whose expressiveness is basically restricted to the subsumption of affirmative monadic predicates. A background theory (or knowledge base) consisting of axioms of this form determines an *information domain*, which is the space of all possible information states that can arise during processing, provided the background theory is true. Information processing can then be seen as a traversal from less specific to more

G. Kern-Isberner, W. Rödder, and F. Kulmann (Eds.): WCII 2002, LNAI 3301, pp. 108–130, 2005.

specific states. Within this framework, it is natural to ask for appropriate ways of employing *default assumptions*. The idea is that if an information state justifies a certain default assumption, one can traverse to a more specific information state by taking this assumption as affirmed.

The paper is organized as follows: Section 2.1 introduces some basic terminology and notation. In addition to standard notions like interpretation, validity, theory, model etc, we introduce the *specialization* relation given by an interpretation and consider *persistence* with respect to this relation. In Section 2.2, we define the key concept of an *information domain* of a theory as the ordered universe of a certain canonical model of that theory. In Section 3.1, it is shown that (explicit) term negation trivializes the information domain. Section 3.2 presents a natural *intuitionistic* interpretation of conditional and negation with respect to the information domain of a theory. Section 4.1 shows how *default predicates* give rise to *extensions* of information states. In Section 4.2, this approach is generalized to defaults in the style of Reiter [17].

2   The Logic of Affirmative Assertions

2.1   Preliminaries

Let Σ be a set of *atomic* one-place predicates. The logical language we employ in the following is essentially a small fragment of first-order predicate logic presented in a variable-free manner. In particular, the Boolean logical connectives are "lifted" to predicate operators; for instance, if ϕ and ψ are one-place predicates, then $\phi \wedge \psi$ is their logical conjunction.[1] We furthermore introduce two predicates V and Λ which are respectively satisfied by everything and nothing in the universe of discourse. In addition, let $\forall\phi$ stand for $\forall x(\phi x)$, where ϕ is a one-place predicate.

Definition 1 (Term/Statement). *A* Boolean predicate *or* term *over* Σ *is inductively built by* $\wedge$, $\vee$, $\neg$, *and* $\rightarrow$ *from members of* Σ *plus* V *and* Λ. *A Boolean term is* affirmative *(or* positive*) if it is free of* $\neg$ *and* $\rightarrow$. *The term algebra of Boolean terms over* Σ *is denoted by* $B[\Sigma]$, *the algebra of affirmative terms by* $A[\Sigma]$. *A* universal statement *over* Σ *is a statement of the form* $\forall\phi$, *with* $\phi \in B[\Sigma]$.

Interpretation and satisfaction are defined as usual in first-order logic:

Definition 2 (Interpretation/Satisfaction). *A* (set-valued) interpretation *of* Σ *consists of a* universe U *and an* interpretation function M *from* Σ *to the power set* $\wp(U)$ *of* U. *An interpretation function* M *from* Σ *to* $\wp(U)$ *uniquely corresponds to a* satisfaction relation $\vDash$ *from* U *to* Σ, *with* $x \vDash p$ *iff* $x \in M(p)$.

An interpretation function M from Σ to $\wp(U)$ can be inductively extended to a function from $B[\Sigma]$ to $\wp(U)$ by defining $M(V) = U$, $M(\Lambda) = \varnothing$, $M(\phi \wedge$

---

[1] Formally, $\phi \wedge \psi$ is short for $\lambda x(\phi x \wedge \psi x)$, where the λ-notation indicates *predicate abstraction*.

$\psi) = M(\phi) \cap M(\psi)$, $M(\phi \vee \psi) = M(\phi) \cup M(\psi)$, $M(\neg\phi) = U \setminus M(\phi)$, and $M(\phi \rightarrow \psi) = M(\neg\phi \vee \psi)$. Correspondingly, the associated satisfaction relation $\vDash$ can be extended to one from U to $B[\Sigma]$ such that $x \vDash \phi$ iff $x \in M(\phi)$.

The set $M(\phi) = \{x \in U \mid x \vDash \phi\}$ is called the *extent* of ϕ and is also denoted by $[\![\phi]\!]$. The set $\iota(x) = \{p \in \Sigma \mid x \vDash p\}$ is referred to as the *intent* of x.

Definition 3 (Canonical Interpretation). *The* canonical interpretation *of Σ has universe $\wp(\Sigma)$ and takes each $p \in \Sigma$ to the set $\{X \subseteq \Sigma \mid p \in X\}$; that is, $X \vDash p$ iff $p \in X$.*

By associating with each subset of Σ its characteristic function from Σ to $2 = \{0, 1\}$, one gets the standard notion of a *valuation of Σ*. From this point of view, the canonical interpretation takes $\phi \in B[\Sigma]$ to the set of all valuations v of Σ with $v(\phi) = 1$.

Definition 4 (Satisfiability/Validity/Truth). *A Boolean predicate ϕ over Σ is* satisfiable *with respect to an interpretation of Σ with universe U if ϕ is satisfied by some member of U; ϕ is* valid *if ϕ satisfied by every member of U; the statement $\forall \phi$ is* true *if ϕ is valid. The predicate ϕ is* logically valid *if it is valid with respect to all interpretations of Σ.*

Implication and *equivalence* are validity of the conditional and the biconditional, respectively. That is, ϕ (logically) *implies* ψ if $\phi \rightarrow \psi$ is (logically) valid.

Proposition 1. *Given an interpretation of Σ with universe U, then $\phi \in B[\Sigma]$ is satisfied by $x \in U$ just in case ϕ is satisfied by the intent of x under the canonical interpretation.*

Proof. In case $\phi \in \Sigma \cup \{\Lambda, V\}$, there is nothing to show. Now term induction: $x \vDash \neg\phi$ iff $x \nvDash \phi$ iff, by induction hypothesis, $\iota(x) \nvDash \phi$ iff $\iota(x) \vDash \neg\phi$. Similar for the other logical connectives.

Corollary 1. *A Boolean predicate is logically valid iff it is valid with respect to the canonical interpretation.*

An interpretation comes along with a natural specialization ordering on its universe:

Definition 5 (Specialization). *Given an interpretation of Σ with universe U and two members x and y of U, then x is* specialized by y*, in symbols, $x \sqsubseteq y$, if y satisfies every member of Σ that is satisfied by x.*

Remark 1. In case Σ is finite, we can define a dyadic specialization predicate $\sqsubseteq$ directly within the language of first-order logic as follows:

$$x \sqsubseteq y \qquad \text{iff} \qquad \bigwedge\{px \rightarrow py \mid p \in \Sigma\}.^2$$

---

[2] Here and in the following $\bigwedge$ takes a nonempty finite set P of predicates (or statements) to the conjunction of the elements P in any order. Similar use is made of $\bigvee$. In addition, we define $\bigwedge \varnothing = V$ and $\bigvee \varnothing = \Lambda$.

Clearly this gives an interpretation of specialization as defined above. A similar treatment for infinite Σ, however, would call for infinitary logic.

The specialization relation $\sqsubseteq$ is reflexive and transitive, i.e. a preorder. A subset S of U is *upwards closed* if $\uparrow S \subseteq S$, where $\uparrow S = \{y \in U \mid \exists x \in S (x \sqsubseteq y)\}$.

Definition 6 (Persistence). *A predicate ϕ over Σ is persistent if its extent under each interpretation of Σ is upwards closed with respect to specialization.*

Lemma 1. *Given an interpretation of Σ with universe U, then for all $x, y \subset U$,*

$$x \sqsubseteq y \qquad \text{iff} \qquad \forall \phi \in A[\Sigma] (x \vDash \phi \rightarrow y \vDash \phi).$$

Proof. Use the definition of $\sqsubseteq$ and straightforward term induction over $A[\Sigma]$.

Proposition 2. *A Boolean predicate is persistent iff it is logically equivalent to an affirmative predicate.*

Proof. Suppose $\phi \in A[\Sigma]$. Choose any interpretation of Σ with universe U. If $x \vDash \phi$ and $x \sqsubseteq y$, with $x, y \in U$, then $y \vDash \phi$, by Lemma 1. Hence ϕ is persistent. Conversely, suppose $\phi \in B[\Sigma]$ is persistent. Then, with Σ interpreted canonically, $[\![\phi]\!]$ is of the form $\uparrow \mathcal{F}$, where $\mathcal{F}$ is a finite set of finite subsets of Σ. Hence ϕ is equivalent to $\bigvee \{\bigwedge X \mid X \in \mathcal{F}\}$. Now apply Corollary 1.

Remark 2. To many readers, the foregoing observation may look more familiar when formulated in terms of Boolean functions. For a Boolean predicate is persistent iff its corresponding Boolean function is monotone.

Proposition 3. *Let Σ be finite. Given an interpretation of Σ with universe U, every upwards closed subset of U is the extent of an affirmative predicate over Σ.*

Proof. Suppose $V \subseteq U$ is upwards closed with respect to $\sqsubseteq$. Then the image $\mathcal{V} \subseteq \wp(\Sigma)$ of V by ι is upwards closed in turn. By assumption, the set $\mathcal{F}$ of minimal elements of $\mathcal{V}$ is finite. So $\mathcal{V}$ is the extent of $\bigvee \{\bigwedge X \mid X \in \mathcal{F}\}$. Now apply Proposition 1.

Definition 7 (Theory/Model). *A theory Γ over Σ is a set of universal statements over Σ. A model of Γ is an interpretation of Σ with respect to which all statements of Γ are true.*

Given two theories Γ and Γ' over Σ, we say that Γ *entails* Γ', in symbols, $\Gamma \vdash \Gamma'$, if every model of Γ is also a model of Γ'. The theories Γ and Γ' are said to be *equivalent* if they entail each other.

Definition 8 (Conditional Form). *A theory over Σ has* conditional *(or bi-conditional) form if its statements are of the form $\forall(\phi \rightarrow \psi)$ (or $\forall(\phi \leftrightarrow \psi)$), with ϕ and ψ affirmative. The conditional form is* normal, *if ϕ is purely conjunctive (or $\vee$) and ψ is purely disjunctive (or $\wedge$).*

For convenience, we introduce two binary term operators $\preceq$ and $\equiv$ such that $\phi \preceq \psi$ and $\phi \equiv \psi$ are $\forall(\phi \to \psi)$ and $\forall(\phi \leftrightarrow \psi)$, respectively.

Proposition 4. *Every theory is equivalent to a theory in conditional (normal) form and to one in biconditional form.*

Proof. By applying the standard transformations of propositional logic, every Boolean predicate over Σ can be brought into conjunctive normal form. Moreover, the statements

$$\neg p_1 \vee \ldots \vee \neg p_m \vee q_1 \vee \ldots \vee q_n \equiv p_1 \wedge \ldots \wedge p_m \to q_1 \vee \ldots \vee q_n,$$

$$q_1 \vee \ldots \vee q_n \equiv V \to q_1 \vee \ldots \vee q_n,$$

$$\neg p_1 \vee \ldots \vee \neg p_m \equiv p_1 \wedge \ldots \wedge p_m \to \Lambda,$$

are logically true, and $\forall(\phi \wedge \psi)$ is equivalent to $\{\forall\phi, \forall\psi\}$. So every universal statement over Σ is equivalent to a finite set of statements of the form $\phi \preceq \psi$, with ϕ, ψ affirmative. Finally recall that $\phi \to \psi$ is equivalent to $\phi \leftrightarrow \phi \wedge \psi$ (and to $\phi \vee \psi \leftrightarrow \psi$).

Remark 3 (Predicates vs. Propositions). The definitions given in this section are intended as a natural formalization of the simple conceptual framework we put forward in the introduction: information about an entity is accumulated by observing certain properties of that entity and by employing background knowledge in form of subsumption statements. Alternatively, one can of course work in propositional logic, where atomic propositions take the place of atomic monadic predicates. From a conceptual point of view this simply means to think of the entity in question as the "world". The above definitions of interpretation and model then resemble a standard possible world semantics. Moreover, one should keep in mind that any universal first-order theory allows "propositional grounding" by taking the elements of the Herbrand base as atomic propositions.

2.2   Information Domains

There is a standard way to associate with each theory a *canonical model* by restricting the universe of the canonical interpretation appropriately:

Definition 9 (Canonical Model). *Suppose Γ is a theory over Σ. Let $C(\Gamma)$ be the set of all $X \subseteq \Sigma$ which, under the canonical interpretation, satisfy ϕ for every statement $\forall\phi$ of Γ. The* canonical model *of Γ takes $p \in \Sigma$ to $\{X \in C(\Gamma) \mid p \in X\}$.*

Specialization on $C(\Gamma)$ is set inclusion and hence a partial order. We refer to the members of $C(\Gamma)$ as *consistently Γ-closed subsets* of Σ. The following immediate consequence of definitions will prove useful:

Proposition 5. *Let Γ and Γ' be theories over Σ. Then $C(\Gamma \cup \Gamma') = C(\Gamma) \cap C(\Gamma')$. In particular, if $\Gamma \subseteq \Gamma'$ then $C(\Gamma') \subseteq C(\Gamma)$.*

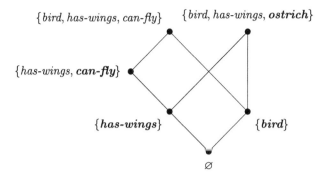

Fig. 1. Information domain of Γ

Notice that if Γ has conditional normal form, i.e. if every statement of Γ is of the form $\bigwedge P \preceq \bigvee Q$, with $P, Q \subseteq \Sigma$ finite, then $X \subseteq \Sigma$ belongs to $C(\Gamma)$ just in case $P \subseteq X \to X \cap Q \neq \varnothing$, for all $(\bigwedge P \preceq \bigvee Q) \in \Gamma$. Moreover notice that the valuations of Σ corresponding to members of $C(\Gamma)$ are precisely the 2-valued models of Γ in the sense of standard propositional logic.

Adapting the terminology of [6], we refer to any ordered set that is order-isomorphic to $C(\Gamma)$ as "the" *information domain* of Γ.

Example 1. Suppose $\Sigma = \{bird, ostrich, can\text{-}fly, has\text{-}wings\}$. Let Γ be the theory over Σ consisting of the statements

$$
\begin{aligned}
can\text{-}fly &\preceq has\text{-}wings, \\
ostrich &\preceq bird \wedge has\text{-}wings, \\
can\text{-}fly \wedge ostrich &\preceq \Lambda, \\
has\text{-}wings \wedge bird &\preceq can\text{-}fly \vee ostrich.
\end{aligned}
$$

The information domain of Γ is depicted in Figure 1. If a member p of a consistently closed set X is given in boldface, X is the *least satisfier* of p. For instance, $\{has\text{-}wings, can\text{-}fly\}$ is the least satisfier of *can-fly*.

A Paradigm for Information Processing. A theory represents knowledge about a certain domain of discourse. During processing this knowledge is applied to affirmative knowledge (or information) a cognitive agent has acquired about a certain object or situation. So the theory provides *background knowledge* and its information domain is the *space of information states* the cognitive agent traverses in the course of accumulating positive information about an entity. Under normal conditions, the traversal is directed from less specific to more specific information states. Consequently, once an affirmative predicate is accepted, more information will not affect this decision. For, by Proposition 1, affirmative predicates are persistent with respect to specialization.

Suppose the cognitive agent affirms ϕ in state x. Under the assumption that her next state only depends on her background knowledge and her affirmative knowledge about the current situation, she will move to the *least satisfier* of ϕ

above x, *if existent*. Put differently, if x is the least satisfier of some affirmative predicate ψ, the target state is the least satisfier of $\phi \wedge \psi$. Of course, least satisfiers do not exist in general. For instance, given the theory of Example 1, the affirmation of *has-wings* in the state $\{bird\}$ does not license a move to another state. Either the agent waits for additional information or she selects one of the minimal satisfiers based on *default assumptions*. The issue of using defaults is taken up in Section 4 in more detail.

Horn Theories. The ordering structure of an information domain depends on the type of its theory. For instance, let Γ be a *Horn theory* over Σ, i.e. a theory whose conditional normal form is free of disjunctions. Then $C(\Gamma)$ is an *inductive intersection system*, i.e. is closed with respect to intersection of nonempty subset systems and union of upwards directed subset systems (and every inductive intersection system arises that way; cf. [15]). In order theoretic terms, the information domains of Horn theories are precisely the so-called *Scott domains*. In particular, every two elements with a common upper bound have a least upper bound.

Since $C(\Gamma)$ is an inductive intersection system, it follows that each subset S of Σ has a unique *closure* $cl_\Gamma(S)$ with respect to Γ, namely $\bigcap\{X \in C(\Gamma) \mid S \subseteq X\}$, provided that S is *consistent* at all, i.e. a subset of some member of $C(\Gamma)$. We may assume that the statements of Γ are either of the form $\bigwedge P \preceq \Lambda$ or $\bigwedge P \preceq q$, with $P \subseteq \Sigma$ finite and $q \in \Sigma$. The closure of a (consistent) subset of Σ can then be determined by repeated application of the consequence operator d_Γ that takes any (consistent) subset S of Σ to

$$d_\Gamma(S) \;=\; S \cup \{q \mid P \subseteq S, \, (\textstyle\bigwedge P \preceq q) \in \Gamma\}.$$

More precisely, we have by standard fixpoint arguments:

Proposition 6. *Let Γ be a Horn theory over Σ. If $S \subseteq \Sigma$ is consistent then*

$$cl_\Gamma(S) \;=\; \bigcap\{X \subseteq \Sigma \mid S \subseteq X = d(X)\} \;=\; \bigcup\{d_\Gamma^i(S) \mid i \geq 0\}.$$

Algebraic Representation. The canonical model of a theory Γ can be seen as a *representation* of the knowledge given by Γ. Indeed, let Γ^* be the theory (over the same set of atoms) that consists of all statements $\forall \phi$ such that ϕ is satisfied by each element of the information domain of Γ. Since first-order logic is (strongly) complete, it follows that Γ^* is the *deductive closure* of Γ. So Γ is determined by its canonical model up to (deductive) equivalence.

It is also possible to come up with an *algebraic* representation of Γ (which, for instance, is favored in [14]). According to a classical result of Birkhoff, there is a categorical equivalence between the finite ordered sets and the finite distributive lattices with zero and unit (cf. [5]). Adapted to the present context, with Σ assumed as finite, the information domain can be constructed from the *Lindenbaum algebra* of a theory, and vice versa.

Definition 10 (Lindenbaum Algebra). *Given a theory Γ over Σ, the Lindenbaum algebra $L(\Gamma)$ (of affirmative terms) of Γ is the quotient $A[\Sigma]/{\simeq_\Gamma}$, with $\phi \simeq_\Gamma \psi$ iff $\Gamma \vdash \phi \equiv \psi$.*

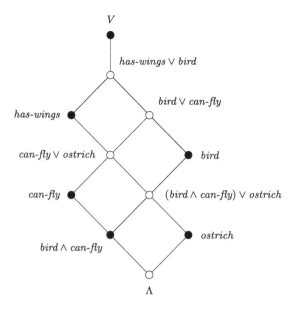

Fig. 2. Lindenbaum algebra of theory of Example 1

Remark 4. Our definition of the Lindenbaum algebra of Γ differs from the standard textbook definition insofar as we consider only affirmative terms and not Boolean terms in general. If $A[\Sigma]$ is replaced by $B[\Sigma]$, the resulting lattice is Boolean, i.e. complemented. Another way to construct this Boolean lattice is to the apply the Lindenbaum construction to the affirmative terms of the Booleanization $\overline{\Gamma}$ of Γ; see Section 3.1 below. Moreover, Booleanization induces an embedding of the distributive lattice $L(\Gamma)$ into the Boolean lattice $L(\overline{\Gamma})$, and this embedding is universal in a certain sense; see [23] and [15] for details.

Example 2. Let Γ be as in Example 1. The Lindenbaum algebra of Γ is depicted in Figure 2. The shaded circles indicate the $\vee$-*irreducible* elements of $L(\Gamma)$, which stand in a one-to-one order-reversing correspondence with the elements of the information domain of Γ; cf. Figure 1. Conversely, the structure of $L(\Gamma)$ can be recovered from the information domain by taking all upwards closed subsets (ordered by set inclusion).

3   Conditional and Negation

3.1   Classical Negation

Let us briefly reflect on the status of negation in the framework of information processing presented so far. On the one hand, we put no restriction on negation in theories. More precisely, the matrix of our universal statements may consist of arbitrary Boolean terms. (But notice that negation *of* universal statements, which gives rise to existential ones, is not allowed.) On the other hand, we are only interested in the contribution of theories to support *positive* assertions.

Consider again Example 1. The statement *ostrich* $\preceq \neg$*can-fly* cannot be used to infer an affirmative predication (at least as long as $\neg$*can-fly* is not regarded as an affirmative predicate on its own – see below). The conditional normal form *can-fly* $\wedge$ *ostrich* $\preceq \Lambda$ reveals that its sole purpose is to cut down the number of possible information states (and to detect inconsistencies in the knowledge base). Similarly, take the statement *has-wings* $\wedge$ *bird* $\wedge \neg$*ostrich* $\preceq$ *can-fly*, which is logically equivalent to the last statement of Example 1. As long as we have no positive interpretation of $\neg$*ostrich*, the statement is of no use for moving to a higher information state. One such positive reading of negation, *term negation*, is presented in the following. Another one, *negation as failure*, will be sketched in Section 4.2.

Predicate Negation and Booleanization. We have argued that though negation may be present in theories, there remains an asymmetry between negated and non-negated predicates if we think of a theory as a knowledge base for accumulating positive information. This asymmetry towards positive information is reflected in the construction of the information domain, whose elements represent sets of affirmative predicates and whose ordering signals the increase of positive information.

The asymmetry can be resolved by regarding negated predicates as taking part in affirmative assertions. That is, predications involving negated predicates are not viewed as denials but as assertions. This is the traditional distinction between *predicate denial* and *term negation*, which Horn [8] traces back to Aristotle; schematically: 'x (is not) A' versus 'x is (not A)'.

To reconcile predicate negation with the viewpoint of affirmative predication one can "Booleanize" a theory by supplementing it with all negated atoms and all statements that instantiate the laws of contradiction and excluded middle for atoms:

Definition 11 (Booleanization). *The* Booleanization *of a theory* Γ *over* Σ *is the theory* $\overline{\Gamma}$ *over* $\Sigma \cup \{-p \,|\, p \in \Sigma\}$, *where*

$$\overline{\Gamma} = \Gamma \cup \{p \wedge -p \preceq \Lambda \,|\, p \in \Sigma\} \cup \{V \preceq p \vee -p \,|\, p \in \Sigma\}.$$

Since each member of $C(\overline{\Gamma})$ contains either p or $-p$, but not both, if follows that no member of $C(\overline{\Gamma})$ is a proper subset of another member of $C(\overline{\Gamma})$, that is:

Proposition 7. *The information domain of the Booleanization of a theory is an antichain.*

Example 3. Suppose $\Sigma = \{a, b, c\}$ and $\Gamma = \{a \wedge b \preceq \Lambda, \ a \vee b \preceq c\}$. Then $C(\Gamma)$ consists of $\varnothing$, $\{c\}$, $\{a, c\}$, and $\{b, c\}$, whereas $C(\overline{\Gamma})$ consists of $\{-a, -b, -c\}$, $\{-a, -b, c\}$, $\{a, -b, c\}$, and $\{-a, b, c\}$; see Figure 3.

The previous example indicates that Booleanization means loss of specialization while keeping all elements of the information domain. Indeed:

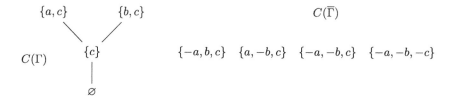

$C(\Gamma)$

Fig. 3. Trivializing effect of Booleanization on specialization

Proposition 8. *There is a one-to-one correspondence between the information domain of a theory and that of its Booleanization.*

Proof. Let f be the function from $C(\overline{\Gamma})$ to $C(\Gamma)$ that takes Y to $Y \cap \Sigma$. For every $X \in C(\Gamma)$, the set $\overline{X} = X \cup \{-p \mid p \in \Sigma \setminus X\}$ is consistently closed with respect to $\overline{\Gamma}$. Clearly, $\overline{X} \cap \Sigma = X$ and $\overline{Y} \cap \Sigma = Y$. Hence f is onto and one-to-one.

3.2   Intuitionistic Conditional and Negation

It is implicit in *intuitionistic logic* that knowledge is persistent. Since this point of view is much in accordance with the principles of information processing presented here, it seems reasonable to adopt an intuitionistic interpretation of negation and conditional.[3] Roughly speaking, this interpretation is obtained by "persistifying" classical negation and conditional.

Recall from Section 2.1 that a predicate ϕ is persistent iff its extent $[\![\phi]\!]$ under every interpretation is upwards closed with respect to specialization. In order to express the property of persistence more succinctly, let us introduce an additional predicate operator $\square$, where, given an interpretation of Σ with universe U, satisfaction of $\square\phi$ by $x \in U$ is defined as follows:

$$x \vDash \square\phi \qquad \text{iff} \qquad \forall y \in U(x \sqsubseteq y \rightarrow y \vDash \phi).$$

In terms of extents, $x \in [\![\square\phi]\!]$ iff $\uparrow\{x\} = \{y \in U \mid x \sqsubseteq y\} \subseteq [\![\phi]\!]$. In the following we speak of predicates that are constructed by the Boolean connectives and $\square$ over $\Sigma \cup \{\Lambda, V\}$ briefly as *predicates over* Σ. The definition of satisfaction, validity, etc can be carried over from the Boolean case without changes.

Remark 5. Recall from Remark 1 that specialization can be defined within first-order logic as long as Σ is finite (and in infinitary logic otherwise). The same can thence be said of $\square$.

With the help of $\square$, the condition of persistence can be formulated as follows: A predicate ϕ over Σ is persistent if $\phi \equiv \square\phi$ is true with respect to every interpretation of Σ.

---

[3] See [22] for an introduction to intuitionistic logic.

Proposition 9. *For every two predicates ϕ, ψ over Σ, the following predicates are valid with respect to all interpretations of Σ:*

(i) $\Box(\phi \to \psi) \to (\Box\phi \to \Box\psi)$,   (ii) $\Box\phi \to \Box\Box\phi$,   (iii) $\Box\phi \to \phi$.

Proof. Either by recourse to the interpretation of $\Box$ or by recognizing that (i), (ii), and (iii) are the axioms of modal logic **S4**, which are known to be valid in all reflexive and transitive Kripke frames.[4]

Corollary 2. *For any predicate ϕ, the predicate $\Box\phi$ is persistent.*

Proof. By Proposition 9(ii) and (iii), $\Box\Box\phi \equiv \Box\phi$ is true under all interpretations.

In particular, it follows by Proposition 3:

Corollary 3. *Suppose Σ is finite and ϕ is a predicate over Σ. Then, for every interpretation of Σ, there is an affirmative predicate over Σ that is equivalent to $\Box\phi$ with respect to that interpretation.*

The following simple example illustrates that finiteness is essential here:

Example 4. Suppose Σ is infinite and Γ is the set of all statements $p \preceq \neg q$, with $p, q \in \Sigma$, $p \neq q$. Then $C(\Gamma) = \{\varnothing\} \cup \{\{p\} \mid p \in \Sigma\}$, and $[\![\Box\neg p]\!] = \{\{q\} \mid q \neq p\}$ is upwards closed but clearly not the extent of an affirmative predicate over Σ.

Another thing to keep in mind when applying Corollary 3 is the dependence on the given interpretation:

Example 5. Suppose $\Sigma = \{a, b\}$. If $\Gamma = \varnothing$ then $C(\Gamma) = \{\varnothing, \{a\}, \{b\}, \{a, b\}\}$; thus $\Box\neg a$ is equivalent to Λ. On the other hand, if $\Gamma = \{a \wedge b \preceq \Lambda\}$ then $C(\Gamma) = \{\varnothing, \{a\}, \{b\}\}$ and $\Box\neg a$ is equivalent to b.

If not otherwise indicated, Σ is henceforth assumed to be finite.

Persistent Conditional and Negation. Classical conditional and negation do not preserve persistence. If we want persistent versions of conditional and negation, we need to "persistify" them by means of $\Box$; that is, we define predicate operators $\Rightarrow$ and $\sim$ such that

$$\phi \Rightarrow \psi = \Box(\phi \to \psi) \qquad \text{and} \qquad \sim\phi = \Box\neg\phi.$$

In particular, $\sim\phi \equiv \phi \Rightarrow \Lambda$. (Defining $\Rightarrow$ and $\sim$ along these lines essentially means to embed intuitionistic logic into modal logic **S4**, which goes back to Gödel.)

In order to determine the extent of $\phi \Rightarrow \psi$, the following reformulation of definitions is sometimes useful:

$$[\![\phi \Rightarrow \psi]\!] = \{x \mid \uparrow\{x\} \cap [\![\phi]\!] \subseteq [\![\psi]\!]\}.$$

---

[4] See e.g. [11] or [1].

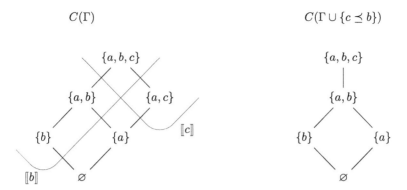

$$C(\Gamma) \qquad\qquad\qquad C(\Gamma \cup \{c \preceq b\})$$

Fig. 4. Black cats and other animals

Example 6. Consider again the theory Γ of Example 1. One can read off easily from Figure 1 that the extent of *bird* $\Rightarrow$ *can-fly* is $\uparrow\{$*has-wings, can-fly*$\}$ with respect to the information domain $C(\Gamma)$ of Γ. Consequently, *bird* $\Rightarrow$ *can-fly* is equivalent to *can-fly*.

In the rest of this section, we look more closely at the interrelation between $\phi \preceq \psi$ and $V \preceq \phi \Rightarrow \psi$ (i.e. between $\forall(\phi \to \psi)$ and $\forall(\Box(\phi \to \psi))$).

Lemma 2. *If ϕ is persistent then $\phi \wedge \psi \preceq \chi$ is equivalent to $\phi \preceq \psi \Rightarrow \chi$.*

Proof. Suppose $\phi \wedge \psi \preceq \chi$ is true. That is, $\phi \wedge \psi \to \chi$ and hence $\phi \to (\psi \to \chi)$ are valid. Consequently, $\Box(\phi \to (\psi \to \chi))$ and thus $\Box\phi \to \Box(\psi \to \chi)$, by Proposition 9(i), are valid too. Since ϕ is persistent, it follows that $\phi \preceq \psi \Rightarrow \chi$ is true. Similar for the reverse direction.

So, given two affirmative predicates ϕ and ψ, we know by Lemma 2 that $\phi \preceq \psi$ is equivalent to $V \preceq \phi \Rightarrow \psi$. Hence one might expect that extending a theory Γ by $\phi \preceq \psi$ has the same effect as extending Γ by $V \preceq \phi \Rightarrow \psi$. However, this is clearly a mistake, because in general, $\phi \to \psi$ does no imply $V \to (\phi \Rightarrow \psi)$ (i.e. $\Box(\phi \to \psi)$). The following example illustrates this fact:

Example 7. Suppose $\Sigma = \{a, b, c\}$ and $\Gamma = \{c \preceq a\}$. To bring some life into Γ, say a is *animal*, b is *black*, and c is *cat*. The information domain of Γ, represented by $C(\Gamma)$, is depicted on the left of Figure 4. Consider the extent of $c \Rightarrow b$ in $C(\Gamma)$. By definition, $c \Rightarrow b$ is satisfied by $X \in C(\Gamma)$ iff $\uparrow\{X\} \cap [\![c]\!] \subseteq [\![b]\!]$. One can easily read off from Figure 4 that $c \Rightarrow b$ is equivalent to b. Let us contrast $c \Rightarrow b$ with the statement $c \preceq b$. The predicate $c \Rightarrow b$ is satisfied by those states of the information domain of which it is "persistently true" that they satisfy *black* if they satisfy *cat*. For example, the least *animal* state $\{a\}$ does not satisfy $c \Rightarrow b$ because it can be specialized to a *cat* state that is not necessarily a *black* state (represented by $\{a, c\}$). The statement $c \preceq b$, on the other hand, does not exclude non-black non-cats; see Figure 4 for the effect of extending Γ by $c \preceq b$.

Remark 6. Since Σ is assumed as finite, $\Rightarrow$ is an operation on the Lindenbaum algebra of any theory over Σ, with $a \Rightarrow b = \bigvee \{c \mid c \wedge a \leqslant b\}$. Hence $\Rightarrow$ gives the Lindenbaum algebra the structure of a *Heyting algebra*; the element $a \Rightarrow b$ is also known as the *relative pseudo-complement* of a relative to b.

4   Defaults

The picture developed so far is that of a cognitive agent whose background theory determines her possible information states, which are ordered by specialization. Triggered by positive information, the agent traverses her space of information states (her information domain) from less specific states to more specific ones. Let us now expand this picture to the effect that the cognitive agent may also jump to more specific states on the basis of her *default assumptions*.

4.1   Poolean Defaults

The approach of Poole [16] to default reasoning builds on the general idea to presume certain assertions as true unless information to the contrary is available.[5] Poole's approach can be adapted to the framework of information domains in the following way. Suppose Γ is a theory over Σ and Φ is a set of affirmative predicates which are regarded as defaults. Let us refer to these data briefly as a *default system*:

Definition 12 (Default System). *A default system over Σ is a pair $\langle \Gamma, \Phi \rangle$, where Γ is a theory and Φ is a set of affirmative predicates over Σ.*

The question is how to associate with each information state of Γ or, more generally, with each subset of Σ, a more specific information state by employing the defaults in Φ. Since the goal is to presume as many defaults true as possible, we may ask for an information state that satisfies a maximal set of defaults. At the same time, we do not want the new state to represent more information than enforced by the defaults. In short, given a subset S of Σ, we ask for a minimally specific information state extending S that satisfies a maximal subset of Φ. The following preference relation allows to encode both conditions into one minimality constraint. Let Φ_X be $\{\phi \in \Phi \mid X \vDash \phi\}$.

Definition 13 (Preference Relation). *Let $\langle \Gamma, \Phi \rangle$ be a default system, and $X, Y \in C(\Gamma)$. We say that X is preferred over Y with respect to Φ, in symbols, $X <_\Phi Y$, if either $\Phi_Y \subset \Phi_X$ or $\Phi_Y = \Phi_X$ and $X \subset Y$.*

The desired information states can then be described as those consistently Γ-closed supersets of S that are minimal with respect to the preference relation induced by Φ:

---

[5] As noted in [10, p. 66], this idea goes back well beyond Poole.

Definition 14 (Preferred Extension). *A* preferred extension *of S with respect to* $\langle \Gamma, \Phi \rangle$ *is a* $<_\Phi$-*minimal element of* $\{X \in C(\Gamma) \mid S \subseteq X\}$.

Remark 7. Our notion of a preferred extension is closely related to that of a *preferred model* in the sense of [20].

Example 8. Suppose $\Sigma = \Phi = \{a, b, c\}$ and $\Gamma = \{a \wedge b \preceq \Lambda,\ a \wedge c \preceq \Lambda\}$. Then $C(\Gamma) = \{\varnothing, \{a\}, \{b\}, \{c\}, \{b, c\}\}$. The empty set has two preferred extensions with respect to $\langle \Gamma, \Phi \rangle$, namely $\{a\}$ and $\{b, c\}$.

In the previous example, the preferred extensions of the empty set coincide with the maximal elements of the information domain. This is just an instance of the following general fact:

Proposition 10. *If* $\langle \Gamma, \Phi \rangle$ *is a default system over* Σ *such that* $\Phi = \Sigma$ *then the preferred extensions of S with respect to* $\langle \Gamma, \Phi \rangle$ *are the maximal elements of* $\{X \in C(\Gamma) \mid S \subseteq X\}$.

Proof. Since $\Phi = \Sigma$, we have that $\Phi_X = X$. Hence $X <_\Phi Y$ iff $Y \subset X$.

We now consider a slightly different approach to defining extensions, which rests on the idea that an extension should satisfy all default predicates it is consistent with and be minimal with respect to this condition.

Definition 15 (Consistency). *With respect to a theory* Γ *over* Σ, *a predicate* ϕ *over* Σ *is said to be* consistent *with a subset S of* Σ *if there is a consistently* Γ-*closed superset of S that satisfies* ϕ.

In terms of information states, the foregoing definition runs as follows: A predicate ϕ over Σ is consistent with an information state X of Γ if X is specialized by a state of Γ satisfying ϕ. In short, ϕ is consistent with X if $X \vDash \Diamond\phi$, where $\Diamond$ stands for $\neg\square\neg$.

Definition 16 (Closed under Defaults). *An information state X is* closed under *a default predicate* ϕ *if X* satisfies ϕ *whenever* ϕ *is consistent with X, that is, if* $X \vDash (\Diamond\phi \rightarrow \phi)$. *A state is* closed *under a set of defaults if it is closed under every default of that set.*

Definition 17 (Minimal Extension). *A* minimal extension *of S with respect to* $\langle \Gamma, \Phi \rangle$ *is a* $\subseteq$-*minimal element of* $\{X \in C(\Gamma) \mid X$ *is* Φ-*closed and* $S \subseteq X\}$.

If $\Phi = \Sigma$ then the Φ-closed states are precisely the maximal elements of $C(\Gamma)$. Hence, by Proposition 10, the minimal extensions coincide with the preferred extensions in this case. In general, it turns out that every preferred extension is a minimal extension but not the other way round.

Lemma 3. *Suppose* $\langle \Gamma, \Phi \rangle$ *is a default system over* Σ. *Then every preferred extension of* $S \subseteq \Sigma$ *with respect to* $\langle \Gamma, \Phi \rangle$ *is* Φ-*closed.*

Proof. Let X be a preferred extension of S with respect to $\langle \Gamma, \Phi \rangle$. Suppose $X \vDash \Diamond \phi$ for some $\phi \in \Phi$, that is, there is a $Y \in C(\Gamma)$ such that $X \subseteq Y \vDash \phi$. Then $\Phi_X \subseteq \Phi_Y$. Hence $\Phi_X = \Phi_Y$, because, by assumption, $Y \not\prec_\Phi X$. So $X \vDash \phi$.

Proposition 11. *Suppose $\langle \Gamma, \Phi \rangle$ is a default system over Σ. Every preferred extension of S is a minimal extension of S.*

Proof. Let X be a preferred extension of S with respect to $\langle \Gamma, \Phi \rangle$. By Lemma 3 we know that X is Φ-closed. Suppose there is a Φ-closed $Y \in C(\Gamma)$ such that $S \subseteq Y \subseteq X$. Then $\Phi_Y = \Phi_X$, because Y is Φ-closed. Since $Y \not\prec_\Phi X$, it follows that $Y = X$.

Example 9. Suppose $\Sigma = \{a, b, c\}$, $\Gamma = \{a \wedge b \preceq \Lambda,\ b \preceq c\}$, and $\Phi = \{b\}$. Then $C(\Gamma)$ consists of the sets $\varnothing$, $\{a\}$, $\{c\}$, $\{a, c\}$, and $\{b, c\}$, of which $\{a\}$, $\{a, c\}$, and $\{b, c\}$ are Φ-closed. Hence the minimal extensions of $\varnothing$ with respect to $\langle \Gamma, \Phi \rangle$ are $\{a\}$ and $\{b, c\}$, whereas the only preferred extension of $\varnothing$ is $\{b, c\}$ (since $\Phi_{\{a\}} = \varnothing$ and $\Phi_{\{b,c\}} = \{b\}$).

Default Conditionals. Typically, defaults appear as conditionals. For instance, it is not assumed by default that everything can fly, but that everything can fly if it is a bird. Since we want default predicates to be persistent,[6] it is thus natural to employ the intuitionistic conditional.

What does it mean to take $\phi \Rightarrow \psi$ as a default assumption? Being in a state x that satisfies $\phi \Rightarrow \psi$ means that every state y at least as specific as x that satisfies ϕ also satisfies ψ. To assume $\phi \Rightarrow \psi$ thus means something like this: Whatever observations we will make about the object or situation we are currently accumulating information about, we are sure that if it satisfies ϕ, it satisfies ψ. This approach, however, has some drawbacks.

Consider once more the theory of Example 1. Suppose we use *bird $\Rightarrow$ can-fly* to express the default that birds typically can fly. Then it seems reasonable to assume *bird $\Rightarrow$ can-fly* to hold as long as no information to the contrary is available. As observed in Example 6, this means that one can jump to state $\{has\text{-}wings, can\text{-}fly\}$ without any additional information at hand. But this commitment appears to be stronger than wanted because the given theory does not exclude entities (say a certain yet unknown species of bats) which are not birds, have wings, and cannot fly. This problem of default conditionals, Poolean style, will be remedied by the more general approach à la Reiter presented in Section 4.2.

Another problem is how to cope with conflicting defaults, witness the following example:

Example 10. Let Γ be the theory over $\{student, adult, working, non\text{-}working\}$ whose statements are

---

[6] To prevent misunderstandings, this does not mean that default assumptions are persistent in the sense that they will be never given up again. Of course, quite the contrary is the case when belief revision comes into play.

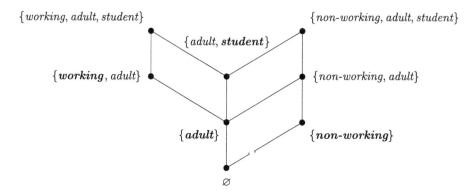

Fig. 5. Information domain of Γ

$$student \preceq adult, \qquad working \preceq adult, \qquad working \wedge non\text{-}working \preceq \Lambda.$$

The information domain of Γ is shown in Figure 5. Consider the two default assumptions that students are not working whereas adults are working. Since students are adults there is no state of the information domain at which both assumptions are persistently true. For $student \Rightarrow non\text{-}working \equiv non\text{-}working$ and $adult \Rightarrow working \equiv working$, but $working \wedge non\text{-}working \equiv \Lambda$.

4.2   Default Theories

Similar to the treatment of Poolean defaults in the previous section, Reiter's [17] well-known approach to default logic can be adapted to the framework of information domains. Again we are interested in default extensions of information states, which are best seen as models and not as theories. Our approach is thus in the same spirit as that of Rounds and Zhang [18, 24], see Remark 9 below, while style of presentation is to some degree inspired by [11].

Definition 18 (Default). *A default over Σ is triple $\langle \phi, \Psi, \chi \rangle$, where ϕ and χ are affirmative predicates and Ψ is a finite set of predicates over Σ.*

It is common to call ϕ the *prerequisite*, χ the *consequent*, and the members of Ψ the *justifications* of the default. If $\Psi = \{\psi_1, \ldots, \psi_n\}$, the default $\langle \phi, \Psi, \chi \rangle$ is also written in the form

$$\frac{\phi : \psi_1, \ldots, \psi_n}{\chi}.$$

Definition 19 (Default Theory). *A default theory over Σ is a pair $\langle \Gamma, \Delta \rangle$ consisting of a theory Γ over Σ and a set Δ of defaults over Σ.*

Definition 20 (Applicability/Reduct). *With respect to a theory Γ over Σ, a default $\langle \phi, \Psi, \chi \rangle$ is applicable to a subset S of Σ if every $\psi \in \Psi$ is consistent with S. Given a default theory $\langle \Gamma, \Delta \rangle$, the reduct Δ^S of Δ with respect to S is the set of all statements $\phi \preceq \chi$ such that $\langle \phi, \Psi, \chi \rangle \in \Delta$ is applicable to S.*

Generalizing Definition 16, we say that an information state X is *closed with respect to* Δ if X belongs to $C(\Delta^X)$, that is, if X satisfies $\phi \to \chi$ whenever $\langle \phi, \Psi, \chi \rangle \in \Delta$ is applicable to X. The concept of a minimal extension can be generalized likewise:

Definition 21 (Minimal Extension). *Given a default theory $\langle \Gamma, \Delta \rangle$ over Σ and a subset S of Σ, a minimal extension of S with respect to $\langle \Gamma, \Delta \rangle$ is a minimal element of $\{X \in C(\Gamma \cup \Delta^X) \mid S \subseteq X\}$.*

Notice that by making these generalizations we assume that a Boolean default predicate ϕ has the same effect as the default $V{:}\phi/\phi$.[7] Indeed, by definition, ϕ is consistent with S iff $V{:}\phi/\phi$ is applicable to S, in which case $V \preceq \phi$ belongs to the reduct with respect to S.

In general, the notion of a minimal extension does not properly capture the intuition of a default extension. For instance, the minimal extension $\{a\}$ of $\varnothing$ in Example 9 should not count as a default extension because neither the statements $a \wedge b \preceq \Lambda$ and $b \preceq c$ of Γ nor the default $V{:}b/b$ provide any reason to assume a by default. Viewed from another angle, the problem is that $\{a\}$ *is not minimal in $C(\Gamma)$ under the condition of being closed with respect to $\varnothing = \Delta^{\{a\}}$.* A promising candidate for defining default extensions then has the following form, which is essentially an adaption of Reiter's definition [17] (see also [18]):

Definition 22 (Default Extension). *Let $\langle \Gamma, \Delta \rangle$ be a default theory over Σ. Then X is a default extension of $S \subseteq \Sigma$ with respect to $\langle \Gamma, \Delta \rangle$ if X is minimal in $\{Y \in C(\Gamma \cup \Delta^X) \mid S \subseteq Y\}$.*

Proposition 12. *With respect to a default theory $\langle \Gamma, \Delta \rangle$ over Σ, every default extension of $S \subseteq \Sigma$ is a minimal extension of S.*

Proof. Suppose X is minimal in $\{Y \in C(\Gamma \cup \Delta^X) \mid S \subseteq Y\}$. Then $X \in C(\Gamma \cup \Delta^X)$, that is, X is closed with respect to Δ. It remains to show that X is minimal in $\{Y \in C(\Gamma \cup \Delta^Y) \mid S \subseteq Y\}$. Suppose $S \subseteq X' \subseteq X$ such that $X' \in C(\Gamma \cup \Delta^{X'})$. Then $\Delta^X \subseteq \Delta^{X'}$ and thus $C(\Delta^{X'}) \subseteq C(\Delta^X)$ (cf. Proposition 5). So $X' \in C(\Gamma \cup \Delta^X)$. Hence, by assumption, $X \subseteq X'$ and, consequently, $X = X'$. $\qquad\blacksquare$

The following two examples show that default extensions need not be unique.

Example 11. Suppose $\Sigma = \{a, b\}$, $\Gamma = \{a \wedge b \preceq \Lambda\}$, and $\Delta = \{V{:}a/a, V{:}b/b\}$. Then $C(\Gamma) = \{\varnothing, \{a\}, \{b\}\}$. Since $\Delta^{\{a\}} = \{V \preceq a\}$ and $\Delta^{\{b\}} = \{V \preceq b\}$, both $\{a\}$ and $\{b\}$ are default extensions of $\varnothing$ with respect to $\langle \Gamma, \Delta \rangle$ (which coincide with the minimal extensions in this case; cf. Proposition 10).

Example 12. Let Σ and Γ be as in Example 10 and

$$\Delta = \left\{ \frac{student : non\text{-}working}{non\text{-}working}, \frac{adult : working}{working} \right\}.$$

---

[7] See also [16–Sect. 4].

Then the least satisfier of *student*, which is {*adult, student*}, has the two default extensions {*working, adult, student*} and {*non-working, adult, student*}, which are incompatible. Clearly it would be preferable in this case if only the latter state counts as an extension because students are specific adults and hence the default assumptions about students should have stronger priority than those about adults.

Remark 8 (Prioritizing Defaults). Some approaches to prioritization allow to impose an arbitrary partial ordering on the set of defaults (see e.g. [11] and the detailed discussion in [2]). Within the framework developed so far, it seems more natural to make use of the base theory Γ in order to impose a priority ordering on defaults, simply by preferring an (applicable) default over another if the prerequisite of the first implies that of the second with respect to Γ.

Horn Default Theories. Let us call a default theory $\langle \Gamma, \Delta \rangle$ *Horn* if Γ is a Horn theory and each member of Δ has purely conjunctive prerequisite and consequent. Then every reduct of Δ is a Horn theory. It follows that X is a default extension of S just in case $X = \bigcap \{ Y \in C(\Gamma \cup \Delta^X) \mid S \subseteq Y \}$, that is, X is the closure $cl_{\Gamma \cup \Delta^X}(S)$ of S with respect to $\Gamma \cup \Delta^X$. Hence, by Proposition 6:

Proposition 13. *Let $\langle \Gamma, \Delta \rangle$ be a Horn default theory over Σ. Then X is a default extension of $S \subseteq \Sigma$ with respect to $\langle \Gamma, \Delta \rangle$ iff*

$$X = cl_{\Gamma \cup \Delta^X}(S) = \bigcup \{ d^i_{\Gamma \cup \Delta^X}(S) \mid i \geq 0 \}.$$

The following reformulation of this result comes close to Reiter's iterative characterization of default extensions and is essentially a notational variant of [18–Corollary 4.1]:

Corollary 4. *Let $\langle \Gamma, \Delta \rangle$ be a Horn default theory over Σ. Then X is a default extension of $S \subseteq \Sigma$ with respect to $\langle \Gamma, \Delta \rangle$ iff $X = \bigcup \{ Y_i \mid i \geq 0 \}$, with $Y_0 = S$ and $Y_{i+1} = cl_\Gamma(Y_i) \cup d_{\Delta^X}(Y_i)$.*

Proof. One shows easily by induction that $d^i_{\Gamma \cup \Delta^X}(S) \subseteq Y_i \subseteq cl_{\Gamma \cup \Delta^X}(S)$, for every i. Now use Proposition 13.

Remark 9 (Default Information Structures and Domain Theory). As mentioned before, the approach presented here bears some resemblance to that of Rounds and Zhang. A mere technical difference is that they employ so-called *sequent structures*, which are essentially theories in conditional normal form that are closed with respect to entailment (see also [4]). In [18, 24], Rounds and Zhang restrict themselves to *deterministic* sequent structures, that is, in our terminology, to Horn theories. Correspondingly, they consider only Scott domains as information domains. (In addition, they introduce a generalization of default extensions called *dilations*.) More recently [19], they proposed a domain theoretic framework for disjunctive logic programming, which goes much beyond Scott domains and which they also apply to default reasoning. A detailed comparison with this latter approach is a topic for future work.

Default Domains. Defaults à la Reiter are essentially metalevel statements. Alternatively, one can aim at expressing defaults by statements within an appropriate logical language. The first approach of this type goes back to McDermott and Doyle [13, 12], who introduced a modal operator to capture the consistency of assumptions.

We have already defined such a consistency operator $\Diamond$ ($= \neg\Box\neg$) in Section 4.1. A default with prerequisite ϕ, justifications $\psi_1, \ldots, \psi_n$, and consequent χ may then be translated into the statement

$$\phi \wedge \Diamond\psi_1 \wedge \ldots \wedge \Diamond\psi_n \preceq \chi. \tag{1}$$

Remark 10. An information domain can be seen as a Kripke frame for interpreting both, the "modal" operator $\Box$ (and $\Diamond$) as well as the intuitionistic operators $\Rightarrow$ and $\sim$. This double use of a Kripke frame closely resembles the semantical side of Gabbay's [7] intuitionistic approach to nonmonotonic logic; see also [3].

Lemma 4. *The statements $\phi \wedge \Diamond\psi \preceq \chi$ and $\phi \preceq \sim\psi \vee \chi$ are equivalent.*

Proof. The predicates $\phi \wedge \Diamond\psi \to \chi$ and $\phi \to \neg\Diamond\psi \vee \chi$ are equivalent, and so are the predicates $\neg\Diamond\psi$, $\Box\neg\psi$, and $\sim\psi$.

It follows that statement (1) is equivalent to

$$\phi \preceq \sim\psi_1 \vee \ldots \vee \sim\psi_n \vee \chi. \tag{2}$$

Given a theory Γ over Σ (with Σ finite), we know by Corollary 3 that with respect to the information domain of Γ each predicate $\sim\psi_i$ is equivalent to some affirmative predicate over Σ. Let Γ_Δ be the theory over Σ that results from a set Δ of defaults by rewriting them into the form (2) and then replacing the ψ_i's by equivalent affirmative predicates.

Definition 23 (Default Domain). *The* default domain *of a default theory $\langle \Gamma, \Delta \rangle$ is the information domain of the theory $\Gamma \cup \Gamma_\Delta$.*

Example 13. Let Γ be the theory of Example 1 and $\Delta = \{bird : can\text{-}fly/can\text{-}fly\}$. The corresponding default statement is $bird \wedge \Diamond can\text{-}fly \preceq can\text{-}fly$, which is equivalent to $bird \preceq \sim can\text{-}fly \vee can\text{-}fly$. Inspection of Figure 1 reveals that $\sim can\text{-}fly$ is equivalent to *ostrich* (since $\sim can\text{-}fly \equiv can\text{-}fly \Rightarrow \Lambda \equiv \bigvee\{\phi \mid \phi \wedge can\text{-}fly \equiv \Lambda\}$). The default domain of $\langle \Gamma, \Delta \rangle$ is hence of the form as depicted by Figure 6.

Recall that an information state $X \in C(\Gamma)$ is a minimal extension of $S \subseteq \Sigma$ iff X is minimal with respect to $S \subseteq X$ and $X \in C(\Delta^X)$. The latter condition says that for each default $\phi : \psi_1, \ldots, \psi_n/\chi$ of Δ, if X satisfies $\Diamond\psi_i$, for every i, then X satisfies $\phi \to \chi$; in other words, X satisfies $\phi \wedge \Diamond\psi_1 \wedge \ldots \wedge \Diamond\psi_n \to \chi$. The relation between minimal extensions and members of the default domain is hence as follows:

Proposition 14. *Let $\langle \Gamma, \Delta \rangle$ be a default theory over Σ and S a subset of Σ. Then the minimal extensions of S with respect to $\langle \Gamma, \Delta \rangle$ are the minimal elements of $\{X \in C(\Gamma \cup \Gamma_\Delta) \mid S \subseteq X\}$.*

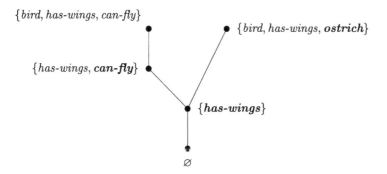

Fig. 6. Default domain of $\langle \Gamma, \Delta \rangle$

Example 14. Suppose $\Sigma = \{p, q\}$, $\Gamma = \varnothing$, and $\Delta = \{V{:}\neg p/q\}$. Recall that $\sim\neg p$ is $\Box\neg\neg p$ and thus equivalent to p, since p is persistent. So we can choose $\Gamma_\Delta = \{V \preceq p \vee q\}$. Hence $C(\Gamma_\Delta) = \{\{p\}, \{q\}, \{p, q\}\}$, whose minimal elements are $\{p\}$ and $\{q\}$. Notice that the only default extension of $\varnothing$ is $\{q\}$.

Since each minimal extension is a default extension of itself, it follows by Proposition 14:

Corollary 5. *A subset X of Σ belongs to $C(\Gamma \cup \Gamma_\Delta)$ just in case X is a default extension of some subset of Σ.*

Remark 11 (Theory Completion). Let Γ be a theory over Σ. If Δ consists of all defaults of the form $V{:}p/p$, with $p \in \Sigma$, then the default domain of $\langle \Gamma, \Delta \rangle$ is the antichain of all maximal elements of $C(\Gamma)$ (cf. Section 4.1). This antichain is thus the information domain of the extension of Γ by the statements $V \preceq p \vee \tilde{p}$, for every $p \in \Sigma$, where $\tilde{p}$ is an affirmative term equivalent to $\sim p$. In [15], such an extension is called a *completion* of Γ. Compare the completion of a theory with its Booleanization; cf. Section 3.1. (Notice in addition that the statements $p \wedge \tilde{p} \preceq \Lambda$ are true by definition.)

Negation as Failure. There is a well-known connection between default logic and logic programs with negation as failure. Adapted to the present context such a program is a theory consisting of statements

$$p_1 \wedge \ldots \wedge p_l \wedge \textbf{not}\, q_1 \wedge \ldots \wedge \textbf{not}\, q_m \preceq r,$$

with $p_i, q_j, r \in \Sigma$. Here, the operator **not** indicates *negation as failure to prove*, whose effect can be characterized as follows (cf. [9]): if **not** q belongs to the premise of a statement, the applicability of the statement is restricted to the case that the theory as a whole does not allow to derive q. One way to make this precise is to adopt the default interpretation of negation as failure: **not** q is interpreted as 'it is consistent to assume $\neg q$'. In short, **not** q is interpreted as $\Diamond\neg q$, which in turn is equivalent to $\neg\Box q$ and hence to $\neg q$, since q is persistent. A *stable model* of a program is then a default extension of the empty set with respect to the program seen as a default theory.

Example 15. Consider the program $\{\mathbf{not}\, p \preceq q,\ \mathbf{not}\, q \preceq p\}$. The corresponding set of defaults is $\{V\!:\!\neg p/q,\ V\!:\!\neg q/p\}$. Represented as default statements: $\Diamond \neg p \preceq q$ and $\Diamond \neg q \preceq p$, which both are equivalent to $V \preceq p \vee q$. So the minimal elements of the default domain are $\{p\}$ and $\{q\}$, which coincide with the default extensions of $\varnothing$ and thus are the stable models of the program in question.

Notice that this approach can be straightforwardly extended to *disjunctive* logic programs, whose statements are of the form

$$p_1 \wedge \ldots \wedge p_l \wedge \mathbf{not}\, q_1 \wedge \ldots \wedge \mathbf{not}\, q_m \preceq r_1 \vee \ldots \vee r_n.$$

4.3   Conclusion and Prospects

We have presented a paradigm for information processing that is centered around the idea of accumulating positive information by taking assertions always as affirmative. To this end, we have associated with each theory, i.e. each set of universally quantified Boolean predicates, a domain of information states that is ordered by specialization. This ordering has turned out to provide a natural interpretation for intuitionistic negation and conditional. Moreover, we have proposed ways to reconcile Poole's and Reiter's approach to default reasoning with our view of processing in information domains.

There are many topics to be further explored. For example, we have neglected questions of inference almost completely. In particular, nothing has been said about nonmonotonic reasoning, revision of default assumptions, etc. This deficiency is reflected by our focus on a *denotational* characterization of extensions. An *operational* account within the presented framework, in contrast, should give an algorithmic description of how the application of defaults gives rise to traversals between information states. Further questions to be addressed are the proper definition of default domains for infinite sets of atoms and the generalization to power domains, which would allow to represent alternatives of information states.[8]

Acknowledgments. The author would like to thank the anonymous reviewers for their helpful comments. One of the reviewers has been particularly supportive in pointing out additional connections to approaches in the literature.

References

1. Patrick Blackburn, Maarten de Rijke, and Yde Venema. *Modal Logic*. Cambridge Tracts in Theoretical Computer Science 53. Cambridge University Press, Cambridge, 2001.
2. Gerhard Brewka and Thomas Eiter. Prioritizing default logic. In Steffen Hölldobler, editor, *Intellectics and Computational Logic*, Applied Logic Series 19. Kluwer, Dordrecht, 2000.

---

[8] Zhang and Rounds [25] present an approach along this line that is based on Scott domains.

3. M. R. B. Clarke. Intuitionistic non-monotonic reasoning – further results. In *Proceedings of the 8th European Conference on Artificial Intelligence (ECAI-88)*, pages 525–527, 1988.

4. Thierry Coquand and Guo-Qiang Zhang. Sequents, Frames, and Completeness. In *Computer Science Logic, 14th Annual Conference of the EACSL*, Lecture Notes in Computer Science 1862, pages 277–291, Berlin, 2000. Springer.

5. Brian A. Davey and Hilary A. Priestley. *Introduction to Lattices and Order*. Cambridge University Press, Cambridge, 2nd edition, 2002.

6. Manfred Droste and Rüdiger Göbel. Non-deterministic information systems and their domains. *Theoretical Computer Science*, 75:289–309, 1990.

7. Dov M. Gabbay. Intuitionistic basis for non-monotonic logic. In *Proceedings of the 6th Conference on Automated Deduction*, Lecture Notes in Computer Science 138, pages 260–273, Berlin, 1982. Springer.

8. Laurence R. Horn. *A Natural History of Negation*. University of Chicago Press, Chicago, 1989.

9. Vladimir Lifschitz. Foundations of logic programming. In Gerhard Brewka, editor, *Principles of Knowledge Representation*, Studies in Logic, Language, and Computation, pages 69–127. CSLI Publications, Stanford, CA, 1996.

10. David Makinson. General patterns in nonmonotonic reasoning. In Dov M. Gabbay, C. J. Hogger, and J. A. Robinson, editors, *Handbook of Logic in Artificial Intelligence and Logic Programming, Vol. 3: Nonmonotonic Reasoning and Uncertain Reasoning*, pages 35–110. Oxford University Press, Oxford, 1994.

11. V. Wiktor Marek and Miroslaw Truszczynski. *Nonmonotonic Logic*. Springer, Berlin, 1993.

12. Drew McDermott. Nonmonotonic logic. II. Nonmonotonic modal theories. *Journal of the Association for Computing Machinery*, 29(1):33–57, 1982.

13. Drew McDermott and John Doyle. Nonmonotonic logic. I. *Artificial Intelligence*, 13(1–2):41–72, 1980.

14. Frank J. Oles. An application of lattice theory to knowledge representation. *Theoretical Computer Science*, 249:163–196, 2000.

15. Rainer Osswald. *A Logic of Classification – with Applications to Linguistic Theory*. PhD thesis, FernUniversität Hagen, Fachbereich Informatik, 2002. http://pi7.fernuni-hagen.de/osswald/papers/osswald-phd.pdf.

16. David Poole. A logical framework for default reasoning. *Artificial Intelligence*, 36:27–47, 1988.

17. Raymond Reiter. A logic for default reasoning. *Artificial Intelligence*, 13:81–132, 1980.

18. William C. Rounds and Guo-Qiang Zhang. Domain theory meets default logic. *Journal of Logic and Computation*, 5(1):1–25, 1995.

19. William C. Rounds and Guo-Qiang Zhang. Clausal Logic and Logic Programming in Algebraic Domains *Information and Computation*, 171(2):183–200, 2001.

20. Yoav Shoham. *Reasoning about Change*. MIT Press, Cambridge, MA, 1988.

21. Michael B. Smyth. Power domains and predicate transformers: A topological view. In Josep Díaz, editor, *Proceedings of ICALP'83*, Lecture Notes in Computer Science 154, pages 662–675, Berlin, 1983. Springer.

22. Dirk van Dalen. Intuitionistic logic. In Dov M. Gabbay and Franz Guenthner, editors, *Handbook of Philosophical Logic, Vol. III: Alternatives to Classical Logic*, pages 225–339. D. Reidel, Dordrecht, 1986.

23. Steven Vickers. *Topology via Logic*. Cambridge Tracts in Theoretical Computer Science 5. Cambridge University Press, Cambridge, 1989.

24. Guo-Qiang Zhang and William C. Rounds. Defaults in domain theory. *Theoretical Computer Science*, 177(1):155–182, 1997.
25. Guo-Qiang Zhang and William C. Rounds. Power defaults. In *Logic Programming and Nonmonotonic Reasoning, 4th International Conference (LPNMR'97)*, Lecture Notes in Computer Science 1265, pages 152–169, Berlin, 1997. Springer.

A Maple Package for Conditional Event Algebras⋆

Piotr Chrząstowski-Wachtel[1,2] and Jerzy Tyszkiewicz[1]

[1] Institute of Informatics, Warsaw University,
Banacha 2, PL-02-097 Warszawa, Poland
[2] Warsaw School of Social Psychology,
Chodakowska 19/31, PL-03-815 Warszawa, Poland
{pch, jty}@mimuw.edu.pl

Abstract. We present a Maple package for experimenting with conditional event algebras of Schay, Adams-Calabrese, Goodman-Nguyen-Walker and Product Space of Goodman and Nguyen.

It includes the implementation of the first polynomial time algorithm for calculating probabilities in the product space conditional event algebra.

Keywords: conditional event theories, systems and implementations, Markov chains, Moore machines, symbolic computation.

1 Introduction

In this paper we describe a Maple package for investigation of conditional event algebras, with the emphasis on the connectives found in Goodman-Nguyen-Walker (GNW) [8, 7, 6], Adams-Calabrese (AC) [1, 2, 3], Schay [15] and Product Space (PS) [5] conditional event algebras. It is intended primarily as a research tool. We have implemented here the main ideas of [18, 17, 16]. Emphasis is put on the computation of probabilities, not on deciding equality or inequality of conditional events. All the formulas are translated into their corresponding three-valued Moore machines, and these into Markov chains, based on a table of probabilities specified by the user. Finally, a polynomial time algorithm is invoked to calculate the probability of the conditional event, which is by the results of [17, 16] equal to ratios of certain limiting probabilities associated with the Markov chain.

Even though the implemented algorithms are theoretically reasonably efficient, more complicated calculations require a lot of time. This is however not a drawback of our system. It is rather a price we pay for a greatly increased functionality. What we gain is that the tool we present is a set of symbolic computation procedures, and it offers the possibility of deriving symbolic results. It is therefore possible to provide symbolic parameters as the values of probabilities and get the probabilities of conditional events in the form of algebraic expressions involving these parameters. Needless to say, it is of fundamental importance in research, even if the computations are slowed down by manipulations on formulas rather than numbers.

⋆ Supported by the Australian Research Council ARC grant A 49800112 (1998–2000) (P.Ch.-W. and J.Ty.) and by the Polish Research Council KBN grant 7 T11C 027 20 (J.Ty.).

G. Kern-Isberner, W. Rödder, and F. Kulmann (Eds.): WCII 2002, LNAI 3301, pp. 131–151, 2005.
© Springer-Verlag Berlin Heidelberg 2005

Furthermore, the intermediate results, the Moore machine and the Markov chain, are also represented in a symbolic form and can be analyzed in detail.

Finally, the procedures we have implemented are general enough to permit operations not present in the original definitions of Schay, AC, GNW and PS. E.g., it is possible to nest the PS conditioning operator, or to apply Calabrese and GNW reconditioning operators to PS expressions. We did not investigate the meaning and properties of such mixed formal systems. However, anybody interested can start making experiments with our tool to see what's in there.

The package has been written for Maple 6 or later. It is available, together with all other materials mentioned in this paper, from URL http://www.mimuw.edu.pl/~jty/Hagen.html.

2 Computer Algebra Systems

In this section we describe the functionality of typical computer algebra systems, called also symbolic computation systems. According to [9],

> Computer Algebra is a subject of science devoted to methods for solving mathematically formulated problems by symbolic algorithms, and to implementation of these algorithms in software and hardware. It is based on exact finite representation of finite and infinite mathematical objects and structures, and allows for symbolic and abstract manipulation by a computer. Structural mathematical knowledge is used during the design as well as for verification and complexity analysis of the respective algorithms. Therefore computer algebra can be effectively employed for answering questions from various areas of computer science and mathematics, as well as natural sciences and engineering, provided they can be expressed in a mathematical model.

The goals described above are exactly the needs we had when we started thinking about an implementation of the ideas described in [18, 17, 16]. It is therefore not surprising that we decided to use one of the computer algebra systems as a vehicle for our implementation. Our choice of *Maple* was mainly based on its availability to us.

There are many general-purpose computer algebra systems, and dozens of specialized ones. The former ones (*Maple* is among them) typically involve hundreds or even thousands of built-in efficient procedures for dealing with a huge number of predefined data structures: integers, rationals, algebraic numbers, polynomials, rational functions, matrices, lists, formulas, equalities, functions, integrals, etc., etc., as well as user-defined ones. Moreover, these data types are *exact*: $1/3$ *is* $1/3$, and not $0.3 \ldots 3$, π is π and not $3.14159 \ldots$, and consequently $e^{\pi i} + 1$ really equals 0 in *Maple* and other computer algebra systems:

```
>   Z:=exp(1)^(Pi*I)+1;
```
$$Z := (e)^{(I\pi)} + 1$$
```
>   simplify(Z);
```
$$0$$

Such systems are very often used as a kind of advanced mathematical calculator, allowing a working mathematician or computer scientist to perform instantly standard calculations he might need — integrate or differentiate a function, check the sum of a series, solve an equation or compute the Jordan form of a matrix.

```
>   f:=sin(x^(1/2));
```
$$f := \sin(\sqrt{x})$$

```
>   int(f,x);
```
$$2\sin(\sqrt{x}) - 2\sqrt{x}\cos(\sqrt{x})$$

```
>   diff(f,x);
```
$$\frac{1}{2}\frac{\cos(\sqrt{x})}{\sqrt{x}}$$

```
>   sum(1/i^6,i=1..infinity);
```
$$\frac{1}{945}\pi^6$$

```
>   solve(a^3+a-2=0,a);
```
$$1, -\frac{1}{2} + \frac{1}{2}I\sqrt{7}, -\frac{1}{2} - \frac{1}{2}I\sqrt{7}$$

```
>   A:=matrix(3,3,[[a,a+1,a+1],[3,4,4],[5,6,6]]);
```
$$A := \begin{bmatrix} a & a+1 & a+1 \\ 3 & 4 & 4 \\ 5 & 6 & 6 \end{bmatrix}$$

```
>   linalg[jordan](A);
```
$$\begin{bmatrix} 0 & 0 & 0 \\ 0 & 5 + \frac{1}{2}a + \frac{1}{2}\sqrt{132 + 12a + a^2} & 0 \\ 0 & 0 & 5 + \frac{1}{2}a - \frac{1}{2}\sqrt{132 + 12a + a^2} \end{bmatrix}$$

Besides that, each such a system has an internal programming language, which allows users to write larger programs, which not only can do what all the typical programming languages can, but also utilize the wealth of built-in functions. This way a good deal of the computational effort of a mathematician can be mechanized.

Our package is a set of *Maple* procedures implementing fragments of the theory of temporal conditionals we developed in the earlier papers [18, 17, 16]. It can be used in the calculator mode: one can check probabilities of certain conditionals, compare their behaviors, see their simplified forms. Of course, all other capabilities of *Maple* can be employed in this analysis, as the results of our procedures are quite normal data objects. Our procedures can also be used in the more advanced programming mode: own procedures can be written on top of what we provide. We did it ourselves, analyzing the Monty Hall problem and the Penney Ante Game.

A large part of the functionality of computer algebra systems are also their visualization tools, allowing one to create diagrams, 3D pictures, animations and other graphical representations of mathematical objects. This adds another way the results of our procedures can be investigated. A small glimpse is shown in this paper, some more advanced examples were presented during the workshop in Hagen. The fully functional lecture worksheet can be downloaded from the URL http://www.mimuw.edu.pl/ ~jty/Hagen.mws.

3   Functionality of the Package

Below we describe the basic functionality of the package, illustrating our descriptions with an example session. In this presentation every line printed in a typewriter font starting with a > sign is an input for Maple, that should be typewritten by the user from the keyboard. Below each such line a response from Maple is displayed, as it is seen inside Maple.

Example 1. First we have to activate the package. We assume all the files to be stored in the default Maple directory, which may be system-dependent.

```
>    read "cea.maple";with(cea);
```

> **module** *cea* ()
> **export** *moore2markov*, *form2moore*, *asymp*;
> · · ·
> **end module**
> [*asymp*, *form2moore*, *moore2markov*]

Maple responds with the list of available procedures.

3.1   Main Data Structures

The following are the main data structures used in the package.

– Formula.
– Moore machine.
– Markov chain.
– Probability table.

3.2   Formulas

Formulas are represented as Maple expressions, using neutral operators (whose names are preceded by & in Maple). It is required that the leafs of the parse tree of the formulas are either unevaluated names or constants &true and &false. The implemented connectives are the following: &not, common to all the formalisms, &orAC, &andAC, &cond form the Adams-Calabrese algebra, &orGNW, &andGNW, &cond form the Goodman, Nguyen and Walker algebra, next &or, &and, &condPS are the connectives of the product space algebra, and, besides that, re-conditioning operators &condC of Calabrese and &condGNW of Goodman, Nguyen and Walker. The operators of Schay are already among those. It is possible to nest these operators in any way, except that &condPS should not be used with argument formulas containing any of &condC and &condGNW. The logical meaning of these connectives is explained in the section on conversion of formulas into Moore machines.

Example 2. The following are example formulas:

```
> F1:= &not(x &cond y) &andAC (z &cond (x &orAC y));
```
$$F1 := \&\text{not}(x \&\text{cond} y) \&\text{andAC} (z \&\text{cond} (x \&\text{orAC} y))$$

This was a pure AC formula. The next one is pure PS.

```
> F2:= &not(x &condPS y) &and (z &condPS (x &or y));
```
$$F2 := \&\text{not}(x \&\text{condPS} y) \&\text{and} (z \&\text{condPS} (x \&\text{or} y))$$

The last one is a mixture of PS and GNW.

```
> F3:= &not(x &cond y) &andGNW (z &condPS (x &or y));
```
$$F3 := \&\text{not}(x \&\text{cond} y) \&\text{andGNW} (z \&\text{condPS} (x \&\text{or} y))$$ □

3.3   Moore Machines

We explain briefly what a Moore machine is, following [11]. It is essentially a deterministic finite automaton, with all the states labeled by letters of an *output alphabet*. A Moore machine has no acceptance condition, as it is supposed to compute functions rather than to accept/reject its input. On a given input sequence a Moore machine computes exactly as finite deterministic automata do, and the output it produces is the sequence of labels of the states visited during the computation.

We assume that the machine has n states. A Moore machine is represented as a list of two components. The first is a square table indexed by pairs of natural numbers. It represents the transition function in the following way: if an event m is stored in the table under the index (i, j), then if the machine is in the state i and m happens, the state changes to j. The second component is a list of the labels of the states, in the order in which they appear in the transition table. The first state is by default the initial one. The possible labels of the states are 0 (false), 1 (true) and 2 (undefined).

The events stored in the table represent elements of the free Boolean algebra, generated by several generators. Elements are represented in DNF (Disjunctive Normal Form), in the following way. Generators are unevaluated names of Maple (e.g., x). Literals are either generators or negated generators (e.g., x or `not(x)`). Clauses (conjunctions of literals) are represented as sets of literals (e.g., `{x,not(y)}`), disjunctions of clauses are represented as sets of clauses (e.g., $(x \wedge \neg y) \vee (\neg x \wedge y)$ is represented by `{{x,not(y)},{not(x),y}}`).

In one disjunction, all clauses must involve precisely the same variables. It is therefore guaranteed that a clause is never a proper subset of another one, if they appear in the same disjunction. This is motivated by the needs of the implementation, as this property significantly simplifies the computation of probabilities. E.g., $x \vee (\neg x \wedge y)$ can not be represented by `{{x},{not(x),y}}`, which is forbidden; an equivalent expression, like `{{x,not(y)},{x,y},{not(x),y}}`, can be used instead.

By this convention `{{}}` is a disjunction consisting of the empty clause `{}` only, and hence is always true. On the other hand, `{}` is always false, when considered as empty disjunction of clauses.

Next, for any state i of the Moore machine, the values in the row corresponding to i must be logically mutually exclusive, and their union must be the full event.

As long as these rules are preserved, the user can hand-make any Moore machine and subsequently use our tools to analyze it. We give an example of this kind later on.

3.4   Conversion of Formulas into Moore Machines

The conversion of a formula into a Moore machine is done by invoking the function `form2moore`. It has one parameter: the formula, and as a result it produces a pair: the transition table and the list of labels of the states of a Moore machine.

When we type a formula in the infix form, Maple automatically creates its parse tree. The details of this process are unimportant, except that we do not have any impact on the binding strength of the operators we define. Therefore, in order to assure the correct result, formulas must be entered with all the parentheses. After the parsing, we have names (of variables) and/or two Boolean constants in the leafs of the tree. The names of the operators are in its internal nodes.

The method used to construct an automaton from an expression is to traverse the parse tree in the natural order — first converting the operand formulas into Moore machines, and only then processing the connective in the current node. This order of processing the tree is called the *postfix order*. At each node the following steps are performed:

- If the node is a variable (0-ary operator), then create the variable automaton using the procedure `var2moore`. The transition table of the automaton for variable x is shown below:

$$\begin{bmatrix} \{\{x\}\} \, \{\{not(x)\}\} \\ \{\{x\}\} \, \{\{not(x)\}\} \end{bmatrix}.$$

  The label of the first state is 1, and of the second one is 0.
- If the node is an operator with one argument, then the operator is not, and we negate the labels of the child automaton, leaving the transition table unaltered. As we are working in three valued logic, we give here the truth table of the negation operator.

&not x	
x	&not x
0	1
1	0
2	2

- If the node is an operator with at least two arguments, then
  - If it is the `&condPS` operator, then we invoke a specially designed procedure `ps_cond2moore`, which differs significantly from all the other procedures for dealing with Moore machines. It creates the Moore machine representation of the product space conditioning operator. Its description can be found in [17, 16]. Below we present the graphical representation of that Moore machine. On the picture below, the Moore machine of x `&condPS` y is drawn according to the standard conventions for drawing finite automata: states are vertices, and arrows correspond to transitions changing the state. The label of an arrow is the event that causes the corresponding change of state. Arrows labeled by an empty event are omitted.
  - Otherwise, it is one of the following eight binary operators: `&or`, `&orAC`, `&orGNW`, `&and`, `&andAC`, `&andGNW`, `&condC`, `&condGNW`. In this case we create the product machine of the two child machines. Conceptually, its

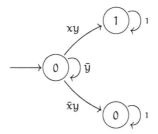

set of states is the Cartesian product of the sets of states of the child machines. In the product machine, the event at an edge from (a, b) to (c, d) is the set-theoretic intersection of the events at the edges from a to c and from b to d in their respective automata. The label of a state (a, b) is the result of the operator taken from the table `Operators` applied to the labels of the states a and b in their respective automata. Next we minimize the resulting Moore machine. It is important to do so immediately, to assure that the subsequent operations will be performed on smaller machines.

The truth tables of the presently implemented operators are as follows:

x &and y			
x\y	0	1	2
0	0	0	
1	0	1	
2			

x &andAC y			
x\y	0	1	2
0	0	0	0
1	0	1	1
2	0	1	2

x &andGNW y			
x\y	0	1	2
0	0	0	0
1	0	1	2
2	0	2	2

x &or y			
x\y	0	1	2
0	0	1	
1	1	1	
2			

x &orAC y			
x\y	0	1	2
0	0	1	0
1	1	1	1
2	0	1	2

x &orGNW y			
x\y	0	1	2
0	0	1	2
1	1	1	1
2	2	1	2

x &cond y			
x\y	0	1	2
0	2	0	
1	2	1	
2			

x &condC y			
x\y	0	1	2
0	2	0	0
1	2	1	1
2	2	2	2

x &condGNW y			
x\y	0	1	2
0	2	0	0
1	2	1	2
2	2	2	2

Space left in some places indicate that the connective is undefined for the corresponding pair of arguments.

Example 3. Translating formulas into Moore machines.

```
>   M1:=form2moore(F1);
```

$$M1 := [M, [2, 0, 1]]$$

```
>   M2:=form2moore(F2);
```

$$M2 := [M, [0, 0, 1, 1]]$$

An example of the true structure of the Moore machine.

```
>    op(M2[1]);
```

$[\{\%1, \{z,\ \textbf{not}\ y,\ \textbf{not}\ x\}\},$
$\{\{y,\ \textbf{not}\ x,\ \textbf{not}\ z\}, \{x,\ \textbf{not}\ y,\ \textbf{not}\ z\}, \{x,\ y,\ \textbf{not}\ z\}, \{x,\ y,\ z\}\}, \{\{x,\ z,\ \textbf{not}\ y\}\},$
$\{\{y,\ z,\ \textbf{not}\ x\}\}]$
$[\{\}, \{\{y,\ \textbf{not}\ x,\ \textbf{not}\ z\}, \%1, \{x,\ \textbf{not}\ y,\ \textbf{not}\ z\}, \{x,\ y,\ \textbf{not}\ z\}, \{z,\ \textbf{not}\ y,\ \textbf{not}\ x\},$
$\{y,\ z,\ \textbf{not}\ x\}, \{x,\ z,\ \textbf{not}\ y\}, \{x,\ y,\ z\}\}, \{\}, \{\}]$
$[\{\}, \{\{x,\ y,\ \textbf{not}\ z\}, \{x,\ y,\ z\}\},$
$\{\%1, \{x,\ \textbf{not}\ y,\ \textbf{not}\ z\}, \{z,\ \textbf{not}\ y,\ \textbf{not}\ x\}, \{x,\ z,\ \textbf{not}\ y\}\},$
$\{\{y,\ \textbf{not}\ x,\ \textbf{not}\ z\}, \{y,\ z,\ \textbf{not}\ x\}\}]$
$[\{\}, \{\}, \{\}, \{\{y,\ \textbf{not}\ x,\ \textbf{not}\ z\}, \%1, \{x,\ \textbf{not}\ y,\ \textbf{not}\ z\}, \{x,\ y,\ \textbf{not}\ z\},$
$\{z,\ \textbf{not}\ y,\ \textbf{not}\ x\}, \{y,\ z,\ \textbf{not}\ x\}, \{x,\ z,\ \textbf{not}\ y\}, \{x,\ y,\ z\}\}]$
$\%1 := \{\ \textbf{not}\ y,\ \textbf{not}\ x,\ \textbf{not}\ z\}$

The "%1" symbol is introduced by Maple to denote a commonly occurring subexpression and is explained by the system below.

The drawing (hand-made) of this Moore machine can be seen below. In each state the number is its label, and the subscript indicates the number of this state in the Maple representation above. The procedure of converting the table representation into a drawing is as follows: the table gives the matrix of transitions row-wise. The entry in row i and column j gives the event causing the machine to change its state from i to j. Entries are given in the DNF, i.e., they are disjunctions of conjunctions of literals, and their particular format is explained in Section 3.3. While converting the table into a drawing, shown below, the events have been simplified where possible and appropriate.

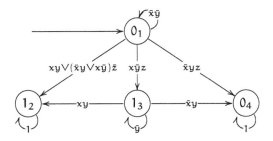

```
>    M3:=form2moore(F3);
```

$$M3 := [M, [0, 0, 2, 1, 0]]$$

3.5   Probability Table

Probability table is a (virtual) table of Maple expressions, indexed by all atoms of the free Boolean algebra $\mathfrak{B}$ generated by all the primitive events. This uniquely determines a discrete probability space whose field of measurable elements is $\mathfrak{B}$.

This table is virtual in the sense that we provide a sophisticated data structure, a table with a custom indexing function `independent`, which allows one to retrieve values from the table, which are not stored explicitly, but inferred from the values which are really stored in it. This permits one to enter a few values only and have the remaining probabilities inferred in an automatic way. E.g., if one wants to have n mutually independent primitive events, it is enough to specify $2n$ probabilities, and all the remaining ones will be computed on demand. Recall that without this functionality, one would always have to enter 2^n values manually. The table assumes implicitly that the unspecified values should be reconstructed according to the principle that all the relevant events are mutually independent.

In the package, the probability space is represented as a product of several smaller probability spaces, called *clusters*. Each cluster must be completely specified by assigning probabilities to all its atoms, with the exception that the elements of probability 0 need not be specified.

The reader should note that there are two levels at which unspecified probabilities may show up and be inferred.

- The first is the cluster level. Values of probabilities missing in the definition of a cluster are assumed to be 0.
- The second is the probability space level. By principle, probabilities of events involving atomic events from different clusters *must be unspecified*. They are reconstructed assuming that all the clusters are mutually independent.

Example 4. A table P of probabilities is created by issuing the command

```
>  P:=table(independent,[]);
```

$$P := \text{table}(independent, [])$$

This command creates an empty P and turns on the machinery reconstructing the missing values. Assuming we have an event x, we can set its probability by issuing

```
>  P[{x}]:=a;  P[{not(x)}]:=1-a;
```

$$P_{\{x\}} := a$$

$$P_{\{ \text{ not } x\}} := 1 - a$$

This creates a cluster, which is the probability space whose atoms are x and $not(x)$. Observe that the probabilities are indeed symbolic values. As explained above, it is mandatory to set both the probability of x and of its complement $not(x)$, even though the latter can be determined from the former (and vice versa). Of course, it would be possible to infer the probability of $not(x)$. However, such an extension would be

- computationally expensive (and one should remember that the procedure of reconstructing missing values is invoked for each entry in the transition table of an automaton),
- difficult to combine with another extension, which we think is more fundamental and which **is** implemented: that all unspecified probabilities are assumed to be 0.

Now we create another cluster

```
>  P[{y,z}]:=1/8;P[{y,not(z)}]:=1/8;P[{not(y),z}]:=1/4;
>  P[{not(y),not(z)}]:=1/2;
```

$$P_{\{z, y\}} := \frac{1}{8}$$

$$P_{\{y, \text{ not } z\}} := \frac{1}{8}$$

$$P_{\{z, \text{ not } y\}} := \frac{1}{4}$$

$$P_{\{\text{ not } y, \text{ not } z\}} := \frac{1}{2}$$

Now y and z are not independent. The system infers probabilities which have not been entered explicitly:

> `P[{y,x}];`

$$\frac{a}{4}$$

□

As we already mentioned, in the implemented solution the inference algorithm is invoked whenever a value from the table is accessed, which causes a lot of computational overhead, if some values are accessed many times.

The other possibility would be to infer the probabilities as the values are entered into P and record them in a table. However, with this solution it would become almost impossible to make any modification of the once entered values, and would require a lot of storage space (exponential in the number of clusters). The present algorithm does not need this.

Lastly, we explain why we didn't decide to go even farther with automation of implicit inference of probabilities not supplied explicitly by the user.

The answer is the following complexity-theoretic result. We use here and in the short discussion which follows standard terminology and results from that field. A good reference book is [11].

Theorem (Pitowsky [14]). *Let* $a_1, \ldots, a_n$ *be events and let* P *be a partial assignment of probabilities to certain elements in* $\mathfrak{B}$, *the Boolean algebra freely generated by* $a_1, \ldots, a_n$.

It is $\mathcal{NP}$-*hard to determine if* P *can be extended to a probability measure defined on all elements of* $\mathfrak{B}$.

In the workshop version of the paper we claimed this result, being at that time unaware of the earlier book of Pitowsky [14], where this theorem has been shown. Below we present our own proof, which differs significantly from the one found in cf. [14], and is therefore of independent interest.

Proof. We reduce the following $\mathcal{NP}$-complete problem to our problem. It is a reformulation of the well-known 3-CNF SAT problem.

Given: A propositional formula φ in 3-DNF form, i.e., a disjunction of clauses, where each clause is a conjunction of at most 3 literals (literals are variables or negated variables).

Question: Is φ not a tautology?

The reduction is as follows: Let a formula φ in 3-DNF form be given, say

$$\bigvee_{i=1}^{n_1} (l_{i,1} \wedge l_{i,2} \wedge l_{i,3}) \vee \bigvee_{i=n_1+1}^{n_2} (l_{i,1} \wedge l_{i,2}) \vee \bigvee_{i=n_2+1}^{n_3} l_{i,1},$$

where each of the $l_{i,j}$ is a literal: either a propositional variable or a negation of a propositional variable. W.l.o.g., in each clause the same variable can appear at most once.

Now from φ we construct (this construction can be easily performed in polynomial time) the following partial assignment of probabilities:

$$P(a_{i,1} \cap a_{i,2} \cap a_{i,3}) = \frac{1}{8 \cdot n_3} \quad \text{for } 1 \leq i \leq n_1,$$

$$P(a_{i,1} \cap a_{i,2}) = \frac{2}{8 \cdot n_3} \quad \text{for } n_1 < i \leq n_2, \tag{1}$$

$$P(a_{i,1}) = \frac{4}{8 \cdot n_3} \quad \text{for } n_2 < i \leq n_3.$$

Our goal is to show that P can be extended to a total probability measure on $\mathfrak{B}$ iff φ is not a tautology.

It is clear that φ is a tautology iff

$$\bigcup_{i=1}^{n_1} (a_{i,1} \wedge a_{i,2} \wedge a_{i,3}) \cup \bigcup_{i=n_1+1}^{n_2} (a_{i,1} \wedge a_{i,2}) \cup \bigcup_{i=n_2+1}^{n_3} a_{i,1} \tag{2}$$

is the full event in $\mathfrak{B}$.

The sum of all values assigned in (1) is less than 1, hence if φ is a tautology, (1) can not be extended to a total probability measure on $\mathfrak{B}$.

In turn, if φ is not a tautology, there is at least one atom α of $\mathfrak{B}$ which is not included in (2). Now, we assign equal probability $p > 0$ to all the atoms not equal to α, and the rest of the probability to α. This assignment clearly extends to a total probability measure on $\mathfrak{B}$, as every assignment of nonnegative values to atoms does, provided that the sum of the assigned values is 1. Under this probability measure

- all of the events in the first line of (1) have the same probability (because all of them are unions of an equal number of atoms, and α is not among them);
- all of the events in the second line have equal probability (which is twice as large as that of the events from the first line);
- all of the events in the third line have equal probability (which is four times as large as that of the events from the first line).

Now changing p (and therefore the probability of α as well) we can adjust the so constructed probability measure to agree with the assignment (1). Indeed, the probabilities are, as it is easy to verify, continuous functions of p, hence by varying p we can obtain any value of their probabilities between the maximal and the minimal achievable value. It remains to be seen that the values given in (1) are indeed between these extreme values.

The minimal value is 0 for all the events from (1); this happens when $p = 0$. The maximal value is achieved when $P(\alpha) = 0$; under such a probability measure the events in the first line of (1) are assigned a probability slightly larger than $1/8$. Indeed, each of these events contains exactly $1/8$ of the total number of atoms, and each atom has probability slightly larger than the reciprocal of the number of atoms (because of the zero probability of α). The events from the second and the third line will have probabilities respectively twice and four times larger. At the same time the values from (1) are positive, but not larger than $1/8$, $1/4$ and $1/2$, respectively, in the first, second and third line, i.e., they are indeed located between the extreme achievable values.

The demonstrated equivalence shows that it is $\mathcal{NP}$-hard to determine if a partial assignment of probabilities can be extended to a full probability measure on $\mathfrak{B}$.   □

It is commonly believed that no $\mathcal{NP}$-hard problem can be solved by a deterministic or randomized sequential algorithm of polynomial time complexity. As a consequence, the same applies to the question, given a partial assignment of probabilities to certain events, whether there exists a probability space which extends this partial assignment. Certainly, the computational problem of determining the remaining probabilities can only be harder than the decision problem.

In contrast, our algorithms implemented both for the reconstruction of probabilities and for the verification of the table are essentially linear time in the number of values explicitly stored in the table. This shows the advantage of using probability spaces decomposed into clusters.

Another remark is that we indeed do not know if the decision problem considered in the Theorem is in $\mathcal{NP}$. It might appear to be even harder. We have a simple $\mathcal{PSPACE}$ algorithm solving this problem, hence its true complexity is between $\mathcal{NP}$ and $\mathcal{PSPACE}$. However, from the purely practical standpoint, $\mathcal{NP}$-hardness is already enough to rule out the possibility of implementing any efficient algorithm to test if the entered values can be extended to a probability measure at all, let alone to reconstruct it, if we do not impose the cluster structure.

3.6   Markov Chain

A Markov chain is represented as a list of two components: the first is a square table of Maple expressions, indexed by pairs of natural numbers, which represents (or perhaps is) the table of transition probabilities, and the second is a list of labels of the states, in the order in which they appear in the transition table (and is a reminiscence of the Moore machine from which the Markov chain stems). The initial probability assignment in the Markov chain assigns 1 to the state number 0, and 0 to all other states of the Markov chain. Again, the user can hand-make any Markov chain and use it subsequently in the session.

The implementation requires that the Markov chain is a list made of two components:

1. A square table with symbolic and/or numeric entires, where the sum of the entries in each row simplifies to 1.
2. A list of labels of length equal to the dimension of the table, with all entries 0, 1 or 2.

3.7   Conversion of a Moore Machine into a Markov Chain

The conversion of a Moore machine into a Markov chain looks very simple. It is an application of the function calculating the probability of a disjunction to every item in the automaton table. Having a Boolean polynomial as an entry in the table, we replace that polynomial by its probability. The table is organized as a non-typed structure, so that we can act on it in both cases: when it contains numeric values, and when it contains symbolic data. We take the needed values from the table of probabilities, where the probabilities of clusters (see Section 3.5) are stored. The table is *not checked* for correctness in this procedure. Therefore, if the table is incomplete or contradictory,[1] the value returned from the table may be a wrong one.

The conversion function is invoked by `moore2markov(M,P)`, where M is the Moore machine and P is the probability table.

Example 5. Translating Moore machines into Markov chains.

> `T1:=moore2markov(M1,P);`

$$
T1 := \left[\begin{bmatrix} \frac{3}{4} - \frac{3}{4}a & \frac{1}{8} + \frac{5}{8}a & \frac{1}{8} + \frac{1}{8}a \\[6pt] \frac{3}{4} - \frac{3}{4}a & \frac{1}{8} + \frac{5}{8}a & \frac{1}{8} + \frac{1}{8}a \\[6pt] \frac{3}{4} - \frac{3}{4}a & \frac{1}{8} + \frac{5}{8}a & \frac{1}{8} + \frac{1}{8}a \end{bmatrix}, [2, 0, 1] \right]
$$

> `T2:=moore2markov(M2,P);`

$$
T2 := \left[\begin{bmatrix} \frac{3}{4} - \frac{3}{4}a & \frac{1}{8} + \frac{5}{8}a & \frac{1}{4}a & \frac{1}{8} - \frac{1}{8}a \\[6pt] 0 & 1 & 0 & 0 \\[6pt] 0 & \frac{1}{4}a & \frac{3}{4} & \frac{1}{4} - \frac{1}{4}a \\[6pt] 0 & 0 & 0 & 1 \end{bmatrix}, [0, 0, 1, 1] \right]
$$

> `T3:=moore2markov(M3,P);`

$$
T3 := \left[\begin{bmatrix} \frac{3}{4} - \frac{3}{4}a & \frac{1}{8} + \frac{1}{2}a & \frac{1}{4}a & \frac{1}{8} - \frac{1}{8}a & \frac{1}{8}a \\[6pt] 0 & 1 & 0 & 0 & 0 \\[6pt] 0 & 0 & \frac{3}{4} & \frac{1}{4} - \frac{1}{4}a & \frac{1}{4}a \\[6pt] 0 & 0 & \frac{3}{4} & \frac{1}{4} - \frac{1}{4}a & \frac{1}{4}a \\[6pt] 0 & 0 & \frac{3}{4} & \frac{1}{4} - \frac{1}{4}a & \frac{1}{4}a \end{bmatrix}, [0, 0, 2, 1, 0] \right]
$$

□

3.8   Calculating Asymptotic Probabilities in the Markov Chain

To facilitate the computation, we split the procedure into 3 algorithms, permitting the user to indicate which one to use, depending on the type of the Markov chain to be analyzed.

---

[1] The former happens when the entries of the table can be extended in more than one way to a probability distribution with independent clusters, the latter when there is no way to extend them to a probability distribution.

AC and GNW. If we have a 3-or-less-state Markov chain resulting from a AC and/or GNW formula, then we use the procedure `asymp` with 2 arguments: the Markov chain and switch `three`, indicating that we are working with a three-valued algebra. This permits invoking a very efficient algorithm. The algorithm is extremely simple and fast: it just reads the value of the probability from the automaton transition table. It is the probability of getting into the state with label 1 (from an arbitrary other state). This algorithm *does not work* (may give false results) for Markov chains not created from AC and/or GNW formulas.

Product space. If the Markov chain is an absorbing one and results from a PS formula, then such Markov chain has only states labeled 0 or 1. We use then the procedure `asymp` with another 2 parameters: the Markov chain and switch `PS`, indicating the type of the algebra. This permits invoking another, quite efficient algorithm (although not as fast as the one for AC and GNW). Again, this algorithm *does not work* (fails or gives false results) for Markov chains created from non-PS formulas, like those using AC and/or GNW connectives, nested uses of `&condPS`, etc.

It has been proved in [17, 16] that any Markov chain resulting from a PS formula is absorbing. Absorbing states are those where there is 1 on the diagonal of the Markov chain's table. There is at most one such state of each label 1 and 0. If the first (i.e., the starting) state is absorbing, then the probability is 1 when the label is 1, and 0 otherwise.

In all other cases, the algorithm uses a method taken from [13]. Denote by P the matrix $(p(i,j))$ of transition probabilities, by Q the submatrix of rows and columns corresponding to transient states, and by R the submatrix of rows corresponding to transient states and columns corresponding to absorbing states. Let Id be a diagonal matrix with 1's on the diagonal and 0's elsewhere. Let $B = (Id - Q)^{-1} * R$. Then the probability is the entry in B in the row corresponding to the initial state in the column corresponding to the (only) absorbing state labeled by 1 in the Markov chain.

Other Markov chains. If we have a Markov chain resulting from mixing the connectives of AC, GNW and PS within one formula, or nesting PS conditioning operators, or a hand-made one, the third algorithm must be used. The procedure is general, so it works also for the previous two cases, but as always with general procedures, it has an overhead, which is quite big. In this case the switch should be `general`, or can be omitted.

The result will be computed in all cases except for these, where the limit is of the type $\frac{0}{0}$. In such cases we are not able to give any sensible result, and the system produces an error message.

At present an easy but rather inefficient algorithm, taken from [12], is implemented. It has the following description: denote again by P the matrix $(p(i,j))$ of transition probabilities, and let $f(x)$ be its characteristic polynomial $\det(x * I - P)$, where I is an identity matrix of the same dimensions as P. Let $f(x) = (x-1)^r \phi(x)$, where $\phi(1) \neq 0$. Then let $\Phi = \phi(P)/\phi(1)$ (where multiplications and additions are now understood as matrix multiplications and additions). Φ is the table of limit transition probabilities, i.e., $\Phi(i,j)$ is the limit for $n \to \infty$ of the probability of getting from i to j in n steps. Now the value we are looking for is

$$\frac{\sum_{\text{label of i is 1}} \Phi(0, i)}{\sum_{\text{label of i is 0 or 1}} \Phi(0, i)}.$$

Example 6. Calculating the probabilities and analysing the results. (The drawing is Maple-made this time.)

> `lim1:=asymp(T1,three);`

$$lim1 := \frac{1}{2} \frac{1+a}{1+3a}$$

> `lim2:=asymp(T2,PS);`

$$lim2 := -\frac{1}{2} \frac{-1-a+2a^2}{1+3a}$$

> `lim3:=asymp(T3,general);`

$$lim3 := -\frac{-1-a+2a^2}{5+18a}$$

> `plot([lim1,lim2,lim3],a=0..1,linestyle=[2,4,1],`

> `legend=["AC","PS","PS with GNW"], color=[red,blue,green]);`

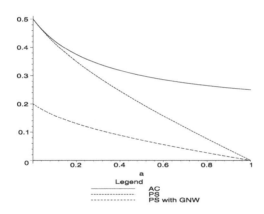

4   Hand-Made Moore Machines — Monty Hall

The *Monty Hall problem* involves a classical game show situation and is named after Monty Hall, the long time host of the the TV game show "Let's Make a Deal".

There are three doors labeled A, B, and C. A car is put behind one of the doors, while goats are put behind the other two. The rules are as follows:

– The player selects a door.
– The host selects a different door with a goat behind it and opens it.
– The host gives the player the option to change his/her original choice to the remaining closed door.
– The door finally selected by the player is opened and he/she wins either a car or a goat.

The Monty Hall problem, "to change or not to change", became the subject of intense controversy.

We show here that the Monty Hall game can be implemented in our package, and thus can be analyzed in a mechanical way, avoiding all the problems with understanding conditional probabilities. We assume the following:

1. First the car is placed behind on of the doors, and the events are accordingly denoted A, B and C.
2. The contestant chooses one of the doors, and the events corresponding to his/her choice are a, b and c, respectively.
3. The host opens one of the doors without the car. If the door chosen by the contestant has a car behind, the host has two choices, and makes a random choice, tossing a coin, with possible outcomes m and not(m). Otherwise the host has no choice and must open the only remaining door with a goat.
4. The contestant either changes the choice or not, tossing another coin, whose outcomes are s and not(s). This, together with the previous choices, determines finally whether the contestant wins the car or not.

These rules are implemented in the following procedure MontyHall, which creates a model of the complete game in the form of a Moore machine. The Maple programming language details are not so important, and space limitations do not permit us to discuss them in every detail anyway. The point we want to make is that the machine has initially 94 states (not all of which are really used), and still it is possible to write quite a short and structuralized description of it. The greatest amount of work is necessary to assure correct numbering of states, which is responsible for the long terms appearing in the table indices.

```
MontyHall:=proc()
local M,L,Val1,Val2,Val3,Val4,i,j,k,1;

M:=table(emptyset);

Val1:=proc(j);[{A},{B},{C}][j+1] end proc;

Val2:=proc(j);[{a},{b},{c}][j+1] end proc;

Val3 := proc(j, k)
  if j = 1 then if k = 2 then {{m}} else {{not m}} end if
  else if k = 1 then {{m}} else {{not m}} end if
  end if
end proc;

Val4:=proc(j);[{s},{not(s)}][j+1] end proc;

L:=array(1..94,[seq(2,i=1..94)]);

for i from 0 to 2 do M[1,2+i]:={Val1(i)};
    for j from 0 to 2 do
      M[2+i,5+i*3+j]:={Val2(j)};
```

```
         for k from 0 to 2 do
          if i=j and k<>i then
            M[5+i*3+j,14+i*9+j*3+k]:={Val3(j,k)};
            for l from 0 to 1 do
               L[41+i*18+j*6+k*2+l]:=1;
               M[14+i*9+j*3+k,41+i*18+j*6+k*2+l]:={Val4(l)};
               M[41+i*18+j*6+k*2+l,41+i*18+j*6+k*2+l]:={{}};
            end do
          elif i<>j and k<>i and k<>j then
            M[5+i*3+j,14+i*9+j*3+k]:={{}};
            for l from 0 to 1 do
               L[41+i*18+j*6+k*2+l]:=1-l;
               M[14+i*9+j*3+k,41+i*18+j*6+k*2+l]:={Val4(l)};
               M[41+i*18+j*6+k*2+l,41+i*18+j*6+k*2+l]:={{}};
            end do
          end if
         end do
     end do
end do;

L:=convert(L,list);

[M,L]

end proc;
```

Next we save the procedure in a file `MH.maple` and henceforth analyse it using our package. We suppress almost all the output in the following Maple session (by using colons : instead of semicolons ; as separators), because it does not provide new information.

```
>   restart;
>   read "cea.maple": with(cea): read "MH.maple":
```

We have loaded the package and the Monty Hall procedure.

```
>   MH:=MontyHall():
```

The Moore machine implementing the Monty Hall problem has been created. It is only now when we decide what the probabilities of events m and s are. All the previous modeling could be done on a purely logical level. In the following lines we create the probability table. Note that we use symbolic parameters as the probabilities of events m and s; thus our analysis covers a whole class of possible strategies for the player and the host.

```
>   P:=table(independent):
>   P[{A,not(B),not(C)}]:=1/3:
>   P[{not(A),B,not(C)}]:=1/3:
>   P[{not(A),not(B),C}]:=1/3:
>   P[{a,not(b),not(c)}]:=1/3:
>   P[{not(a),b,not(c)}]:=1/3:
```

```
>  P[{not(a),not(b),c}]:=1/3:
>  P[{m}]:=beta:P[{not(m)}]:=1-beta:
>  P[{s}]:=alpha:P[{not(s)}]:=1-alpha:
>  T:=moore2markov(MH,P):
```

The Markov chain has been created from the Moore machine. The following two lines are data type conversions, which permitted us not to specify many transitions of the initial Moore machine.

```
>  T:=subsop([1,1,1]=sparse,T):
>  T:=simplify([Matrix(nops(T[2]),T[1]),T[2]]):
```

And now we run the asymptotic probability evaluation.

```
>  WIN:=asymp(T,general);
```

$$WIN := \frac{1}{3}\alpha + \frac{1}{3}$$

Altogether, we see that changing the door increases the probability of wining the car, and indeed we see how much it does. If the probability of changing is α, then the probability of win is $1/3 + \alpha/3$. At the same time, it is independent of the strategy of the host, as long as the player chooses the door at random with equal probabilities. This is manifested by the fact that β does not occur in the winning probability expression. This expression could have been determined "manually" using standard symmetry arguments. However, symmetry fails in the case when the probabilities that car is located behind each of the doors are allowed to differ. However, we can still perform the analysis using Maple. If we modify the probability table using

```
>  P[{a,not(b),not(c)}]:=Gamma:
>  P[{not(a),b,not(c)}]:=Delta:
>  P[{not(a),not(b),c}]:=1-Gamma-Delta:
```

overriding the old values of $1/3$, we can still compute the probability of winning the car, by redoing the previous computations.

```
>  T:=moore2markov(MH,P):
>  T:=subsop([1,1,1]=sparse,T):
>  T:=simplify([Matrix(nops(T[2]),T[1]),T[2]]):
>  st:=time():SOL:=asymp(T,general);time()-st;
```

$$SOL := 1 - 2\gamma\Delta\alpha - 2\delta\Gamma\alpha - 4\gamma\Gamma\alpha - 4\delta\Delta\alpha + \delta\Gamma + \gamma\Delta + 2\gamma\Gamma + 2\gamma\alpha + 2\delta\Delta$$
$$+ 2\delta\alpha + 2\Gamma\alpha + 2\Delta\alpha - \gamma - \delta - \Gamma - \Delta - \alpha$$

$$9.607$$

In this case we measured the time necessary to perform the computations, and it is below 10 seconds. We doubt if this result could be reproduced in reasonable time by manual computations. Note that β is still absent in the formula.

5 An Extension: Penney Ante Game

There is an interesting game, called Penney Ante. It is described nicely in the famous book [10], so we give only a short account here. There are two players I and II, each of whom has a pattern (word) of length n over the alphabet $\{H, T\}$. Let the patterns be w_I and w_{II}. Now the players repeatedly toss a coin, until the first moment when one of w_I and w_{II} is a suffix of the sequence of toss results. The question is what are the probabilities of win for each of the players. Unexpectedly it turns out that this game is unfair, i.e., the win probabilities need not be $1/2$ and $1/2$. Moreover, Penney Ante is intransitive: for every $m > 2$ there are three patterns x, y, z of that length such that player I has a higher chance to win in each of the following three cases:

- $w_I = x$ and $w_{II} = y$;
- $w_I = y$ and $w_{II} = z$;
- $w_I = z$ and $w_{II} = x$.

This very interesting phenomenon is certainly due to conditioning. Indeed, every pattern eventually shows up in a sequence of coin tosses with probability 1. The win probability for player I in Penney Ante can be defined precisely as

$$\lim_{n \to \infty} \Pr_n (w_I \text{ has occurred} | w_{II} \text{ has not occurred})$$

where $\Pr_n$ is the probability for n tosses. We suspect that several real life situations can be, at least partially, described in a way very similar to the mechanism of Penney Ante. Indeed, the way stock market investors decide to buy and/or sell based on technical analysis resembles very much Penney Ante — they are waiting for a certain pattern of prices of the shares they are interested in. However, who sells/buys first, changes the market price and can prevent the other players from doing so.

Penney Ante game can be expressed in the temporal calculus of [18, 16], and we did not resist the desire to implement it. For that we created a new package cpa which, besides all the functions of cea, has an additional function PA, which, called with two patterns w_I and w_{II} over an arbitrary alphabet, creates a Moore machine corresponding to the win of player I. This machine can be later on analysed as we did before.

```
>   restart;read "cpa.maple": with(cpa);
            [PA, asymp, form2moore, moore2markov]
>   T:=not(H):
>   R[{H}]:=1/2;R[{T}]:=1/2:
```

This says that "tails" is the complement of "heads" and that the coin is fair.

Below we demonstrate that Penney Ante is nontransitive. We do so by computing all the possible win probabilities for three patterns.

```
>   Pat:=[[H,T,T,H],[T,H,T,T],[T,T,H,T]]:
>   S:=array(1..3,1..3,[]):
>   for W1 from 1 to 3 do
>   for W2 from 1 to 3 do
```

```
> S[W1,W2]:=asymp(moore2markov(PA(Pat[W1],Pat[W2]),R))
> end do
> end do;
> op(S);
```

$$\begin{bmatrix} \dfrac{1}{2} & \dfrac{5}{12} & \dfrac{7}{12} \\[2mm] \dfrac{7}{12} & \dfrac{1}{2} & \dfrac{2}{5} \\[2mm] \dfrac{5}{12} & \dfrac{3}{5} & \dfrac{1}{2} \end{bmatrix}$$

As we see, $[H,T,T,H]$ is better than $[T,T,H,T]$, the latter gives better chances than $[T,H,T,T]$, which in turn is better than $[H,T,T,H]$.

5.1   Further Work

We intend to add more functions to the package. Among them we think of an independence test for conditional events. We also consider implementing alternative embeddings of PS into the temporal calculus of [18, 16], which yield different Moore machines and different Markov chains than the embedding used here. However, they are guaranteed to give precisely the same probability evaluations. Also, if there is enough interest in investigating nonstandard Markov chains, we can implement a more efficient general Markov chain algorithm for calculating asymptotic probabilities.

Finally, we consider implementing GNW to AC and AC to GNW translators (the latter restricted to the cases when the translation is possible), following the method described in [4].

References

1. Ernest W. Adams. Probability and the logic of conditionals. In P. Suppes and J. Hintikka, editors, *Aspects of inductive logic*, pages 265–316. North-Holland, Amsterdam, 1966.
2. Philip G. Calabrese. An algebraic synthesis of the foundations of logic and probability. *Inform. Sci.*, 42(3):187–237, 1987.
3. Philip G. Calabrese. Reasoning with uncertainity using conditional logic and probability. In *Proc. First International Symposium on Uncertainity Modeling and Analysis by IEEE Computer Society*, pages 682–688. IEEE, 1990.
4. Piotr Chrząstowski-Wachtel, Jerzy Tyszkiewicz, Arthur Ramer, and Achim Hoffmann. Mutual definability of connectives in conditional event algebras of Schay-Adams-Calabrese and Goodman-Nguyen-Walker. *Information Processing Letters*, 79(4):155–160, 2001.
5. I. R. Goodman. Toward a comprehensive theory of linguistic and probabilistic evidence: two new approaches to conditional event algebra. *IEEE Trans. Systems Man Cybernet.*, 24(12):1685–1698, 1994. Special issue on conditional event algebra (San Diego, CA, 1992).
6. I. R. Goodman and H. T. Nguyen. Mathematical foundations of conditionals and their probabilistic assignments. *Internat. J. Uncertain. Fuzziness Knowledge-Based Systems*, 3(3):247–339, 1995.

7. I. R. Goodman, H. T. Nguyen, and E. A. Walker. *Conditional inference and logic for intelligent systems*. North-Holland Publishing Co., Amsterdam, 1991. A theory of measure-free conditioning.

8. Irwin R. Goodman. A measure-free approach to conditioning. In *Proc. Third AAAI Workshop on on Uncert. in AI*, pages 270–277. University of Washington, Seattle, 1987.

9. Johannes Grabmeier, Erich Kaltofen, and Volker Weispfenning, editors. *Computer algebra handbook*. Springer Verlag, 2003.

10. Ronald L. Graham, Donald E. Knuth, and Oren Patashnik. *Concrete mathematics*. Addison-Wesley Publishing Company, Reading, MA, second edition, 1994. A foundation for computer science.

11. John E. Hopcroft and Jeffrey D. Ullman. *Introduction to automata theory, languages, and computation*. Addison-Wesley Publishing Co., Reading, Mass., 1979. Addison-Wesley Series in Computer Science.

12. Marius Iosifescu. *Finite Markov processes and their applications*. John Wiley & Sons Ltd., Chichester, 1980. Wiley Series in Probability and Mathematical Statistics.

13. John G. Kemeny and J. Laurie Snell. *Finite Markov chains*. Springer-Verlag, New York-Heidelberg, 1976. Reprinting of the 1960 original, Undergraduate Texts in Mathematics.

14. Itamar Pitowsky. *Quantum probability—quantum logic*, volume 321 of *Lecture Notes in Physics*. Springer-Verlag, Berlin, 1989.

15. Geza Schay. An algebra of conditional events. *J. Math. Anal. Appl.*, 24:334–344, 1968.

16. Jerzy Tyszkiewicz, Piotr Chrząstowski-Wachtel, Arthur Ramer, and Achim Hoffman. *The theory of temporal conditionals*. in perparation.

17. Jerzy Tyszkiewicz, Achim Hoffmann, and Arthur Ramer. Embedding conditional event algebras into temporal calculus of conditionals. arXiv:cs.AI/0110004, 2001.

18. Jerzy Tyszkiewicz, Arthur Ramer, and Achim Hoffmann. Temporal calculus of conditional objects and conditional events. arXiv:cs.AI/0110003, 2001.

Conditional Independences in Gaussian Vectors and Rings of Polynomials

František Matúš

Institute of Information Theory and Automation,
Pod vodárenskou věží 4, 182 08 Prague, Czech Republic
matus@utia.cas.cz

Abstract. Inference among the conditional independences in nondegenerate Gaussian vectors is studied by algebraic techniques. A general method to prove implications involving the conditional independences is presented. The method relies on computations of a Groebner basis. Examples of the implications are discussed.

1 Introduction

Conditional independence (CI) constraints have become a crucial ingredient in building highly structured probabilistic models of artificial intelligence. They provide a language communicating qualitative features and induce decompositions of probability distributions amenable to practical computations. Though the major part of research has been concerned with CI constraints related to graphs, cf. [14, 2], an effort is devoted also to the general problem of discrete structures describing simultaneous occurence of independences within a random vector, for references see [19, 9]. Challenging problems involving CI constraints have emerged also in statistics [7], information theory [21, 20], secret-sharing schemes of cryptography [1], representations of matroid theory [11] and instances of the probabilistic method in combinatorics [17].

For a random vector $\xi = (\xi_i)_{i \in N}$ indexed by a finite set N, the stochastic conditional independences $\xi_I \perp\!\!\!\perp \xi_J | \xi_K$ among its subvectors $\xi_I = (\xi_i)_{i \in I}$, ξ_J, ξ_K, where $I, J, K \subseteq N$ are disjoint, can be encoded by means of the discrete structure

$$[\xi] = \{(ij|K) \in \mathcal{R};\ \xi_i \perp\!\!\!\perp \xi_j | \xi_K\}.$$

Here, ij denotes a two-element set $\{i, j\}$ and $\mathcal{R}$ is the family of all ordered pairs $(ij|K)$ with $ij, K \subseteq N$ disjoint. The conditional independence $\xi_I \perp\!\!\!\perp \xi_J | \xi_K$ holds if and only if $(ij|L) \in [\xi]$ for all $i \in I$, $j \in J$ and L satisfying $K \subseteq L \subseteq (I \cup J \cup K) \setminus ij$, see [8, Lemma 3].

Manipulations with the CI constraints can be facilitated when understanding which subset $\mathcal{K}$ of $\mathcal{R}$ can be represented as $\mathcal{K} = [\xi]$ by means of some random vector ξ. This difficult problem remains open even in the discrete setting when N has more than four elements and ξ takes an arbitrary finite number of values, see [10, 9]. The analogous problem of representability by Gaussian vectors has

G. Kern-Isberner, W. Rödder, and F. Kulmann (Eds.): WCII 2002, LNAI 3301, pp. 152–161, 2005.

not been yet investigated in depth. This contribution aims at first steps in this direction, restricting attention to nondegenerate vectors. The degenerate case seems to differ only by additional technicalities. Systematic study of the structures $[\xi]$ is motivated also by a hope to reveal new classes of statistical models for discrete and especially Gaussian vectors, possibly admitting parsimonious parametrizations and effective computations.

Let $A = (a_{i,j})_{i,j=1}^{n}$ be the covariance matrix of a Gaussian vector $\xi = (\xi_1, \ldots, \xi_n)$, $n \geqslant 1$. When the vector is nondegenerate the matrix A is positive definite, in symbols $A > 0$. It is well-known that the variable ξ_1 is conditionally independent of ξ_2 given $(\xi_3, \ldots, \xi_n)$ if and only if the $(1,2)$-element of the inverse A^{-1} of A vanishes. Denoting submatrices of A by $A_{I,J} = (a_{i,j})_{i \in I, j \in J}$ and determinants by $|A_{I,J}|$, $I, J \subseteq N = \{1, \ldots, n\}$, this element equals $(-1)^{i+j}|A_{N \setminus \{1\}, N \setminus \{2\}}|/|A|$, see for example [15, p. 11]. Therefore, more generally, a conditional independence $\xi_i \perp\!\!\!\perp \xi_j | \xi_K$ for $(ij|K) \in \mathcal{R}$ holds if and only if $|A_{jK,iK}|$, equal to $|A_{iK,jK}|$, vanishes; here iK abbreviates $\{i\} \cup K$ and similarly with jK. Since the independences are determined by the covariance matrix A of ξ one can work directly with positive definite matrices and use the notation

$$\langle\!\langle A \rangle\!\rangle = \{(ij|K) \in \mathcal{R}; \ |A_{iK,jK}| = 0\}$$

in place of $[\xi]$. The main problem of this contribution is now to recognize whether a given subset $\mathcal{K}$ of $\mathcal{R}$ can be written as $\mathcal{K} = \langle\!\langle A \rangle\!\rangle$ for some $A > 0$. An equivalent problem is to find for which families $\mathcal{L}, \mathcal{M} \subseteq \mathcal{R}$ the implication

$$\mathcal{L} \subseteq \langle\!\langle A \rangle\!\rangle \quad \Rightarrow \quad \mathcal{M} \cap \langle\!\langle A \rangle\!\rangle \neq \emptyset \tag{1}$$

holds for all $A > 0$. In fact, $\mathcal{K} = \langle\!\langle A \rangle\!\rangle$ for some $A > 0$ if and only if (1) with $\mathcal{L} = \mathcal{K}$ and $\mathcal{M} = \mathcal{R} \setminus \mathcal{K}$ is violated by some $A > 0$.

By Sylvester's criterion [15, p. 96], the set of positive definite $n \times n$ matrices is a semialgebraic subset of a Euclidean space, that is, a set defined by a finite number of polynomial inequalities. To decide whether for given $\mathcal{L}, \mathcal{M}$ the implication (1) quantified by 'for all $A > 0$' is true or false amounts to decide whether the semialgebraic set

$$\{A > 0; \ |A_{iK,jK}| = 0 \text{ for } (ij|K) \in \mathcal{L}, \text{ and } |A_{iK,jK}| \neq 0 \text{ for } (ij|K) \in \mathcal{M}\}$$

is empty or not, respectively. Algorithms for manipulations with semialgebraic sets are available, see [12], and can recognize whether a given semialgebraic set is empty. Small instances of some statistical problems were attacked by such algorithms in [6], reporting successful treatment of a trivial example of the implication (1), see (5) below. Nevertheless, the general algorithms seem to provide no insight into the structure of this special problem and are of considerable complexity.

Since the construction $A \mapsto \langle\!\langle A \rangle\!\rangle$ makes sense even for a fairly general matrix, for example for any symmetric matrix, the quantification $A > 0$ in the main problem can be replaced by the assumption that A belongs to another set of symmetric matrices. Modified problems obtained in this way approximate the main problem and are of theoretical interest *per se* because they enrich the insight into abstract CI structures. A convenient replacement of $A > 0$ may require A to belong to an algebraic set (a set defined by polynomial equalities) or a constructible set of matrices (finite union of differences of algebraic sets). When matrices are admitted to have the complex numbers as elements numerous methods of the classical algebraic geometry, supported by computational procedures, notably for Groebner bases, are at disposal.

For example, when $A > 0$ the principal minors $|A_{I,I}|$ are positive for $\emptyset \neq I \subseteq N$, and a straightforward relaxation of the quantification $A > 0$ is to require A to be in the set $\mathbb{A}$ of complex symmetric matrices having all proper principal minors nonzero. The discrete structures $\langle\!\langle A \rangle\!\rangle$ for $A \in \mathbb{A}$ keep reasonable properties and remain within the class of semigraphoids, which has been a framework for CI relations, see [14, 7, 9]. When $n = 3$ this relaxation works very well and provides the same class of structures $\langle\!\langle A \rangle\!\rangle$ as if $A > 0$, see Example 1. However, when $n \geqslant 4$ this is not the case, see Example 3. In general, an 'exact' relaxation, of purely theoretical interest, to a constructible set of matrices exists. In fact, the disjoint union of the sets of matrices

$$\{A \in \mathbb{A}; \ |A_{iK,jK}| = 0 \ \text{ for } (ij|K) \in \mathcal{K}, \ \text{ and } |A_{iK,jK}| \neq 0 \ \text{ for } (ij|K) \in \mathcal{R} \setminus \mathcal{K}\}$$

over all $\mathcal{K} \subseteq \mathcal{R}$ admitting the representation $\mathcal{K} = \langle\!\langle A \rangle\!\rangle$ for some $A > 0$ is a constructible set; obviously, (1) is valid for all $A > 0$ if and only if it is valid for all A in the union.

This contribution is organized as follows. In Section 2, examples of the implication (1) are presented for complex matrices. When specialized to positive definite matrices, they summarize basic properties of CI in nondegenerate Gaussian vectors. In Section 3, the implication (1) is analysed when A belongs to the set of matrices that violate each of a given set of polynomial equalities (such sets, for example $\mathbb{A}$ itself, are constructible when working with a finite number of equalities). A necessary and sufficient condition for the validity of such an implication is presented in Proposition 1. The condition can be practically verified by computing a single Groebner basis. Section 4 discusses possible uses of this result: all basic implications of Section 2 can be obtained as consequences of Proposition 1, and two additional implications for CI in Gaussian vectors are derived by a more sophisticated relaxation.

2   Basic Properties of Conditional Independence

In this section, basic properties of CI in nondegenerate Gaussian vectors are derived as consequences of a single determinantal identity. It is convenient to abbreviate $A_{I,I}$ by A_I and take $|A_\emptyset| = 1$.

Lemma 1. *For any complex matrix A, and $L \subseteq N$, $k \in N \setminus L$, and $i, j \in N \setminus kL$*

$$(-1)^{[i,j;k]} |A_{ikL,jkL}||A_L| = |A_{iL,jL}||A_{kL}| - |A_{iL,kL}||A_{kL,jL}| \qquad (2)$$

where $[i, j; k] = 1$ if k is between i and j, and $[i, j; k] = 0$ otherwise.

Proof. When $L = \emptyset$ this is an obvious formula for determinant expansion. When $L \neq \emptyset$ one can suppose that A_L is invertible, the opposite case following then by continuity. First, let all elements of L be smaller than the minimum of the set $ijk = \{i, j, k\}$. The Schur complement of A_L in $A_{ikL,jkL}$ is the matrix $B_{ik,jk} = A_{ik,jk} - A_{ik,L}(A_L)^{-1}A_{L,jk}$. Then $|A_{ikL,jkL}|$ factors into $|A_L||B_{ik,jk}|$ by [15, Theorem 3.1.1a]. Similarly, the four determinants of the right-hand side of (2) can be factored as $|A_{iL,jL}| = |A_L||B_{i,j}|$, etc. Then (2) follows from the expansion formula for $|B_{ik,jk}|$.

If L and i, j, k are in a general position then a unique permutation π of $ijk \cup L$ exists such that $\pi(l_1) < \pi(l_2)$ for $l_1 < l_2$ with l_1, l_2 in L or in ijk, and such that the elements of $\pi(L) = \{\pi(l); l \in L\}$ are smaller than the minimum of $\pi(i)\pi(j)\pi(k)$. Let C be the matrix obtained by applying π to the rows and columns of A simultaneously. By the previous argument

$$(-1)^{[\pi(i),\pi(j);\pi(k)]} |C_{\pi(ikL),\pi(jkL)}||C_{\pi(L)}|$$
$$= |C_{\pi(iL),\pi(jL)}||C_{\pi(kL)}| - |C_{\pi(iL),\pi(kL)}||C_{\pi(kL),\pi(jL)}| \, .$$

Then $[\pi(i), \pi(j); \pi(k)] = [i, j; k]$, $|C_{\pi(L)}| = |A_L|$ and $|C_{\pi(kL)}| = |A_{kL}|$. Denoting by n_i the number of all $l \in L$ greater than i, one can see that $|C_{\pi(ikL),\pi(jkL)}|$ is the $(-1)^{n_i+n_j}$ multiple of $|A_{ikL,jkL}|$. Similar observations concern the expressions $|C_{\pi(iL),\pi(jL)}|$ and $|C_{\pi(iL),\pi(kL)}||C_{\pi(kL),\pi(jL)}|$. Hence, (2) obtains in full generality. $\qquad\square$

Remark 1. The identity (2) can be viewed as a technical refinement of its special case

$$|A_N||A_{N\setminus ij}| = |A_{N\setminus i}||A_{N\setminus j}| - |A_{N\setminus i,N\setminus j}||A_{N\setminus j,N\setminus i}| \tag{3}$$

where $ij = \{1, 2\}$. Thanks to a personal communication of David Mond, University of Warwick, who kindly drove our attention to [13, 5, 4], the identity (3) can be put into the framework of the identity (1) of [4, p. 156]. In fact, the latter rewrites in our notation as

$$\sum_{j \in M \setminus J} (-1)^j |D_{N,jJ}||D_{N,M\setminus jJ}| = 0 \tag{4}$$

where $M = \{1, 2, \ldots, 2n\}$, J is a subset of M of the cardinality $n - 1$ and D is any $n \times 2n$ matrix; (4) is a special case of the straightening laws of [5]. Then, for $J = \{2, \ldots, n\}$ and

$$D = \begin{bmatrix} A & \begin{vmatrix} 1 & 0 \\ 0 & 1 \\ \vdots & \vdots \\ 0 & 0 \end{vmatrix} & A_{N,\{3,\ldots,n\}} \end{bmatrix}$$

a simple computation derives (3) from (4).

Corollary 1. *For any symmetric complex matrix* A

$$\{(ij|kL),(ij|L)\} \subseteq \langle\!\langle A\rangle\!\rangle \Rightarrow (ik|L) \in \langle\!\langle A\rangle\!\rangle \ \text{or} \ (jk|L) \in \langle\!\langle A\rangle\!\rangle, \tag{5}$$

$$\{(ij|kL),(ik|L)\} \subseteq \langle\!\langle A\rangle\!\rangle \Rightarrow (ij|L) \in \langle\!\langle A\rangle\!\rangle, \qquad if \ |A_{kL}| \neq 0, \tag{6}$$

$$\{(ij|kL),(ik|L)\} \subseteq \langle\!\langle A\rangle\!\rangle \Rightarrow (ik|jL) \in \langle\!\langle A\rangle\!\rangle, \qquad if \ |A_L||A_{kL}| \neq 0, \tag{7}$$

$$\{(ij|kL),(ik|jL)\} \subseteq \langle\!\langle A\rangle\!\rangle \Rightarrow (ij|L) \in \langle\!\langle A\rangle\!\rangle, \qquad if \ |A_{jkL}||A_L| \neq 0, \tag{8}$$

$$\{(ij|L),(ik|L)\} \subseteq \langle\!\langle A\rangle\!\rangle \Rightarrow (ij|kL) \in \langle\!\langle A\rangle\!\rangle, \qquad if \ |A_L| \neq 0. \tag{9}$$

Proof. Rewriting (2) for a symmetric matrix A gives

$$(-1)^{[i,j;k]}\,|A_{ikL,jkL}||A_L| - |A_{iL,jL}||A_{kL}| = -|A_{iL,kL}||A_{jL,kL}|\,.$$

Hence, $|A_{ikL,jkL}| = 0$ and $|A_{iL,jL}| = 0$ imply $|A_{iL,kL}||A_{jL,kL}| = 0$ which proves (5).

Along the same guidelines, the identity (2) allows to conclude $(ij|L) \in \langle\!\langle A\rangle\!\rangle$ in (6).

To prove (7), one switches j and k in Lemma (1)

$$(-1)^{[i,k;j]}\,|A_{ijL,jkL}||A_L| = |A_{iL,kL}||A_{jL}| - |A_{iL,jL}||A_{jL,kL}|\,. \tag{10}$$

and combines this equality with (6).

When $|A_{ikL,jkL}| = 0$ and $|A_{ijL,jkL}| = 0$ the left-hand sides in (2) and (10) vanish. Solving the two homogeneous equations for $|A_{iL,jL}|$ and $|A_{iL,kL}|$, their determinant is equal to $|A_{jL}||A_{kL}| - |A_{jL,kL}||A_{kL,jL}|$. Using Lemma 1 with $i = j$, this determinant equals $|A_{jkL}||A_L|$ and is nonzero by assumption. Hence, $|A_{iL,jL}| = 0$ and (8) follows.

The implication (9) follows directly from (2). □

A subset $\mathcal{K}$ of $\mathcal{R}$ is called a *semigraphoid* if it fulfills

$$\{(ij|kL),(ik|L)\} \subseteq \mathcal{K} \iff \{(ik|jL),(ij|L)\} \subseteq \mathcal{K}\,, \tag{11}$$

cf. (6) and (7), and a *pseudographoid* if it fulfills the implication $\Rightarrow$ in

$$\{(ij|kL),(ik|jL)\} \subseteq \mathcal{K} \iff \{(ij|L),(ik|L)\} \subseteq \mathcal{K}\,, \tag{12}$$

cf. (8). A semigraphoid which is a pseudographoid is called a *graphoid*. (For the pseudographoid axiom compare [14, (3.6e) and (3.6b)] and [9, p. 107]. For (5)see [14, (3.34f)].)

By Corollary 1, to ensure that $\langle\!\langle A\rangle\!\rangle$ satisfies (11) and (12) it suffices to assume that $|A_I|$ is nonzero for every nonempty proper subset of N or, equivalently to the above notation, $A \in \mathbb{A}$. Since all positive definite matrices belong to $\mathbb{A}$ the conditional independences in nondegenerate Gaussian vectors enjoy (11), (12) and (5) with $\mathcal{K}$ or $\langle\!\langle A\rangle\!\rangle$ replaced by $[\xi]$.

Example 1. For $N = \{1,2,3\}$, $n = 3$, Corollary 1 implies that if $A \in \mathbb{A}$ then $\langle\!\langle A\rangle\!\rangle$ is equal to one of the families

$$\emptyset, \{(ij|\emptyset)\}, \{(ij|k)\}, \{(ij|\emptyset),(ik|\emptyset),(ij|k)\} \ \text{and} \ \mathcal{R}\,.$$

The five families are representable by the following positive definite matrices

$$
\begin{bmatrix} 1 & \epsilon & \epsilon^2 \\ \epsilon & 1 & \epsilon^4 \\ \epsilon^2 & \epsilon^4 & 1 \end{bmatrix}
\quad
\begin{bmatrix} 1 & 0 & \epsilon \\ 0 & 1 & \epsilon \\ \epsilon & \epsilon & 1 \end{bmatrix}
\quad
\begin{bmatrix} 1 & \epsilon^3 & \epsilon^2 \\ \epsilon^3 & 1 & \epsilon \\ \epsilon^2 & \epsilon & 1 \end{bmatrix}
\quad
\begin{bmatrix} 1 & 0 & 0 \\ 0 & 1 & \epsilon \\ 0 & \epsilon & 1 \end{bmatrix}
$$

($\epsilon \neq 0$ sufficiently small) and the identity matrix, respectively. Therefore, if $n = 3$ the class of structures $\{\langle\!\langle A \rangle\!\rangle; \ A > 0\}$ coincides with the class $\{\langle\!\langle A \rangle\!\rangle; \ A \in \mathbb{A}\}$. This is no longer true when $n > 3$, see Example 3.

3   Implications and Polynomial Rings

This section focuses on a necessary and sufficient condition for validity of (1) when A is in a certain constructible set of matrices. The condition involves some elementary material on rings of polynomials which is available in an accessible form in [3,12].

When studying $\langle\!\langle A \rangle\!\rangle$ for a symmetric matrix A with nonzero diagonal elements there is no loss of generality in assuming that all diagonal elements are equal to 1. In fact, multiplying each i-th row and i-th column of A by a number b_i satisfying $b_i^2 = a_{i,i}^{-1}$ one obtains a matrix A' rendering $\langle\!\langle A \rangle\!\rangle = \langle\!\langle A' \rangle\!\rangle$; in addition, A' can be made positive definite or belongs to $\mathbb{A}$ if A was positive definite or belonged to $\mathbb{A}$, respectively. From now on all matrices are supposed to be symmetric and to have the ones on their diagonals.

Let x_{ij} be an indeterminate indexed by a two-element set ij, $1 \leqslant i < j \leqslant n$, and $\mathfrak{R}$ be the ring of polynomials in these indeterminates with coefficients in the field of complex numbers. By little abuse of language we will substitute complex matrices A to polynomials, having in mind that the indeterminates x_{ij} are replaced by the off-diagonal elements $a_{i,j} = a_{j,i}$. For $(ij|K) \in \mathcal{R}$ the determinant $|A_{iK,jK}|$ expands as the sum of products of elements of $A_{iK,jK}$ and this expansion gives rise to a polynomial of $\mathfrak{R}$. Denoting the polynomial by $f_{(ij|K)}$, one has $f_{(ij|K)}(A) = |A_{iK,jK}|$. The notation $f_{(ij|K)}$ is extended to $f_{\mathcal{M}} = \prod_{(ij|K)\in\mathcal{M}} f_{(ij|K)}$ for $\mathcal{M} \subseteq \mathcal{R}$. Similarly, the expansion of $|A_I|$ gives rise to a polynomial denoted by f_I. For example, $f_{(12|3)} = x_{12} - x_{13}x_{23}$, $f_{(13|2)} = x_{12}x_{23} - x_{13}$ and $f_{12} = 1 - x_{12}^2$.

Let p_r, $1 \leqslant r \leqslant t$, be some polynomials in $\mathfrak{R}$ and $\mathbb{B}$ be the set of complex matrices A (symmetric, with ones on diagonals) such that $p_r(A) \neq 0$ for all $1 \leqslant r \leqslant t$. If y_r are additional indeterminates, the polynomials $1 - y_r p_r$ belong to the ring $\mathfrak{T} = \mathfrak{R}[y_1, \ldots, y_t]$ of polynomials in both x_{ij}'s and y_r's. Obviously, a matrix A together with complex numbers $b_1, \ldots, b_t$ solve the polynomial equations $1 - y_r p_r = 0$ if and only if $A \in \mathbb{B}$ and $b_r = p_r(A)^{-1}$. Thus $\mathbb{B}$ is a coordinate projection of an algebraic set and is therefore constructible by [3, pp. 258–9]. In the following assertion another indeterminate z and the ring $\mathfrak{T}[z]$ of polynomials with one more indeterminate come into play.

Proposition 1. *Let $\mathcal{L}, \mathcal{M}$ be subfamilies of $\mathcal{R}$, p_r be a polynomial in $\mathfrak{R}$, $1 \leqslant r \leqslant t$, and $\mathbb{B}$ be the corresponding set of complex matrices. The implication*

$$\mathcal{L} \subseteq \langle\!\langle A \rangle\!\rangle \quad \Rightarrow \quad \mathcal{M} \cap \langle\!\langle A \rangle\!\rangle \neq \emptyset \,, \qquad A \in \mathbb{B} \,, \tag{13}$$

is true if and only if the ring $\mathfrak{T}[z]$ is generated by the polynomials $f_{(ij|K)}$ for $(ij|K) \in \mathcal{L}$, $1 - y_r p_r$ for $1 \leqslant r \leqslant t$, and $1 - z f_{\mathcal{M}}$. This is equivalent to the incidence of $f_{\mathcal{M}}$ to the radical $\sqrt{I_{\mathcal{L}}}$ of the ideal $I_{\mathcal{L}}$ generated by $f_{(ij|K)}$ for $(ij|K) \in \mathcal{L}$, and $1 - y_r p_r$ for $1 \leqslant r \leqslant t$.

Proof. If $\mathfrak{T}[z]$ is generated as above then the constant polynomial 1 can be combined from the listed polynomials. Thus, polynomials $h_{(ij|K)}$, h_r, and h in $\mathfrak{T}[z]$ exist such that

$$\sum_{(ij|K) \in \mathcal{L}} h_{(ij|K)} f_{(ij|K)} + \sum_{r=1}^{t} h_r [1 - y_r p_r] + h[1 - z f_{\mathcal{M}}] = 1 \,.$$

Substituting $A \in \mathbb{B}$ for x_{ij}'s and the nonzero numbers $p_r(A)^{-1}$ for y_r's the sum over r equals zero. Then

$$\sum_{(ij|K) \in \mathcal{L}} h^*_{(ij|K)} |A_{iK,jK}| + h^* \left[1 - z \prod_{(ij|K) \in \mathcal{M}} |A_{iK,jK}| \right] = 1$$

where $h^*_{(ij|K)}$ and h^* are polynomials involving only the indeterminate z. If $\langle\!\langle A \rangle\!\rangle \subseteq \mathcal{L}$ the sum over $\mathcal{L}$ vanishes. One concludes that the product must be equal to zero, that is, $\mathcal{M} \cap \langle\!\langle A \rangle\!\rangle \neq \emptyset$ and (13) holds.

Suppose (13) is true. When the polynomials $f_{(ij|K)}$ for $(ij|K)$ in $\mathcal{L}$, and $1 - y_r p_r$ for $1 \leqslant r \leqslant t$ vanish after substitution of complex numbers a_{ij} for x_{ij} and b_r for y_r then the symmetric matrix A with the off-diagonal elements $a_{i,j} = a_{ij}$ (and ones on the diagonal) belongs to $\mathbb{B}$. Now, $0 = f_{(ij|K)}(A) = |A_{iK,jK}|$ for $(ij|K)$ in $\mathcal{L}$ and thus $\langle\!\langle A \rangle\!\rangle \subseteq \mathcal{L}$. Due to (13), one has $(ij|K) \in \langle\!\langle A \rangle\!\rangle$ for some $(ij|K) \in \mathcal{M}$. Hence, $\prod_{(ij|K) \in \mathcal{M}} |A_{iK,jK}|$ equals zero and the polynomial $f_{\mathcal{M}}$ vanishes by the same substitution, $f_{\mathcal{M}}(A) = 0$. Hilbert's Nullstellensatz [3, p. 170] implies $f_{\mathcal{M}} \in \sqrt{I_{\mathcal{L}}}$.

If $f_{\mathcal{M}}$ belongs to the radical then $f_{\mathcal{M}}^m$ belongs to $I_{\mathcal{L}}$ for an integer $m \geqslant 1$. Rewriting the constant polynomial 1 as $z^m f_{\mathcal{M}}^m + [1 - z^m f_{\mathcal{M}}^m]$ and further as $z^m f_{\mathcal{M}}^m + h^*[1 - z f_{\mathcal{M}}]$ where $h^* \in \mathfrak{T}[z]$ it combines from a polynomial from $I_{\mathcal{L}}$ and $1 - z f_{\mathcal{M}}$. Thus, the ring $\mathfrak{T}[z]$ is generated by $I_{\mathcal{L}}$ and $1 - z f_{\mathcal{M}}$, cf. [3, p. 176]. □

4   Discussion and Examples

When examining an instance of (1) for $A > 0$ Proposition 1 can be useful. Indeed, if each polynomial p_r is chosen so that $p_r(A) \neq 0$ for every $A > 0$ with ones on its diagonal, Proposition 1 provides a sufficient condition for validity of the instance.

For example, since (2) implies

$$-(-1)^{[i,j;k]} z f_L f_{(ij|kL)} + z f_{kL} f_{(ij|L)} + \left[1 - z f_{(ik|L)} f_{(jk|L)}\right] = 1$$

Proposition 1 with $\mathcal{L} = \{(ij|kL), (ij|L)\}$, $\mathcal{M} = \{(ik|L), (jk|L)\}$ and $t = 0$ proves validity of (5) for the symmetric matrices with ones on diagonals, and then obviously for the symmetric matrices with nonzero diagonal elements. It is not difficult to see (and we do not give details) that all implications of Corollary 1 quantified by $A \in \mathbb{A}$ can be obtained in this way from Proposition 1 with $t = 1, 2$ and f_I's in the role of p_r's. Hence, all basic properties of CI in nondegenerate Gaussian vectors can be viewed as consequences of Proposition 1.

A more sophisticated relaxation of the quantification $A > 0$ works as follows. Given a matrix A and a permutation π of the set N, the matrix $A^\pi = (a_{\pi(i),\pi(j)})_{i,j=1}^n$ is obtained by permuting rows and columns of A simultaneously. Obviously, $A > 0$ if and only if $A^\pi > 0$. By [15, Theorem 36.2.1] the Hadamard product $A \circ A^\pi = (a_{i,j}\, a_{\pi(i),\pi(j)})_{i,j=1}^n$ of $A > 0$ and A^π is positive definite. Hence, the principal minors of Hadamard products $A^{\pi_1} \circ \ldots \circ A^{\pi_k}$ are positive for $A > 0$, $k \geqslant 1$ and arbitrary permutations $\pi_1, \ldots, \pi_k$ of N. Such minors are of interest when substituted for p_r's in Proposition (1).

Example 2. To prove the implication

$$\{(12|3), (13|4), (14|2)\} \subseteq \langle\!\langle A \rangle\!\rangle \quad \Rightarrow \quad (12|\emptyset) \in \langle\!\langle A \rangle\!\rangle \tag{14}$$

for $A > 0$ observe that $1 - a_{2,3}^2 a_{2,4}^2 a_{3,4}^2$ is a principal minor of the Hadamard product $A \circ A^{\pi_1} \circ A^{\pi_2}$ of three matrices and the polynomials

$$f_{(12|3)}, \; f_{(13|4)}, \; f_{(14|2)}, \; 1 - y_1(1 - x_{23}^2 x_{24}^2 x_{34}^2) \; \text{and} \; 1 - z x_{12}$$

generate $\mathfrak{T}[z]$ because

$$1 = (1 - z x_{12}) + z x_{12}\left(1 - y_1(1 - x_{23}^2 x_{24}^2 x_{34}^2)\right) + y_1 z x_{12}\left[1 - x_{23}^2 x_{24}^2 x_{34}^2\right]$$

and

$$x_{12}\left[1 - x_{23} x_{24} x_{34}\right] = f_{(12|3)} + x_{23} f_{(13|4)} - x_{23} x_{34} f_{(14|2)} .$$

Hence, Proposition 1 with $\mathcal{L} = \{(12|3), (13|4), (14|2)\}$, $\mathcal{M} = \{(12|\emptyset)\}$, $t = 1$ and p_1 equal to $(1 - x_{23}^2 x_{24}^2 x_{34}^2)$ implies (14) for the symmetric matrices A with ones on diagonals and $1 - a_{2,3}^2 a_{2,4}^2 a_{3,4}^2 \neq 0$, in turn, for the positive definite matrices.

Analogously,

$$\{(12|3), (23|4), (34|1), (41|2)\} \subseteq \langle\!\langle A \rangle\!\rangle \quad \Rightarrow \quad (12|\emptyset) \in \langle\!\langle A \rangle\!\rangle \tag{15}$$

holds for $A > 0$, using the principal minor $1 - a_{1,3}^2 a_{2,4}^2$ of the Hadamard product $A \circ A^\pi$ of two matrices and the identity

$$x_{12}\left[1 - x_{13}^2 x_{24}^2\right] = f_{(12|3)} + x_{13} f_{(23|4)} + x_{13} x_{24} f_{(34|1)} - x_{13}^2 x_{24} f_{(14|2)} .$$

Remark 2. From the computational point of view, a ring of polynomials is generated by a finite list of polynomials if and only if any Groebner basis of the ideal generated by the list contains a constant (nonzero) polynomial. Therefore, a verification whether $\mathfrak{T}[z]$ is generated as in Proposition 1 has the same complexity as a computation of a single Groebner basis.*

Example 3. For ϵ small and nonzero the matrices A given by

$$
\begin{bmatrix} 1 & \epsilon^2 & \epsilon^4 & \epsilon^3 \\ \epsilon^2 & 1 & \epsilon^{-2} & \epsilon \\ \epsilon^4 & \epsilon^{-2} & 1 & \epsilon \\ \epsilon^3 & \epsilon & \epsilon & 1 \end{bmatrix} \quad \text{and} \quad \begin{bmatrix} 1 & \epsilon^2 & \epsilon & \epsilon \\ \epsilon^2 & 1 & \epsilon & \epsilon^{-1} \\ \epsilon & \epsilon & 1 & \epsilon^2 \\ \epsilon & \epsilon^{-1} & \epsilon^2 & 1 \end{bmatrix}
$$

belong to $\mathbb{A}$ and render $\langle\!\langle A \rangle\!\rangle$ equal to

$$
\{(12|3),(13|4),(14|2)\} \quad \text{and} \quad \{(12|3),(23|4),(34|1),(41|2)\},
$$

respectively. By (14) and (15), these two subsets of $\mathcal{R}$ are not of the form $\langle\!\langle A \rangle\!\rangle$ for some $A > 0$. (Note that the above two matrices are not positive definite.) Therefore, to exclude only vanishing of proper minors of covariance matrices is not sufficient for the algebraic study of CI in Gaussian vectors.

Due to the following assertion, it is not difficult to lift the implications (14) and (15) of Example 2 to a general form.

Lemma 2. *Let* $\mathcal{L},\mathcal{M} \subseteq \mathcal{R}$ *and* $L \subseteq N$ *such that* $L \cap (ij \cup K) = \emptyset$ *for every* $(ij|K)$ *in* $\mathcal{L} \cup \mathcal{M}$. *If (1) is true for all* $A > 0$ *or* $A \in \mathbb{A}$ *then it remains true also when* $\mathcal{L}$ *and* $\mathcal{M}$ *are replaced by* $\mathcal{L}_L = \{(ij|K \cup L); (ij|K) \in \mathcal{L}\}$ *and* $\mathcal{M}_L$, *respectively.*

A proof is simple and therefore omitted.

Corollary 2. *For any positive definite matrix* A, $L \subseteq N$ *and* $i,j,k,l \in N \setminus L$ *different*

$$
\{(ij|kL),(ik|lL),(il|jL)\} \subseteq \langle\!\langle A \rangle\!\rangle \quad \Rightarrow \quad (ij|L) \in \langle\!\langle A \rangle\!\rangle, \tag{16}
$$

$$
\{(ij|kL),(jk|lL),(kl|iL),(li|jL)\} \subseteq \langle\!\langle A \rangle\!\rangle \quad \Rightarrow \quad (ij|L) \in \langle\!\langle A \rangle\!\rangle. \tag{17}
$$

A prototype of the implication (16) with $(ij|L)$ replaced by $(ik|jL)$ goes back to [16], dealing with relational databases. Its validity for random vectors taking finite number of values was established in [18].

Acknowledgement. This research was supported by the grant A 1075104 of GA AV and the grant 201/01/1482 of GA ČR.

---

* For example, in the software package MATHEMATICA, the following input
`GroebnerBasis[{x₁₂−x₁₃x₂₃, x₁₃−x₁₄x₃₄, x₁₂x₂₄−x₁₄, 1−y₁[1−x²₂₃x²₂₄x²₃₄], 1−z x₁₂},`
$\{x_{12}, x_{13}, x_{14}, x_{23}, x_{24}, x_{34}, y_1, z\}]$ is evaluated as the list $\{1\}$ consisting of the constant polynomial 1, cf. Example 2.

References

1. L. Csirmaz (1996) The dealer's random bits in perfect secret-sharing schemes. *Studia Scient. Math. Hungarica* **32** 429–437.
2. R.G. Cowell, A.P. Dawid, S.L. Lauritzen and D.J. Spiegelhalter (1999) *Probabilistic Networks and Expert Systems*. Springer: New York.
3. D. Cox, J. Little and D. O'Shea (1997) *Ideals, Varieties, and Algorithms*. Undergraduate Texts in Mathematics, Springer: New York.
4. C. DeConcini, D. Eisenbud and D. Procesi (1980) Young diagrams and determinantal varieties, *Inventiones math.* **56** 129-165
5. P. Doubilet, G.C. Rota and J. Stein (1974) On the foundations of the combinatorial theory IX: combinatorial methods in invariant theory, *Studies in Applied Math.* **53** 185-216.
6. D. Geiger and Ch. Meek (1999) Quantifier Elimination for statistical problems. *Proceedings of Fifteenth Conference on Uncertainty in Artificial Intelligence* (Eds. K. Laskey and H. Prade), Morgan Kaufmann, 226-235.
7. S.L. Lauritzen (1996) *Graphical Models*. Oxford University Press: Oxford.
8. F. Matúš (1992) On equivalence of Markov properties over undirected graphs. *Journal of Applied Probability* **29** 745–749.
9. F. Matúš (1997) Conditional independence structures examined via minors. *Annals of Mathematics and Artificial Intelligence* **21** 99–128.
10. F. Matúš (1999) Conditional independences among four random variables III: final conclusion. *Combinatorics, Probability and Computing* **8** 269–276.
11. F. Matúš (1999) Matroid representations by partitions. *Discrete Mathematics* **203** 169–194.
12. B. Mishra (1993) *Algorithmic Algebra*. Texts and monographs in Computer Science, Vol. XIV, Springer: New York.
13. D. Mond and R. Pellikaan (1989) Fitting ideals and multiple points of analytic mappings. In: Algebraic and Analytic Geometry, Patzcuaro 1987, E. Ramirez de Arellano ed., Springer Lecture Notes in Mathematics 1414, pp. 107–161.
14. J. Pearl (1988) *Probabilistic Reasoning in Intelligent Systems*. Morgan Kaufman: San Meteo, California.
15. V.V. Prasolov (1994) *Problems and Theorems in Linear Algebra*. AMS Translations of Mathematical Monographs, Vol. 134, AMS: Providence, Rhode Island.
16. Y. Sagiv and S.F. Walecka (1988) Subset dependencies and a completeness result for a subclass of embedded multivalued dependencies. *Journal of ACM* **29** 103-117.
17. J. Spencer (1994) *Ten Lectures on the Probabilistic Method*. SIAM: Philadelphia, Pennsylvania.
18. M. Studený (1991) Conditional independence relations have no finite complete characterization. *Transactions of the 11-th Prague Conference on Information Theory, Statistical Decision Functions and Random Processes, Vol.B*, Academia, Prague, 377-396 (also Kluwer, Dordrecht).
19. M. Studený (2002) On stochastic conditional independence: the problems of characterization and description. *Annals of Mathematics and Artificial Intelligence* **35** 323–341.
20. R.W. Yeung (2002) *A First Course in Information Theory*. Kluwer Academic/Plenum Publishers: New York.
21. Z. Zhang and R.W. Yeung (1997) A non-Shannon-type conditional inequality of information quantities. *IEEE Trans. Information Theory* **43** 1982–1986.

Looking at Probabilistic Conditionals from an Institutional Point of View[*]

Christoph Beierle and Gabriele Kern-Isberner

Praktische Informatik VIII - Wissensbasierte Systeme,
Fachbereich Informatik, FernUniversität Hagen, D-58084 Hagen, Germany
{christoph.beierle, gabriele.kern-isberner}@fernuni-hagen.de

Abstract. We show how probabilistic logic and probabilistic conditional logic can be formalized in the framework of institutions, thereby supporting the study of structural properties of both syntax and semantics of these logics. By using the notions of institution morphism and institution embedding, the relationships between probabilistic propositional logic, probabilistic conditional logic, and the underlying two-valued propositional logic are investigated in detail, telling us, for instance, precisely how to interpret probabilistic conditionals as probabilistic facts or in a propositional setting and vice versa.

1 Introduction

Probabilistic logic has played a major role in the modelling of and reasoning about uncertain knowledge [Pea88, Par94, N.R69]. Whereas in boolean-valued classical logic, everything is essentially just *true* or *false*, assigning probability values to sentences opens up a whole universe of possibilities to express the (un)certainty of a formula. But what are the exact relationships between probabilistic logic and the underlying two-valued logic, or between probabilistic logic and the logic of probabilistic conditionals?

To answer these questions, we use the notion of an institution which Goguen and Burstall introduced as a general framework for logical systems [GB92]. Institutions have been used not only for the general study of logics, but also as a basis for specification and development languages, see e.g. [BG80, ST97, Tar96, GR02].

An institution formalizes the informal notion of a logical system, including syntax, semantics, and the relation of satisfaction between them. The latter poses the major requirement for an institution: that the satisfaction relation is consistent under the change of notation.

Whereas in [GB92] the examples for an institution are mostly based on classical logics, this paper presents a thorough investigation of probabilistic logic and the logic of probabilistic conditionals within the institution framework, thus

---

[*] The research reported here was partially supported by the DFG – Deutsche Forschungsgemeinschaft (grant BE 1700/5-1).

G. Kern-Isberner, W. Rödder, and F. Kulmann (Eds.): WCII 2002, LNAI 3301, pp. 162–179, 2005.

opening up a novel view on these non-classical logics. Considering probabilistic logic and probabilistic conditionals as institutions provides new tools and aspects for the study of structural properties of both syntax and semantics of these logics. It will allow us to immediately apply various general results for institutions to our probabilistic logics, giving us for free e.g. the presentation lemma, various closure properties, or the notions and results about theories and their morphisms. By using the notion of institution morphism [GB92, GR02] and a variant thereof which we call institution embedding, we investigate in detail the relationships between propositional, probabilistic propositional, and probabilistic conditional logic. This will tell us precisely how to interpret e.g. probabilistic conditionals as probabilistic facts or in a propositional setting and vice versa.

This paper is organized as follows: In Section 2, we formalize propositional, probabilistic propositional, and probabilistic conditional logic as institutions. In Section 3, we introduce institution embeddings, and develop a complete roadmap of institution morphisms and embeddings between these institutions. Section 4 contains some conclusions and points out further work.

2   Institutions and Probabilistic Logics

After recalling the definition of an institution and fixing some basic notation, we first present propositional logic in the institution framework. We then formalize in two steps probabilistic propositional logic and the logic of probabilistic conditionals as institutions, give some examples, and demonstrate how the well-known concept of marginal distributions occurs within the institution context.

2.1   Preliminaries: Basic Definitions and Notations

If C is a category, $|C|$ denotes the objects of C and $/C/$ its morphisms; for both objects $c \in |C|$ and morphisms $\varphi \in /C/$, we also write just $c \in C$ and $\varphi \in C$, respectively. C^{op} is the opposite category of C, with the direction of all morphisms reversed. The composition of two functors $F : C \to C'$ and $G : C' \to C''$ is denoted by $G \circ F$ (first apply F, then G). For functors $F, G : C \to C'$, a natural transformation η from F to G, denoted by $\eta : F \implies G$, assigns to each object $c \in |C|$ a morphism $\eta_c : F(C) \to G(C) \in /C'/$ such that for every morphism $\varphi : c \to d \in /C/$ we have $\eta_d \circ F(\varphi) = G(\varphi) \circ \eta_c$. $\mathcal{SET}$ and $\mathcal{CAT}$ denote the categories of sets and of categories, respectively. (For more information about categories, see e.g. [HS73] or [Mac72].)

The central definition of an institution [GB92] is the following (cf. Figure 1 that visualizes the relationships within an institution):

Definition 1. *An institution is a quadruple $Inst = \langle\, Sig,\, Mod,\, Sen,\, \models\, \rangle$ with a category Sig of signatures as objects, a functor $Mod : Sig \to \mathcal{CAT}^{op}$ yielding the category of Σ-models for each signature Σ, a functor $Sen : Sig \to \mathcal{SET}$ yielding the sentences over a signature, and a $|Sig|$-indexed relation $\models_\Sigma \subseteq |Mod(\Sigma)| \times Sen(\Sigma)$ such that for each signature morphism $\varphi : \Sigma \to \Sigma' \in /Sig/,$*

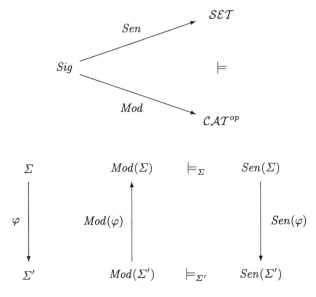

Fig. 1. Relationships within an institution $Inst = \langle Sig, Mod, Sen, \models \rangle$ [GB92]

for each $m' \in |Mod(\Sigma')|$, and for each $f \in Sen(\Sigma)$ the following satisfaction
condition *holds:*

$$m' \models_{\Sigma'} Sen(\varphi)(f) \quad \textit{iff} \quad Mod(\varphi)(m') \models_{\Sigma} f$$

For sets F, G of Σ-sentences and a Σ-model m we write $m \models_{\Sigma} F$ iff $m \models_{\Sigma} f$
for all $f \in F$. The satisfaction relation is lifted to semantical entailment $\models_{\Sigma}$
between sentences by defining $F \models_{\Sigma} G$ iff for all Σ-models m with $m \models_{\Sigma} F$ we
have $m \models_{\Sigma} G$. $F^{\bullet} = \{f \in Sen(\Sigma) \mid F \models_{\Sigma} f\}$ is called the *closure* of F, and F
is *closed* if $F = F^{\bullet}$. The closure operator fulfils the *closure lemma* $\varphi(F^{\bullet}) \subseteq$
$\varphi(F)^{\bullet}$ and various other nice properties like $\varphi(F^{\bullet})^{\bullet} = \varphi(F)^{\bullet}$ or $(F^{\bullet} \cup G)^{\bullet} =$
$(F \cup G)^{\bullet}$. A consequence of the closure lemma is that entailment is preserved
under change of notation carried out by a signature morphism, i.e. $F \models_{\Sigma} G$
implies $\varphi(F) \models_{\varphi(\Sigma)} \varphi(G)$ (but not vice versa).

2.2   The Institution of Propositional Logic

In all circumstances, propositional logic seems to be the most basic logic. The
components of its institution $Inst_{\mathcal{B}} = \langle Sig_{\mathcal{B}}, Mod_{\mathcal{B}}, Sen_{\mathcal{B}}, \models_{\mathcal{B}} \rangle$ will be defined
as follows:

Signatures: $Sig_{\mathcal{B}}$ is the category of propositional signatures. A propositional
signature $\Sigma \in |Sig_{\mathcal{B}}|$ is a (finite) set of propositional variables, $\Sigma = \{a_1, a_2, \ldots\}$.
A propositional signature morphism $\varphi : \Sigma \to \Sigma' \in /Sig_{\mathcal{B}}/$ is a function mapping
propositional variables to propositional variables.

Models: For each signature $\Sigma \in Sig_{\mathcal{B}}$, $Mod_{\mathcal{B}}(\Sigma)$ contains the set of all propositional interpretations for Σ, i.e.

$$|Mod_{\mathcal{B}}(\Sigma)| \;=\; \{I \mid I : \Sigma \to Bool\}$$

where $Bool = \{true, false\}$. Due to its simple structure, the only morphisms in $Mod_{\mathcal{B}}(\Sigma)$ are the identity morphisms. For each signature morphism $\varphi :$ $\Sigma \to \Sigma' \in Sig_{\mathcal{B}}$, we define the functor $Mod_{\mathcal{B}}(\varphi) : Mod_{\mathcal{B}}(\Sigma') \to Mod_{\mathcal{B}}(\Sigma)$ by $(Mod_{\mathcal{B}}(\varphi)(I'))(a_i) := I'(\varphi(a_i))$ where $I' \in Mod_{\mathcal{B}}(\Sigma')$ and $a_i \in \Sigma$.

Sentences: For each signature $\Sigma \in Sig_{\mathcal{B}}$, the set $Sen_{\mathcal{B}}(\Sigma)$ contains the usual propositional formulas constructed from the propositional variables in Σ and the logical connectives $\wedge$ (and), $\vee$ (or), and $\neg$ (not). Additionally, the classical (material) implication $A \Rightarrow B$ is used as a syntactic variant for $\neg A \vee B$. The symbols $\top$ and $\bot$ denote a tautology (like $a \vee \neg a$) and a contradiction (like $a \wedge \neg a$), respectively.

For each signature morphism $\varphi : \Sigma \to \Sigma' \in Sig_{\mathcal{B}}$, the function $Sen_{\mathcal{B}}(\varphi) :$ $Sen_{\mathcal{B}}(\Sigma) \to Sen_{\mathcal{B}}(\Sigma')$ is defined by straightforward inductive extension on the structure of the formulas; e.g., $Sen_{\mathcal{B}}(\varphi)(a_i) = \varphi(a_i)$ and $Sen_{\mathcal{B}}(\varphi)(A \wedge B) = Sen_{\mathcal{B}}(\varphi)(A) \wedge Sen_{\mathcal{B}}(\varphi)(B)$. In the following, we will abbreviate $Sen_{\mathcal{B}}(\varphi)(A)$ by just writing $\varphi(A)$.

In order to simplify notations, we will often replace conjunction by juxtaposition and indicate negation of a formula by barring it, i.e. $AB = A \wedge B$ and $\overline{A} = \neg A$. An *atomic formula* is a formula consisting of just a propositional variable, a *literal* is a positive or a negated atomic formula, an *elementary conjunction* is a conjunction of literals, and a *complete conjunction* is an elementary conjunction containing each atomic formula either in positive or in negated form. Ω_{Σ} denotes the set of all complete conjunctions over a signature Σ; if Σ is clear from the context, we may drop the index Σ.

Note that there is an obvious bijection between $|Mod_{\mathcal{B}}(\Sigma)|$ and Ω_{Σ}, associating with $I \in |Mod_{\mathcal{B}}(\Sigma)|$ the complete conjunction $\omega_I \in \Omega_{\Sigma}$ in which an atomic formula $a_i \in \Sigma$ occurs in positive form iff $I(a_i) = true$.

Satisfaction Relation: For any $\Sigma \in |Sig_{\mathcal{B}}|$, the satisfaction relation

$$\models_{\mathcal{B},\Sigma} \;\subseteq\; |Mod_{\mathcal{B}}(\Sigma)| \times Sen_{\mathcal{B}}(\Sigma)$$

is defined as expected for propositional logic, e.g. $I \models_{\mathcal{B},\Sigma} a_i$ iff $I(a_i) = true$ and $I \models_{\mathcal{B},\Sigma} A \wedge B$ iff $I \models_{\mathcal{B},\Sigma} A$ and $I \models_{\mathcal{B},\Sigma} B$ for $a_i \in \Sigma$ and $A, B \in Sen_{\mathcal{B}}(\Sigma)$.

Proposition 1. *$Inst_{\mathcal{B}} \;=\; \langle\, Sig_{\mathcal{B}}, Mod_{\mathcal{B}}, Sen_{\mathcal{B}}, \models_{\mathcal{B}} \,\rangle$ is an institution.*

It is easy to prove this proposition since the satisfaction condition

$$I' \models_{\mathcal{B},\Sigma'} \varphi(A) \quad \text{iff} \quad Mod_{\mathcal{B}}(\varphi)(I') \models_{\mathcal{B},\Sigma} A$$

holds by straightforward induction on the structure of A. E.g., for a propositional variable a_i, we have $I' \models_{\mathcal{B},\Sigma'} \varphi(a_i)$ iff $I'(\varphi(a_i)) = true$ iff $(Mod_{\mathcal{B}}(\varphi)(I'))(a_i) = true$ iff $Mod_{\mathcal{B}}(\varphi)(I') \models_{\mathcal{B},\Sigma} a_i$.

Example 1. Let $\Sigma = \{s, t, u\}$ and $\Sigma' = \{a, b, c\}$ be two propositional signatures with the atomic propositions s – *being a scholar*, t – *being not married*, u – *being single* and a – *being a student*, b – *being young*, c – *being unmarried*. Let I' be the Σ'-model with $I'(a) = true$, $I'(b) = true$, $I'(c) = false$. Let $\varphi : \Sigma \to \Sigma' \in Sig_\mathcal{B}$ be the signature morphism with $\varphi(s) = a$, $\varphi(t) = c$, $\varphi(u) = c$. The functor $Mod_\mathcal{B}(\varphi)$ takes I' to the Σ-model $I := Mod_\mathcal{B}(\varphi)(I')$, yielding $I(s) = I'(a) = true$, $I(t) = I'(c) = false$, $I(u) = I'(c) = false$.

Note that in the example, φ is neither surjective nor injective. φ not being surjective makes the functor $Mod_\mathcal{B}(\varphi)$ a forgetful functor – any information about b (being young) in I' is forgotten in I. φ not being injective implies that any two propositional variables from Σ mapped to the same element in Σ' are both being identified with the same proposition; thus, under the forgetful functor $Mod_\mathcal{B}(\varphi)$, the interpretation of t (*being not married*) and u (*being single*) will always be equivalent since $\varphi(t) = \varphi(u)$.

2.3   The Institution of Probabilistic Propositional Logic

Based on $Inst_\mathcal{B}$, we can now define the institution of probabilistic propositional logic $Inst_\mathcal{P} = \langle Sig_\mathcal{P}, Mod_\mathcal{P}, Sen_\mathcal{P}, \models_\mathcal{P} \rangle$. We will first give a very short introduction to probabilistics as far as it is needed here.

Let $\Sigma \in |Sig_\mathcal{B}|$ be a propositional signature. A *probability distribution* (or *probability function*) *over* Σ is a function $P : Sen_\mathcal{B}(\Sigma) \to [0, 1]$ such that $P(\top) = 1$, $P(\bot) = 0$, and $P(A \vee B) = P(A) + P(B)$ for any formulas $A, B \in Sen_\mathcal{B}(\Sigma)$ with $AB = \bot$. Each probability distribution P is determined uniquely by its values on the complete conjunctions $\omega \in \Omega_\Sigma$, since $P(A) = \sum\limits_{\omega \in \Omega_\Sigma, \omega \models_{\mathcal{B},\Sigma} A} P(\omega)$.

For two propositional formulas $A, B \in Sen_\mathcal{B}(\Sigma)$ with $P(A) > 0$, the *conditional probability of B given A* is $P(B|A) := \frac{P(AB)}{P(A)}$. Any subset $\Sigma_1 \subseteq \Sigma$ gives rise to a distribution $P_{\Sigma_1} : Sen_\mathcal{B}(\Sigma_1) \to [0, 1]$ by virtue of defining $P_{\Sigma_1}(\omega_1) = \sum\limits_{\omega \in \Omega_\Sigma, \omega \models_{\mathcal{B},\Sigma} \omega_1} P(\omega)$ for all $\omega_1 \in \Omega_{\Sigma_1}$; P_{Σ_1} is called the *marginal distribution of P on Σ_1.*

Signatures: $Sig_\mathcal{P}$ is identical to the category of propositional signatures, i.e. $Sig_\mathcal{P} = Sig_\mathcal{B}$.

Models: For each signature Σ, the objects of $Mod_\mathcal{P}(\Sigma)$ are probability distributions over the propositional variables, i.e.

$$|Mod_\mathcal{P}(\Sigma)| = \{P \mid P \text{ is a probability distribution over } \Sigma\}$$

As for $Mod_\mathcal{B}(\Sigma)$, we assume in this paper that the only morphisms in $Mod_\mathcal{P}(\Sigma)$ are the identity morphisms.

For each signature morphism $\varphi : \Sigma \to \Sigma'$, we define a functor $Mod_\mathcal{P}(\varphi) : Mod_\mathcal{P}(\Sigma') \to Mod_\mathcal{P}(\Sigma)$ by mapping each distribution P' over Σ' to a distribution $Mod_\mathcal{P}(\varphi)(P')$ over Σ. $Mod_\mathcal{P}(\varphi)(P')$ is defined by giving its value for all complete conjunctions over Σ:

$$(Mod_\mathcal{P}(\varphi)(P'))(\omega) := P'(\varphi(\omega)) = \sum_{\omega':\omega' \models_{\mathcal{B},\Sigma'} \varphi(\omega)} P'(\omega')$$

where ω and ω' are complete conjunctions over Σ and Σ', respectively. We still have to ensure:

Proposition 2. $Mod_\mathcal{P}(\varphi)(P')$ *is a* Σ-*model.*

It is straightforward to see that the equation used to define $Mod_\mathcal{P}(\varphi)(P')$ easily generalizes to propositional formulas $A \in Sen_\mathcal{B}(\Sigma)$, that is to say it holds that $(Mod_\mathcal{P}(\varphi)(P'))(A) = P'(\varphi(A))$.

Another immediate consequence is the following:

Proposition 3. *Let* $\varphi : \Sigma \to \Sigma'$ *and* $P := Mod_\mathcal{P}(\varphi)(P')$ *as above. If* φ *is injective, then the marginal distribution* $P'_{\varphi(\Sigma)}$ *of* P' *over* $\varphi(\Sigma) := \{\varphi(x) \mid x \in \Sigma\} \subseteq \Sigma'$ *is identical to* P *with variables* $\varphi(a_i)$ *in* P' *renamed to* a_i *in* P, *i.e.* $P'_{\varphi(\Sigma)}(\omega') = P(\varphi^{-1}(\omega'))$ *for all complete conjunctions* ω' *over* $\varphi(\Sigma)$.

This proposition shows that the well-known (semantical) concept of a marginal distribution coincides with the forgetful functor induced by an injective signature morphism, forgetting the propositional variables not reached by a non-surjective φ. If φ is non-injective, two propositional variables mapped to the same element in Σ' are identified. Therefore, any conjunction containing a negated and a non-negated propositional variable both being identified under φ gets the probability 0. Thus, in the general case for non-surjective and non-injective φ, $Mod_\mathcal{P}(\varphi)(P')$ is defined by 'collapsing' all propositional variables identified under φ and taking the marginal distribution over the remaining variables reached by φ.

Example 2. Let Σ, Σ' and φ be as in Example 1. We define a Σ'-model P' by assigning a probability $P'(\omega')$ to every complete conjunction over Σ':

ω'	$P'(\omega')$	ω'	$P'(\omega')$	ω'	$P'(\omega')$	ω'	$P'(\omega')$
abc	0.1950	$a\overline{b}c$	0.1758	$\overline{a}bc$	0.0408	$a\overline{b}\overline{c}$	0.0519
$\overline{a}bc$	0.1528	$\overline{a}b\overline{c}$	0.1378	$\overline{a}\overline{b}c$	0.1081	$\overline{a}\overline{b}\overline{c}$	0.1378

Thus, for instance, the probability $P'(abc)$ of *being a student, being young,* and *being unmarried* is 0.1950, and the probability $P'(a\overline{b}c)$ of *being a student, not being young,* and *being unmarried* is 0.0408.

The functor $Mod_\mathcal{P}(\varphi)$ transforms P' into the following Σ-model P:

ω	$P(\omega)$	ω	$P(\omega)$	ω	$P(\omega)$	ω	$P(\omega)$
stu	0.2358	$st\overline{u}$	0.0000	$s\overline{t}u$	0.0000	$s\overline{t}\overline{u}$	0.2277
$\overline{s}tu$	0.2609	$\overline{s}t\overline{u}$	0.0000	$\overline{s}\overline{t}u$	0.0000	$\overline{s}\overline{t}\overline{u}$	0.2756

For instance, the probability $P(stu)$ of *being a scholar, being not married,* and *being single* is 0.2358.

Sentences: For each signature Σ, the set $Sen_\mathcal{P}(\Sigma)$ contains *probabilistic facts* of the form

$$A[x]$$

where $A \in Sen_\mathcal{B}(\Sigma)$ is a propositional formula from $Inst_\mathcal{B}$. $x \in [0,1]$ is a probability value indicating the degree of certainty for the occurrence of A.

For each signature morphism $\varphi : \Sigma \to \Sigma'$, the extension $Sen_\mathcal{P}(\varphi) : Sen_\mathcal{P}(\Sigma) \to Sen_\mathcal{P}(\Sigma')$ is defined by $Sen_\mathcal{P}(\varphi)(A[x]) = \varphi(A)[x]$.

Satisfaction Relation: The satisfaction relation $\models_{\mathcal{P},\Sigma} \subseteq |Mod_\mathcal{P}(\Sigma)| \times Sen_\mathcal{P}(\Sigma)$ is defined, for any $\Sigma \in |Sig_\mathcal{P}|$, by

$$P \models_{\mathcal{P},\Sigma} A[x] \quad \text{iff} \quad P(A) = x$$

Note that, since $P(\overline{A}) = 1 - P(A)$ for each formula $A \in Sen_\mathcal{B}(\Sigma)$, it holds that $P \models_{\mathcal{P},\Sigma} A[x]$ iff $P \models_{\mathcal{P},\Sigma} \overline{A}[1-x]$.

Proposition 4. $Inst_\mathcal{P} = \langle Sig_\mathcal{P}, Mod_\mathcal{P}, Sen_\mathcal{P}, \models_\mathcal{P} \rangle$ *is an institution.*

Proof. We are just left to show that the satisfaction condition

$$P' \models_{\mathcal{P},\Sigma'} Sen_\mathcal{P}(\varphi)(A[x]) \quad \text{iff} \quad Mod_\mathcal{P}(\varphi)(P') \models_{\mathcal{P},\Sigma} A[x]$$

holds for each signature morphism $\varphi : \Sigma \to \Sigma' \in Sig_\mathcal{P}$, for each $P' \in Mod_\mathcal{P}(\Sigma')$, and for each $A[x] \in Sen_\mathcal{P}(\Sigma)$:

$$P' \models_{\mathcal{P},\Sigma'} Sen_\mathcal{P}(\varphi)(A[x]) \text{ iff } P' \models_{\mathcal{P},\Sigma'} (\varphi(A)[x]) \qquad \text{iff } P'(\varphi(A)) = x$$
$$\text{iff } (Mod_\mathcal{P}(\varphi)(P'))(A) = x \text{ iff } Mod_\mathcal{P}(\varphi)(P') \models_{\mathcal{P},\Sigma} A[x] \qquad \square$$

Example 3. Let Σ, Σ', P, P' and φ be as in Example 2. Then $P' \models_{\mathcal{P},\Sigma'} b[0.6614]$ since the probability of *being young* is $P'(b) = 0.6614$.

Similarly, $P \models_{\mathcal{P},\Sigma} u[0.4967]$ since under P, the probability of *being single* is $P(u) = 0.4967$. This immediately implies $P' \models_{\mathcal{P},\Sigma'} c[0.4967]$ (i.e. under P', the probability of *being unmarried* is 0.4967) due to the satisfaction condition since $\varphi(u) = c$.

2.4 The Institution of Probabilistic Conditional Logic

We now use $Inst_\mathcal{P}$ to define the institution of probabilistic conditionals $Inst_\mathcal{C} = \langle Sig_\mathcal{C}, Mod_\mathcal{C}, Sen_\mathcal{C}, \models_\mathcal{C} \rangle$.

Signatures: $Sig_\mathcal{C}$ is identical to the category of propositional signatures, i.e. $Sig_\mathcal{C} = Sig_\mathcal{P} = Sig_\mathcal{B}$.

Models: The models for probabilistic conditional logic are again probability distributions over the propositional variables. Therefore, the model functor can be taken directly from probabilistic propositional logic, giving us $Mod_\mathcal{C} = Mod_\mathcal{P}$.

Sentences: For each signature Σ, the set $Sen_\mathcal{C}(\Sigma)$ contains *probabilistic conditionals* (sometimes also called *probabilistic rules*) of the form

$$(B|A)[x]$$

where $A, B \in Sen_\mathcal{B}(\Sigma)$ are propositional formulas from $Inst_\mathcal{B}$. $x \in [0,1]$ is a probability value indicating the degree of certainty for the occurrence of B under the condition A.

Note that the sentences from $Inst_\mathcal{P}$ are included implicitly since a probabilistic fact of the form $B[x]$ can easily be expressed as a conditional $(B|\top)[x]$ with a tautology as trivial antecedent.

For each signature morphism $\varphi : \Sigma \rightarrow \Sigma'$, the extension $Sen_\mathcal{C}(\varphi) : Sen_\mathcal{C}(\Sigma) \rightarrow Sen_\mathcal{C}(\Sigma')$ is defined by straightforward inductive extension on the structure of the formulas: $Sen_\mathcal{C}(\varphi)((B|A)[x]) = (\varphi(B)|\varphi(A))[x]$.

Satisfaction Relation: The satisfaction relation $\models_{\mathcal{C},\Sigma} \subseteq |Mod_\mathcal{C}(\Sigma)| \times Sen_\mathcal{C}(\Sigma)$ is defined, for any $\Sigma \in |Sig_\mathcal{C}|$, by

$$P \models_{\mathcal{C},\Sigma} (B|A)[x] \quad \text{iff} \quad P(A) > 0 \text{ and } P(B \mid A) = \frac{P(AB)}{P(A)} = x$$

Note that for probabilistic facts we have $P \models_{\mathcal{C},\Sigma} (B|\top)[x]$ iff $P(B) = x$ from the definition of the satisfaction relation since $P(\top) = 1$. Thus, $P \models_{\mathcal{P},\Sigma} B[x]$ iff $P \models_{\mathcal{C},\Sigma} (B|\top)[x]$.

Proposition 5. $Inst_\mathcal{C} = \langle Sig_\mathcal{C}, Mod_\mathcal{C}, Sen_\mathcal{C}, \models_\mathcal{C} \rangle$ *is an institution.*

Example 4. Let Σ, Σ', P, P' and φ be as in Example 2. Just as $P' \models_{\mathcal{P},\Sigma'} b[0.6614]$ we now have $P' \models_{\mathcal{C},\Sigma'} (b|\top)[0.6614]$ for the trivial antecedent $\top$. Moreover, $P' \models_{\mathcal{C},\Sigma'} (b|a)[0.8]$ since the probability of *being young* under the condition of *being a student* is $P'(b \mid a) = 0.8$. Corresponding to Example 3, we also have both $P \models_{\mathcal{C},\Sigma} (u|\top)[0.4967]$ and $P' \models_{\mathcal{C},\Sigma'} (c|\top)[0.4967]$.

3 Relating Propositional, Probabilistic, and Probabilistic Conditional Logic

Having stepwise developed propositional, probabilistic, and probabilistic conditional logic, we now turn to study the interrelationships between these logics. There is an obvious translation of sentences mapping A to $A[1.0]$ and mapping $A[x]$ to $(A|\top)[x]$. Furthermore, there is a similar obvious transformation of a propositional interpretation I to the probabilistic model P_I that assigns probability 1 to the single complete conjunction ω with $I(\omega) = true$, and probability 0 to every other complete conjunction. E.g. for the Σ-interpretation I from Example 1, we have $P_I(s\bar{t}\bar{u}) = 1$ and $P_I(st\bar{u}) = 0$.

What happens to satisfaction and entailment when using these standard translations? In order to make these questions more precise, we use the notion of institution morphisms introduced in [GB92] (see also [GR02]) and a variant thereof which we call institution embedding.

3.1   Institution Morphisms

An institution morphism Φ expresses a relation between two institutions $Inst$ und $Inst'$ such that the satisfaction condition of $Inst$ may be computed by the satisfaction condition of $Inst'$ if we translate it according to Φ. The translation is done by relating every $Inst$-signature Σ to an $Inst'$-signature Σ', each Σ'-sentence to a Σ-sentence, and each Σ-model to a Σ'-model.

Definition 2. *Let* $Inst = \langle Sig, Mod, Sen, \models \rangle$ *and* $Inst' = \langle Sig', Mod', Sen', \models' \rangle$ *be two institutions. An* institution morphism Φ *from* $Inst$ *to* $Inst'$ *is a triple* $\langle \phi, \alpha, \beta \rangle$ *with a functor* $\phi : Sig \rightarrow Sig'$, *a natural transformation* $\alpha : Sen' \circ \phi \Longrightarrow Sen$, *and a natural transformation* $\beta : Mod \Longrightarrow Mod' \circ \phi$ *such that for each* $\Sigma \in |Sig|$, *for each* $m \in |Mod(\Sigma)|$, *and for each* $f' \in Sen'(\phi(\Sigma))$ *the following satisfaction condition (for institution morphisms) holds:*

$$m \models_\Sigma \alpha_\Sigma(f') \quad \text{iff} \quad \beta_\Sigma(m) \models'_{\phi(\Sigma)} f' \tag{1}$$

Figure 2 illustrates the relationships within an institution morphism, while Figure 3 visualizes the natural transformation conditions for the sentence and model translation components α and β.

3.2   Relating Propositional and Probabilistic Logic

Since $Inst_\mathcal{B}$, $Inst_\mathcal{P}$, and $Inst_\mathcal{C}$ all have the same category $Sig_\mathcal{B}$ of signatures, a natural choice for the signature translation component ϕ in any morphism between these institutions is the identity $id_{Sig_\mathcal{B}}$ which we will use in this and also the following subsections. Furthermore, it is easy to check that the standard translations discussed above

$$\alpha_{\mathcal{B}/\mathcal{P}} : Sen_\mathcal{B} \Longrightarrow Sen_\mathcal{P} \qquad \alpha_{\mathcal{B}/\mathcal{P},\Sigma}(A) = A[1.0]$$

$$\beta_{\mathcal{B}/\mathcal{P}} : Mod_\mathcal{B} \Longrightarrow Mod_\mathcal{P} \qquad \beta_{\mathcal{B}/\mathcal{P},\Sigma}(I) = P_I \qquad P_I(\omega) = \begin{cases} 1 & \text{if } I(\omega) = true \\ 0 & \text{otherwise} \end{cases}$$

(where $\Sigma \in Sig_\mathcal{B}$ and ω is a complete conjunction over Σ) are natural transformations. Do these give rise to an institution morphism between $Inst_\mathcal{B}$ and $Inst_\mathcal{P}$?

Note that both standard translations, $\alpha_{\mathcal{B}/\mathcal{P}}$ and $\beta_{\mathcal{B}/\mathcal{P}}$, go in the same direction, from propositional to probabilistic logic. Thus, when trying to establish an institution morphism between these two institutions, we take either $\alpha_{\mathcal{B}/\mathcal{P}}$ or $\beta_{\mathcal{B}/\mathcal{P}}$ and try to find a corresponding natural transformation going in the other direction.

First, we look for an institution morphism going from probabilistic to propositional logic where we have the intuitive standard translation mapping A to $A[1]$. However, possibly a little surprising at first sight, no institution morphism using this sentence translation exists as the following proposition shows.

Proposition 6. *There is no* β *such that* $\langle id_{Sig_\mathcal{B}}, \alpha_{\mathcal{B}/\mathcal{P}}, \beta \rangle : Inst_\mathcal{P} \longrightarrow Inst_\mathcal{B}$ *is an institution morphism.*

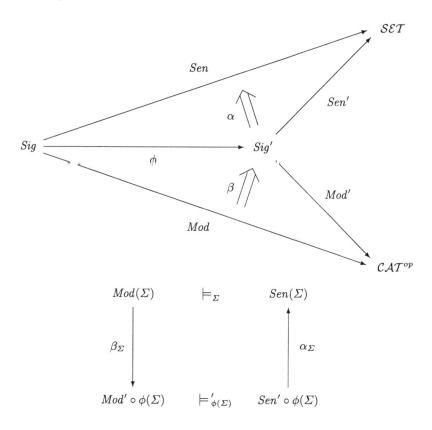

Fig. 2. Relationshipswithinaninstitutionmorphism⟨ ϕ, α, β ⟩ : ⟨ *Sig*, *Mod*, *Sen*, $\models$ ⟩ ⟶ ⟨ *Sig'*, *Mod'*, *Sen'*, $\models'$ ⟩

Proof. Let β : $Mod_\mathcal{P} \implies Mod_\mathcal{B}$ be a natural transformation. We show that the satisfaction condition (1) does *not* hold: Let $\Sigma \in Sig_\mathcal{B}$, $a \in \Sigma$, $x \in [0,1]$ with $x \neq 0$, $x \neq 1$, and $P \in Mod_\mathcal{P}(\Sigma)$ such that $P \models_{\mathcal{P},\Sigma} a[x]$ holds. (Obviously, such a P can be chosen.) Under the propositional interpretation $\beta_\Sigma(P)$, a must be either *true* or *false*. If $\beta_\Sigma(P)(a) = true$, the satisfaction condition would require $P \models_{\mathcal{P},\Sigma} a[1.0]$ which does not hold since $x \neq 1$. On the other hand, if $\beta_\Sigma(P)(a) = false$, we have $\beta_\Sigma(P)(\bar{a}) = true$ and the satisfaction condition would require $P \models_{\mathcal{P},\Sigma} \bar{a}[1.0]$. This can not be the case since $P \models_{\mathcal{P},\Sigma} \bar{a}[1-x]$ and $x \neq 0$.   □

When trying to construct an institution morphism from propositional to probabilistic logic, we must find a natural transformation between the respective sentence functors mapping a probabilistic formula $A[x]$ to a propositional formula. An important question is how we can map sentences with probabilities other than 0 or 1, e.g. something like *being young with probability 0.98*, to sentences which are simply *true* or *false*. The next proposition gives an answer to this question when using the standard model translation yielding P_I.

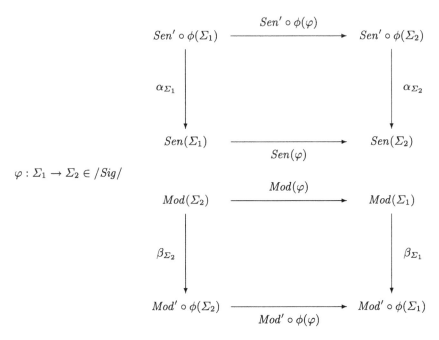

$\varphi : \Sigma_1 \to \Sigma_2 \in /Sig/$

Fig. 3. Natural transformation conditions for sentence and model translations in an institution morphism $\langle \phi, \alpha, \beta \rangle : \langle Sig, Mod, Sen, \models \rangle \longrightarrow \langle Sig', Mod', Sen', \models' \rangle$

Proposition 7. *If $\langle id_{Sig_{\mathcal{B}}}, \alpha, \beta_{\mathcal{B}/\mathcal{P}} \rangle : Inst_{\mathcal{B}} \longrightarrow Inst_{\mathcal{P}}$ is an institution morphism, then for any $\Sigma \in Sig_{\mathcal{B}}$, α_{Σ} maps every sentence $A[x]$ with $x \neq 0$ and $x \neq 1$ to $\bot$.*

Proof. Let $\alpha : Sen_{\mathcal{P}} \Longrightarrow Sen_{\mathcal{B}}$ be a natural transformation, $\Sigma \in Sig_{\mathcal{B}}$, and $A[x] \in Sen_{\mathcal{P}}(\Sigma)$ with $x \in (0,1)$ and thus $x \neq 0$, $x \neq 1$. If $\alpha_{\Sigma}(A[x])$ is not a contradiction, then there exists an interpretation $I \in Mod_{\mathcal{B}}(\Sigma)$ satisfying $\alpha_{\Sigma}(A[x])$. For $I \models_{\mathcal{B}, \Sigma} \alpha_{\Sigma}(A[x])$, the satisfaction relation requires $P_I \models_{\mathcal{P}, \Sigma} A[x]$. Now either $I \models_{\mathcal{B}, \Sigma} A$ or $I \models_{\mathcal{B}, \Sigma} \overline{A}$. For $I \models_{\mathcal{B}, \Sigma} A$, we have $P_I(A) = 1$ and therefore $P_I \models_{\mathcal{P}, \Sigma} A[1]$. For $I \models_{\mathcal{B}, \Sigma} \overline{A}$, we have $P_I(\overline{A}) = 1$ and $P_I(A) = 0$ and therefore $P_I \models_{\mathcal{P}, \Sigma} A[0]$. Thus, in both cases $P_I \models_{\mathcal{P}, \Sigma} A[x]$ can not hold since x is neither 1 nor 0. Therefore, $\alpha_{\Sigma}(A[x])$ must be a contradiction.   □

Proposition 7 leaves little freedom for the translation of sentences in an institution morphism from $Inst_{\mathcal{B}}$ to $Inst_{\mathcal{P}}$: All sentences with non-trivial probabilities must be viewed as contradictions from the propositional point of view. For the remaining cases, an obvious choice is to send $A[1]$ to A and $A[0]$ to $\overline{A}$. Therefore, we define the natural tranformation

$$\alpha_{\mathcal{P}/\mathcal{B}} : Sen_{\mathcal{P}} \Longrightarrow Sen_{\mathcal{B}} \qquad \text{by} \qquad \alpha_{\mathcal{P}/\mathcal{B}, \Sigma}(A[x]) = \begin{cases} A & \text{if } x = 1 \\ \overline{A} & \text{if } x = 0 \\ \bot & \text{otherwise} \end{cases}$$

where $\Sigma \in Sig_{\mathcal{B}}$. Now we have:

Proposition 8. $\langle id_{Sig_B}, \alpha_{P/B}, \beta_{B/P} \rangle : Inst_B \longrightarrow Inst_P$ *is an institution morphism.*

3.3   Relating Probabilistic and Probabilistic Conditional Logic

After looking for morphisms in both directions between $Inst_B$ and $Inst_P$, we now consider the relationships between probabilistic and probabilistic conditional logic. For the two institutions $Inst_P$ and $Inst_C$, we have already mentioned the obvious standard translations given by the natural transformations

$$\alpha_{P/C} : \quad Sen_P \implies Sen_C \qquad \alpha_{P/C,\Sigma}(A[x]) = (A|\top)[x]$$
$$id_{Mod_P} : Mod_P \implies Mod_P$$

where a probabilistic fact is mapped to a probabilistic conditional with $\top$ as trivial condition. Since the model functor in both involved institutions is the same, the standard model translation is of course the identity on $Mod_P = Mod_C$. Now the following is easy to prove:

Proposition 9. $\langle id_{Sig_B}, \alpha_{P/C}, id_{Mod_P} \rangle : Inst_C \longrightarrow Inst_P$ *is an institution morphism.*

Essentially, $\langle id_{Sig_B}, \alpha_{P/C}, id_{Mod_P} \rangle$ is the only institution morphism going from $Inst_C$ to $Inst_P$ by making use of id_{Sig_B} and id_{Mod_P}. This is the case since the satisfaction condition (1) claims for any such institution morphism $\langle id_{Sig_B}, \alpha, id_{Mod_P} \rangle : Inst_C \longrightarrow Inst_P$, that for all $A[x] \in Sen_P(\Sigma)$, $P \models_{C,\Sigma} \alpha_\Sigma(A[x])$ iff $P \models_{P,\Sigma} A[x]$ holds for any $P \in Mod_P(\Sigma)$, that is $\alpha_\Sigma(A[x])$ and $A[x]$ have to be probabilistically equivalent (cf. [PV98, Par94]).

Since the model translation is the identity, the institution morphism in Proposition 9 is obviously *sound* in the sense of [GB92] (all component functors of the natural model transformation must be surjective on objects). Thus, for all $F, G \subseteq Sen_P(\Sigma)$ it holds that $F \models_{P,\Sigma} G$ iff $\alpha_{P/C}(F) \models_{C,\Sigma} \alpha_{P/C}(G)$ (cf. Proposition 35 in [GB92]).

What about the other direction: Is there an institution morphism from $Inst_P$ to $Inst_C$ where the model functor is again the identity? Here, we are looking for a natural transformation α such that $\langle id_{Sig_B}, \alpha, id_{Mod_P} \rangle : Inst_P \longrightarrow Inst_C$ is an institution morphism. α_Σ must map a probabilistic conditional sentence $(B|A)[x]$ to a probabilistic fact $\alpha_\Sigma((B|A)[x])$. Here, we could try to make use of the material implication $A \Rightarrow B$ (or equivalently $\overline{A} \vee B$), i.e., $\alpha_\Sigma((B|A)[x]) = \overline{A} \vee B[x]$ could be considered a possible candidate. α is a natural transformation, but the satisfaction condition for institution morphisms requiring $P \models_{C,\Sigma} (B|A)[x]$ iff $P \models_{P,\Sigma} \overline{A} \vee B[x]$ is *not* satisfied, since $P(\overline{A} \vee B) \geqslant P(B|A)$ for any distribution P, and in general equality does not hold. Actually, there can be no institution morphism from $Inst_P$ to $Inst_C$ which leaves the models unchanged, since no probabilistic fact can be probabilistically equivalent to a probabilistic conditional (cf. the *triviality results* of [Lew76]). In particular, the probabilistic material conditional $\overline{A} \vee B[x]$ cannot be taken as an appropriate counterpart to the genuine probabilistic conditional $(B|A)[x]$ in a propositional probabilistic environment.

3.4   Relating Propositional and Probabilistic Conditional Logic

By now, we have found institution morphisms from both $Inst_B$ and $Inst_C$ to $Inst_P$, but none going from $Inst_P$ to $Inst_B$ or $Inst_C$. What is the situation between propositional logic $Inst_B$ and probabilistic conditional logic $Inst_C$? Here, the obvious standard translations are

$$\alpha_{B/C} : Sen_B \implies Sen_C \qquad \alpha_{B/C,\Sigma}(A) = (A|\top)[1]$$
$$\beta_{B/C} : Mod_B \implies Mod_C \qquad \beta_{B/C,\Sigma}(I) = P_I$$

Thus, $\alpha_{B/C}$ is the (vertical) composition of the standard sentence translations $\alpha_{B/P}$ and $\alpha_{P/C}$, sending A first to $A[1]$ and then to $(A|\top)[1]$. Correspondingly, $\beta_{B/C}$ is the composition of the standard model translation $\beta_{B/P}$ and the identity id_{Mod_P} used when going from $Inst_P$ to $Inst_C$.

When going from probabilistic conditional to propositional logic we have the intuitive sentence mapping A to $(A|\top)[1]$. However, a modification of Proposition 6 shows that no institution morphism using this sentence translation exists.

Proposition 10. *There is no β such that $\langle id_{Sig_B}, \alpha_{B/C}, \beta \rangle : Inst_C \longrightarrow Inst_B$ is an institution morphism.*

Proof. The same argumentation as in the proof of Proposition 6 using $(a|\top)[x]$, $\alpha_{B/C}$, and $\models_C$ instead of $a[x]$, $\alpha_{B/P}$, and $\models_P$, respectively, shows that the satisfaction condition does not hold. $\qquad\square$

When going from propositional logic to probabilistic conditional logic using the standard model translation $\beta_{B/C}$ sending I to P_I, we have to map probabilistic conditionals to propositional sentences. In Section 3.2, we showed in Proposition 7 that when moving from $Inst_B$ to $Inst_P$, probabilistic facts with probabilities other than 0 and 1 must be viewed as contradictory propositions. Not surprisingly, the same holds now for probabilistic conditionals when going from $Inst_B$ to $Inst_C$.

Proposition 11. *If $\langle id_{Sig_B}, \alpha, \beta_{B/C} \rangle : Inst_B \longrightarrow Inst_C$ is an institution morphism, then for any $\Sigma \in Sig_B$, α_Σ maps every sentence $(B|A)[x]$ with $x \neq 0$ and $x \neq 1$ to $\bot$.*

Proof. Let $\alpha : Sen_C \implies Sen_B$ be a natural transformation, $\Sigma \in Sig_B$, and $(B|A)[x] \in Sen_C(\Sigma)$ with $x \in (0,1)$ and thus $x \neq 0$, $x \neq 1$. If $\alpha_\Sigma((B|A)[x]) \in Sen_B(\Sigma)$ is not a contradiction, then there exists an interpretation $I \in Mod_B(\Sigma)$ satisfying $\alpha_\Sigma((B|A)[x])$. For $I \models_{B,\Sigma} \alpha_\Sigma((B|A)[x])$, the satisfaction relation requires $P_I \models_{C,\Sigma} (B|A)[x]$.

Now either $I \models_{B,\Sigma} A\overline{B}$ or $I \models_{B,\Sigma} \overline{A} \vee B$. If $I \models_{B,\Sigma} A\overline{B}$, then $P_I(A) = 1$ and $P_I(B) = 0$ and thus $P_I \models_{C,\Sigma} (B|A)[0]$. If $I \models_{B,\Sigma} \overline{A} \vee B$ and $I \models_{B,\Sigma} \overline{A}$, then $P_I(A) = 0$ and P_I can not satisfy $(B|A)[y]$ for any $y \in [0,1]$. If $I \models_{B,\Sigma} \overline{A} \vee B$ and $I \models_{B,\Sigma} A$, then $I \models_{B,\Sigma} B$ and thus $P_I(A) = 1$ and $P_I(B) = 1$; therefore, $P_I \models_{C,\Sigma} (B|A)[1]$.

Thus, in all cases $P_I \models_{C,\Sigma} (B|A)[x]$ can not hold since x is neither 1 nor 0. Therefore, $\alpha_\Sigma((B|A)[x])$ must be a contradiction. $\qquad\square$

As in Section 3.2, the only choices left for the translation of probabilistic conditionals to propositional sentences is thus the translation of conditionals with the trivial probabilities 1 and 0 since everything else must be interpreted as a contradiction. Since $(B|A)[1]$ represents a conditional *if A then B with probability 1* we might try to map this to the classical (material) implication $A \Rightarrow B$. Likewise, $(B|A)[0]$ could be mapped to the negated implication $\neg(A \Rightarrow B)$. However, a simple counterexample shows that

$$\alpha : Sen_{\mathcal{C}} \Longrightarrow Sen_{\mathcal{B}} \qquad \text{with} \qquad \alpha_{\Sigma}((B|A)[x]) = \begin{cases} A \Rightarrow B & \text{if } x = 1 \\ \neg(A \Rightarrow B) & \text{if } x = 0 \\ \bot & \text{otherwise} \end{cases}$$

does not yield an institution morphism.

Example 5. Let Σ be as in Example 1 and consider the conditional $(u|s)[1]$ (*being single* under the condition of *being a scholar*). Let I be the (unique) interpretation with $I(\overline{s}tu) = true$. Obviously, $I \models_{\mathcal{B},\Sigma} s \Rightarrow u$ and thus $I \models_{\mathcal{B},\Sigma} \alpha_{\Sigma}((u|s)[1])$ since $\alpha_{\Sigma}((u|s)[1]) = s \Rightarrow u$. The satisfaction condition then requires $P_I \models_{\mathcal{C},\Sigma} (u|s)[1]$ which is not the case since $P_I(s) = 0$. Therefore, $\langle id_{Sig_{\mathcal{B}}}, \alpha, \beta_{\mathcal{B}/\mathcal{C}} \rangle$ is not an institution morphism from $Inst_{\mathcal{B}}$ to $Inst_{\mathcal{C}}$.

The example shows that in $Inst_{\mathcal{C}}$, even probabilistic conditionals with probability 1 do not correspond to material implications in $Inst_{\mathcal{B}}$ under an institution morphism with the standard model translation using P_I. In the example, the antecedent s is not satisfied in I; nevertheless, as we know from classical logic, I satisfies $s \Rightarrow u$. On the other hand, P_I does not satisfy $(u|s)[1]$ since the antecedent in $(u|s)[1]$ has probability 0 under P_I, although $P_I(u) = 1$ and therefore, both $P_I \models_{\mathcal{P},\Sigma} u[1]$ and also $P_I \models_{\mathcal{C},\Sigma} (u|\top)[1]$.

Therefore, we modify the translation above by taking the antecedent as a context into account, thus mapping $(B|A)[1]$ not to $A \Rightarrow B$, but to $A \wedge (A \Rightarrow B)$, or equivalently, to AB. $(B|A)[0]$ can still be mapped to $\neg(A \Rightarrow B)$, or equivalently, to $A\overline{B}$, since $A \wedge \neg(A \Rightarrow B) = \neg(A \Rightarrow B)$. This yields the natural tranformation

$$\alpha_{\mathcal{C}/\mathcal{B}} : Sen_{\mathcal{C}} \Longrightarrow Sen_{\mathcal{B}} \qquad \text{with} \qquad \alpha_{\mathcal{C}/\mathcal{B},\Sigma}((B|A)[x]) = \begin{cases} AB & \text{if } x = 1 \\ A\overline{B} & \text{if } x = 0 \\ \bot & \text{otherwise} \end{cases}$$

for every $\Sigma \in Sig_{\mathcal{B}}$. Now we have:

Proposition 12. $\langle id_{Sig_{\mathcal{B}}}, \alpha_{\mathcal{C}/\mathcal{B}}, \beta_{\mathcal{B}/\mathcal{C}} \rangle : Inst_{\mathcal{B}} \longrightarrow Inst_{\mathcal{C}}$ *is an institution morphism.*

Note that $\alpha_{\mathcal{C}/\mathcal{B},\Sigma}$ reflects the three-valued semantics of conditionals, identifying the verifying part AB and the falsifying part $A\overline{B}$ as most important components of conditional information (cf. [DeF74, Cal91]).

Figure 4 summarizes our findings with respect to institution morphisms between the three institutions $Inst_{\mathcal{B}}$, $Inst_{\mathcal{P}}$, and $Inst_{\mathcal{C}}$. Note that using the intuitive standard translations, we have (essentially) exactly one institution morphism between any pair of the three institutions, but none going in the respective opposite direction.

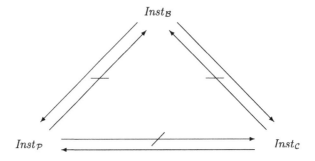

Morphism	Sentence translation		Model translation
$Inst_B \longrightarrow Inst_P$	$A[x] \quad \mapsto$	$\begin{cases} A & \text{if } x = 1 \\ \overline{A} & \text{if } x = 0 \\ \bot & \text{otherwise} \end{cases}$	$I \mapsto P_I$
$Inst_C \longrightarrow Inst_P$	$A[x] \quad \mapsto (A\mid\top)[x]$		$P \mapsto P$
$Inst_B \longrightarrow Inst_C$	$(B\mid A)[x] \mapsto$	$\begin{cases} AB & \text{if } x = 1 \\ A\overline{B} & \text{if } x = 0 \\ \bot & \text{otherwise} \end{cases}$	$I \mapsto P_I$

Fig. 4. Institution morphisms between $Inst_B$, $Inst_P$, and $Inst_C$

3.5   Institution Embeddings

Whereas the translations of sentences and models in an institution morphism go in opposite directions, there are other ways of relating logics. For instance, many-sorted first order logic can be translated to unsorted first order logic by introducing special predicates for each sort and transforming sorted models to unsorted ones. Such translations are often called relativations or embeddings where the translations of both sentences and models go in the same direction. Using the standard translations introduced in the previous subsections, we could step e.g. from $Inst_B$ to $Inst_P$ by sending A to $A[1]$ and I to P_I. We therefore propose the following concept for general institutions.

Definition 3. *Let* $Inst = \langle Sig, Mod, Sen, \models \rangle$ *and* $Inst' = \langle Sig', Mod', Sen', \models' \rangle$ *be two institutions. An* institution embedding Φ *from Inst to Inst' is a triple* $\langle \phi, \alpha, \beta \rangle$ *with a functor* $\phi : Sig \to Sig'$*, a natural transformation* $\alpha : Sen \Longrightarrow Sen' \circ \phi$*, and a natural transformation* $\beta : Mod \Longrightarrow Mod' \circ \phi$ *such that for each* $\Sigma \in |Sig|$*, for each* $m \in |Mod(\Sigma)|$*, and for each* $f \in Sen(\Sigma)$ *the following* embedding condition *holds:*

$$m \models_\Sigma f \quad iff \quad \beta_\Sigma(m) \models'_{\phi(\Sigma)} \alpha_\Sigma(f)$$

Please note that compared to the situation for an institution morphism, both α and β go from *Inst* to *Inst'*.

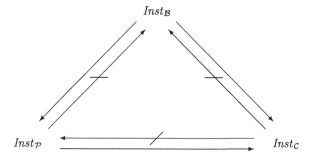

Embedding	Sentence translation	Model translation
$Inst_B \longrightarrow Inst_P$	$A \quad \mapsto \quad A[1]$	$I \mapsto P_I$
$Inst_P \longrightarrow Inst_C$	$A[x] \mapsto (A\vert\top)[x]$	$P \mapsto P$
$Inst_B \longrightarrow Inst_C$	$A \quad \mapsto \quad (A\vert\top)[1]$	$I \mapsto P_I$

Fig. 5. Institution embeddings between $Inst_B$, $Inst_P$, and $Inst_C$

It is easy to check that the standard translations given above yield institution embeddings:

Proposition 13.
$$\langle\, id_{Sig_B},\ \alpha_{B/P},\ \beta_{B/P}\,\rangle:\quad Inst_B \longrightarrow Inst_P$$
$$\langle\, id_{Sig_B},\ \alpha_{P/C},\ id_{Mod_P}\,\rangle:\quad Inst_P \longrightarrow Inst_C$$
$$\langle\, id_{Sig_B},\ \alpha_{B/C},\ \beta_{B/C}\,\rangle:\quad Inst_B \longrightarrow Inst_C$$

are three institution embeddings where the third is the composition of the first and the second one.

All intuitive standard translations introduced previously go in the directions of the institution embeddings in Proposition 13, and all these translations occur in Proposition 13 where we did not have to construct any additional translation. Thus, these are also the only institution embeddings between the three institutions going in the given directions that are based on the standard translations. Furthermore, it is rather obvious that there are no (reasonable) institution embeddings going in the other directions. For instance, in order to show that there is no embedding from $Inst_C$ to $Inst_P$, we can use arguments similar to those given in Section 3.3.

Figure 5 summarizes this embeddings situation between the three institutions $Inst_B$, $Inst_P$, and $Inst_C$.

When comparing institution morphisms and institution embeddings, one should observe that an institution morphism ties the two involved institutions together much closer than an institution embedding. In order to satisfy the embedding condition, all the models and sentences belonging to one of the institutions (i.e. $Inst$) have to be considered. For the satisfaction condition of an institution morphism, the quantification involves both institutions (i.e. models from $Inst$

and sentences from $Inst'$). Therefore, it comes to no surprise that it was easy to find the three embeddings, all going form the easier institution to the more complex one. Such embeddings, however, allow us only to see e.g. how the sentences of the less complex institution act in the more complex one. On the other hand, an institution morphism like $\langle id_{Sig_B}, \alpha_{C/B}, \beta_{B/C} \rangle : Inst_B \longrightarrow Inst_C$ gives us quite detailed information how to interpret sentences from the more complex institution (probabilistic conditionals) in the easier institution (a propositional setting).

4   Conclusions and Further Work

In this paper, we showed how probabilistic logic and probabilistic conditional logic can be formalized as institutions. The use of the general institution framework supports the study of structural properties of both syntax and semantics of these logics, telling us, e.g., how probabilistic facts and conditionals behave under change of notation. By using the notions of institution morphism and institution embedding, we studied in detail the relationships between probabilistic propositional logic, probabilistic conditional logic, and the underlying two-valued propositional logic.

There are various directions in which the work reported here should be extended. For instance, other probabilistic semantics could be considered, in particular: What happens if the presupposition of having antecedents with positive probability is omitted or modified? Moreover, in a forthcoming paper, we will investigate the logic of measure-free conditionals $(B|A)$ with qualitative or ordinal semantics and its relationships to the logics studied here; part of this work has already been done and published in [BKI02b]. In [BKI02a], the institution framework for probabilistic logics developed here is used to provide a new formal approach to knowledge discovery in statistical data.

Acknowledgements. We thank the anonymous referees of this paper for their valuable and helpful comments. The research reported here was partially supported by the DFG – Deutsche Forschungsgemeinschaft within the CONDOR-project under grant BE 1700/5-1.

References

[BG80]   R. Burstall and J. Goguen. The semantics of Clear, a specification language. In *Proceedings of the 1979 Copenhagen Winterschool on Abstract Software Specification*, volume 86 of *LNCS*, pages 292–332, Berlin, Heidelberg, New York, 1980. Springer-Verlag.

[BKI02a] C. Beierle and G. Kern-Isberner. Footprints of conditionals. In D. Hutter and W. Stephan, editors, *Festschrift in Honor of Jörg H. Siekmann*. Springer-Verlag, Berlin, Heidelberg, New York, 2002. (to appear).

[BKI02b] C. Beierle and G. Kern-Isberner. Using institutions for the study of qualitative and quantitative conditional logics. In *Proceedings of the 8th European Conference on Logics in Artificial Intelligence, JELIA'02*, volume LNCS 2424, pages 161–172, Berlin Heidelberg New York, 2002. Springer.

[Cal91]   P.G. Calabrese. Deduction and inference using conditional logic and probability. In I.R. Goodman, M.M. Gupta, H.T. Nguyen, and G.S. Rogers, editors, *Conditional Logic in Expert Systems*, pages 71–100. Elsevier, North Holland, 1991.

[DeF74]   B. DeFinetti. *Theory of Probability*, volume 1,2. John Wiley and Sons, New York, 1974.

[GB92]   J. Goguen and R. Burstall. Institutions: Abstract model theory for specification and programming. *Journal of the ACM*, 39(1):95–146, January 1992.

[GR02]   J. A. Goguen and G. Rosu. Institution morphisms. *Formal Aspects of Computing*, 13(3–5):274–307, 2002.

[HS73]   H. Herrlich and G. E. Strecker. *Category theory*. Allyn and Bacon, Boston, 1973.

[Lew76]   D. Lewis. Probabilities of conditionals and conditional probabilities. *The Philosophical Review*, 85:297–315, 1976.

[Mac72]   S. Mac Lane. *Categories for the Working Mathematician*. Springer-Verlag, New York, 1972.

[N.R69]   N.Rescher. *Many-Valued Logic*. McGraw-Hill, New York, 1969.

[Par94]   J.B. Paris. *The uncertain reasoner's companion – A mathematical perspective*. Cambridge University Press, 1994.

[Pea88]   J. Pearl. *Probabilistic Reasoning in Intelligent Systems*. Morgan Kaufmann, San Mateo, Ca., 1988.

[PV98]   J.B. Paris and A. Vencovska. Proof systems for probabilistic uncertain reasoning. *Journal of Symbolic Logic*, 63(3):1007–1039, 1998.

[ST97]   D. Sannella and A. Tarlecki. Essential comcepts for algebraic specification and program development. *Formal Aspects of Computing*, 9:229–269, 1997.

[Tar96]   A. Tarlecki. Moving between logical systems. In M. Haveraaen, O. Owe, and O.-J. Dahl, editors, *Recent Trends in Data Type Specifications*, volume 1130 of *Lecture Notes in Computer Science*, pages 478–502, Berlin, Heidelberg, New York, 1996. Springer-Verlag.

There Is a Reason for Everything (Probably): On the Application of Maxent to Induction

Jeff B. Paris and Alena Vencovská*

Department of Mathematics,
University of Manchester, Manchester M13 9PL, UK
{jeff, alena}@maths.man.ac.uk

Abstract. In this paper we show how the maxent paradigm may be used to produce an inductive method (in the sense of Carnap) applicable to a wide class of problems in inductive logic. A surprising consequence of this method is that the answers it gives are consistent with, or explicable by, the existence of underlying reasons for the given knowledge base, even when no such reasons are explicitly present. We would conjecture that the same result holds for the full class of problems of this type.

1 Introduction and Notation

In this paper we continue a line of research originally suggested in [24] and completed for the base case ($s = 1$) in [29], [28] and [35].

To motivate the underlying 'problem of induction' to which this research attempts to provide an answer, consider the following problem. I am queuing in a traffic jam on my way to work in the morning and I start to wonder what the chances are that the car I see approaching in my off-side mirror is in the wrong lane and will shortly be attempting to nudge into my lane. Past experience has taught me that if the driver is an elderly woman then it is rather unlikely, if the driver is a middle aged man driving a Range Rover it is quite likely, if there is a tractor further up the road in my lane then it is almost certain, and so on. In addition to this I might have some general knowledge about the situation, for example that about a quarter of the drivers using the road at that time are women, that rather few tractors use this particular route, and so on.

The problem we have here is one of *induction* in the philosophical sense. On the basis of some information about a general population of individuals, and perhaps some particular individuals within it, one wants to assign a measure of belief, which for this paper we shall identify with subjective probability, to properties holding of other individuals from the population. For example, in this case the property that the driver I currently see in my off-side mirror is a queue jumper.

---

* Supported by the UK Engineering and Physical Sciences Research Council (EPSRC) research grant GR/M91396.

G. Kern-Isberner, W. Rödder, and F. Kulmann (Eds.): WCII 2002, LNAI 3301, pp. 180–199, 2005.

In the philosophical debate one commonly thinks of the individuals as 'experiments' and their properties of interest as 'outcomes' of the experiment. Viewed in this way the problem is one of prediction, how much belief to give a future event on the basis of information about some past events, and perhaps, as here, some summary general knowledge. As a real world problem is has provoked considerable debate in philosophical circles, not simply about *what* answer to give but even whether *any* answer can reasonably be given, see for example [8], [9], [12], [14], [17], [21], [22], [23], [30], [31], [32], [33], [34], [36], [37], or for a single source state of the art explication [13]. One reason we would suggest for this lack of consensus is that in any real world example there is always just so much relevant background knowledge particular to that example that it is difficult, if not impossible, to draw any generic conclusions about the problem as a whole.

An alternative, analytic, approach, with which the name of Carnap[1] is generally associated, is to move out of the real world and in today's parlance express the problem as one of *predicate uncertain reasoning*. That is, to formalize the problem of induction as one of specifying a probability function (thought of as capturing an agent's beliefs) on sentences of some predicate language given knowledge in the form of constraints on this probability function. The intention here is that the knowledge should have no further content beyond what is expressed by these constraints, thus avoiding, 'by definition', the open endedness of the real world problem[2] and allowing the precise formulation and critical examination of various seemingly appropriate principles of uncertain reasoning just in the way the formal predicate calculus might be viewed as formalizing an important part of everyday 'exact', i.e. true/false, reasoning. For a selection of influential papers in the subject see [18], [2], [3], [4], [5], [6], [7], [19], [20], [38]. To our knowledge this more philosophical development (to which we would actually see this paper contributing in the main) has concentrated on the case where there is *no background general knowledge* beyond simply the outcomes of some finite number of previous experiments and the interest (as in this paper) was assigning a probability to the outcomes of some future runs of the experiment. On the other hand more recent lines of research in 'knowledge engineering' certainly can be viewed as addressing this more general problem (though it is sometimes unclear whether the intended goal is, as in this paper, the *creation* of personal, or subjective, probabilities, or the statistical estimation of objective, but presumably unknown, probabilities). Whilst, as far as we are aware, none of that work is directly related to this paper we would refer the reader in particular to the themes development by Fagin, Halpern et al (for a sample see [10], [11], [1], [15], [16]).

Returning now to our particular traffic queue example, in this case we certainly do have what we have been referring to as general knowledge. For the

---

[1] A similar development by Johnson, see [18], a decade earlier was somehow overlooked.

[2] Given this severe restriction one might now question why this approach show be at all relevant to the real world problem. To attempt to answer that question here would unfortunately take us too far away from the goal of this paper.

purposes of this example such knowledge might be represented as a set of probabilistic constraints on my personal belief, equivalently subjective probability, function Bel^3. For example the above statements might be formalized by the (infinitely many, as the a_i vary) instances of the constraints

$$Bel(P(a_i)|W(a_i)) = 0.1$$
$$Bel(P(a_i)|R(a_i)) = 0.75$$
$$Bel(P(a_i)|T) = 0.9$$
$$Bel(W(a_i)) = 0.25$$
$$Bel(T) = 0.05$$

$$\cdots\cdots\ \cdot\cdot\ \cdots\cdot$$

where $P(a_i)$ stands for 'car a_i nudges in', $W(a_i)$ stands for 'driver of car a_i is an elderly woman', T stands for 'tractor ahead', etc. and $a_1, a_2, a_3, \ldots$ stand for the potentially infinite stream of cars in the off-side lane, though not necessarily in that order. Notice that is infinite, since there are infinitely many a_i. However, since we shall limit ourselves throughout to only finitely many predicates $P(x), W(x), ..$ and propositions $T, \ldots$, it is finitely specifiable (and in that sense comprehensible) as the instances of finitely many linear constraint schemata.

The sort of question we are interested in here is what belief, as a subjective probability, should be assigned to various assertions, for example that any 2 particular cars will both nudge in, i.e. $P(a_i) \wedge P(a_j)$, on the basis of this knowledge, *and this knowledge alone* (the so called Watt's Assumption of [24]).

Whilst in this case any 'right answers' can hardly be said to be immediately obvious there is one analogous situation in which the answers to some such questions do seem clear. Namely when the knowledge base, K say, has the form

(i) $Bel(P(a_i) \wedge Q_j) - \beta_j Bel(Q_j) = 0, \quad j = 1, 2, ..., q,$

(ii) $Bel(Q_j \wedge Q_k) = 0, \quad 1 \le j < k \le q,$

(iii) $Bel(Q_j) = \lambda_j, \quad j = 1, 2, ..., q,$ (1)

where $\sum_{j=1}^{q} \lambda_j = 1$ and $0 \le \lambda_1, \lambda_2, ..., \lambda_q, \beta_1, \beta_2, ..., \beta_q \le 1^4$.
In this case the Q_j form a *complete set of reasons*[5], in that they are (i) 'reasons' (*for* $P(a_i)$ if $\beta_j > 1/2$, *against* if $\beta_j < 1/2$), (ii) disjoint, and (iii) exhaustive. Given a knowledge base of this special form there is an evident solution based

---

[3] For the purposes of this paper we need only consider probability functions on quantifier free sentences, effectively then the sentences of a propositional language. In this case, for SL the set of sentences of a propositional language L $Bel : SL \to [0, 1]$ is defined to be a *probability function* if it satisfies that for all $\theta, \phi \in SL$ if $\models \theta$ then $Bel(\theta) = 1$ and if $\models \neg(\theta \wedge \phi)$ then $Bel(\theta \vee \phi) = Bel(\theta) + Bel(\phi)$. For more details see [24].

[4] We assume these bounds on the λ_j, β_j without further mention in what follows.

[5] The word 'reason' here should be understood more as an *explanation* rather than a *cause*. For example in the way its being December is a reason for its being cold.

on the implicit assumption that the $P(a_i)$ are, modulo the knowledge base, stochastically independent of each other. In other words,

$$Bel(P(a_i)) = \sum_{j=1}^{q} \lambda_j \beta_j,$$

and more generally

$$Bel(\bigwedge_{i=1}^{m} P^{\epsilon_i}(a_i)) = \sum_{j=1}^{q} \lambda_j \beta_j^k (1 - \beta_j)^{m-k},$$

where $k = \sum \epsilon_i$ and $P^\epsilon = P$ if $\epsilon = 1$ and $P^\epsilon = \neg P$ if $\epsilon = 0$. We call this the *canonical solution* based on this complete set of reasons.

More generally, if instead of just a single predicate $P(x)$ we had a finite set of predicates $P_1(x), P_2(x), ..., P_n(x)$ then we shall say that a knowledge base is a *complete set of reasons* if it has the form

(i) $Bel(P^\epsilon(a_i) \wedge Q_j) - \beta_j^\epsilon Bel(Q_j) = 0, \quad j = 1, 2, ..., q,$

(ii) $Bel(Q_j \wedge Q_k) = 0, \quad 1 \le j < k \le q,$

(iii) $Bel(Q_j) = \lambda_j, \quad j = 1, 2, ..., q,$ (2)

where $\sum_{j=1}^{q} \lambda_j = 1$, $\epsilon \in \{0, 1\}^n$, $\sum_\epsilon \beta_j^\epsilon = 1$ for $j = 1, 2, ..., q$, and

$$P^\epsilon(a_i) = P_1^{\epsilon_1}(a_i) \wedge P_2^{\epsilon_2}(a_i) \wedge ... \wedge P_n^{\epsilon_n}(a_i),$$

the *canonical solution* in this case being giving by

$$Bel(\bigwedge_{i=1}^{m} P^{\epsilon_i}(a_i)) = \sum_{j=1}^{q} \lambda_j \prod_{i=1}^{m} \beta^{\epsilon_i}.$$

This (apparently) provides an acceptable answer in the case of a complete set of reasons. But how should we proceed in the general case? In the case of a *finite* knowledge base K we have argued extensively (see for example [25], [26], [27]) that the *maximum entropy solution*, $ME(K)$[6] provides the only solution to K consistent with the requirements of common sense.

Given these arguments a natural extension of this paradigm to infinite knowledge bases of the above form was suggested in [24] (p197) and developed in [29] and [28]. Namely, that since in reality the a_i only constitute a *potentially* infinite set of individuals, we should take the restrictions of our knowledge base to the sublanguages with only the finitely many $a_1, a_2, ..., a_r$ and consider the limit of

---

[6] For L a finite propositional language and K a finite set of constraints on a probability function $Bel : SL \to [0, 1]$ $ME(K)$ is that probability function Bel satisfying K for which the entropy, $-\sum_\alpha Bel(\alpha) \log Bel(\alpha)$ is maximal, where α ranges over the atoms of SL, that is the sentences of L of the form $p_1^{\epsilon_1} \wedge p_2^{\epsilon_2} \wedge ... \wedge p_n^{\epsilon_n}$ where the p_i are the propositional variables in L and the $\epsilon_i \in \{0, 1\}$.

the maximum entropy solutions to these knowledge bases as r tends to infinity (assuming the limit exists of course).

Formally, let L^r be the propositional language with propositional variables $Q_1, Q_2, ..., Q_q$ and $P_1(a_i), P_2(a_i), ..., P_n(a_i)$ for $i = 1, 2, ..., r$. Let SL^r be the set of sentences of L^r (using the connectives $\neg, \vee, \wedge$, say) and let $K = K(a_1)$ be a finite set of linear constraints

$$\sum_{i=1}^{k} c_i Bel(\theta_i) = d, \tag{3}$$

on a probability function $Bel : SL^1 \to [0, 1]$ where the c_i, d are reals and the $\theta_i \in SL^1$. Let K^r be the combined set of constraints on a probability function $Bel : SL^r \to [0, 1]$ formed by taking the union of the $K(a_i)$ (i.e. the result of replacing a_1 everywhere in $K(a_1)$ by a_i for $i = 1, 2, ..., r$). Let $\mathcal{C}_1$ be the set of such $K(a_1)$ for which K^r is satisfiable (by some probability function $Bel : SL^r \to [0, 1]$) for each $r \geq 1$. It is easy to see that if $K(a_1)$ corresponds to a complete set of reasons then $K(a_1) \in \mathcal{C}_1$.

The following theorem, which is proved in the key case in [29] (and in full generality in [28]), addresses the question of the limiting value of the maximum entropy solutions mentioned above.

Theorem 1. *If $K(a_1) \in \mathcal{C}_1$ and $\epsilon_1, \epsilon_2, ..., \epsilon_m \in \{0, 1\}^n$ then*

$$\lim_{r \to \infty} ME(K^r)(\bigwedge_{i=1}^{m} P^{\epsilon_i}(a_i))$$

exists. Furthermore these limit values agree with the values given by a certain complete set of reasons[7].

The proof of this theorem provides some demystification. The 'reasons' turn out to simply to be the atoms[8] $Q_1^{\delta_1} \wedge Q_2^{\delta_2} \wedge ... \wedge Q_q^{\delta_q}$ from the original language L^1.

Furthermore if $K(a_1)$ starts off as the complete set of reasons (2) (with the obvious extension of this notion to the present formalization) then maximum entropy settles on these reasons (even before taking limits) in the sense that for this $K(a_1)$ (and $r \geq m$)

$$ME(K^r)(\bigwedge_{i=1}^{m} P^{\epsilon_i}(a_i)) = \sum_{j=1}^{q} \lambda_j \prod_{i=1}^{m} \beta^{\epsilon_i}.$$

---

[7] It is interesting to note that for a wide range of $K(a_1)$, indeed probably all $K(a_1) \in \mathcal{C}_1$ although that still remains to be proved in full generality, the result we have here for the maximum entropy inference process also holds for the so called minimum distance inference process, MD, and the limiting centre of mass inference process, CM^∞, see [28].

[8] This may hardly seem like much of a demystification, saying as it does that the culprit is the 'way the rest of the world is outside of the a_i'. However at least in this case we can put a name to the reasons within our existing language.

We shall refer to the general result given by Theorem 1 as the '$s = 1$ case' for reasons which will become apparent in the next section where we consider the '$s = 2$ case'.

2   Pairwise Interactions Between Individuals

In terms of our original example Theorem 1 is pleasing in the sense that it confirms what most of us probably expect of the world, that things happen, in this case motorists nudge in, because there are reasons for them happening. Indeed in this case the reasons are just the 'states of the world', the atoms $Q_1^{\delta_1} \wedge Q_2^{\delta_2} \wedge ... \wedge Q_q^{\delta_q}$.

However there are some happenings in this world whose causes, or reasons, are rather less obvious. For example, in the case of the aforementioned traffic queue I have noticed that when it comes to inconsiderate behavior on the road, in particular nudging in, offending vehicles seem *to me* to somehow infect each other, remotely, with the disease. Thus given that vehicle a_i has nudged in it seems to me far more likely that some other vehicle, a_j say, will also try it on[9]. In this case then I might be inclined to add to my knowledge base some conditional belief constraints

$$Bel(P(a_j)|P(a_i)) = 0.6,$$

or more conveniently for our purposes,

$$Bel(P(a_j) \wedge P(a_i)) - 0.6Bel(P(a_i)) = 0$$

for $i \neq j$. But given such a knowledge base what belief should now be assigned to, say,

$$\bigwedge_{i=1}^{m} P^{\epsilon_i}(a_i) \ ?$$

Clearly the 'common sense' justification which applied in the earlier, simple case, is equally applicable here. The main novelty of this paper is concerned with investigating the answers that this leads us to.

Before doing so however we need to generalize our notation. Let $K = K(a_1, a_2)$ be a finite set of constraints of the same form as (3) but with the $\theta_i \in SL^2$. Let K^r be the set of all constraints on $Bel : SL^r \to [0,1]$ ($r \geq 2$) formed by taking the union of the $K(a_i, a_j)$ for $i, j = 1, 2, ..., r$, $i \neq j$. Let $\mathcal{C}_2$ be the set of such $K(a_1, a_2)$ for which K^r is satisfiable (by some probability function $Bel : SL^r \to [0,1]$) for each $r \geq 2$.

By analogy with Theorem 1 we make the following:

Conjecture. *If $K(a_1, a_2) \in \mathcal{C}_2$ and $\epsilon_1, \epsilon_2, ..., \epsilon_m \in \{0,1\}^n$ then*

$$\lim_{r \to \infty} ME(K^r)(\bigwedge_{i=1}^{m} P^{\epsilon_i}(a_i))$$

---

[9] Perhaps a more widely appreciated example is the phenomenon whereby once one thing has gone well/badly everything else seems to follow suit.

exists. Furthermore these limit values agree with the values given by a certain complete set of reasons.

In a forthcoming technical report we shall give a proof of this conjecture in the case of a single property P, limiting ourselves in the appendix of this paper to a proof of the 'base case' of this conjecture when there is just a single property P and $K(a_1, a_2) \in C_2$ is of the form

$$\{ \, Bel(P(a_1)) = a, \; Bel(P(a_2)) = a, \; Bel(P(a_1) \wedge P(a_2)) = b \, \}.$$

[This is a 'base case' in the sense that the general problem for a single property essentially reduces to a combination of such cases.]
Precisely:

Theorem 2. *If* $K(a_1, a_2) = \{ \, Bel(P(a_1)) = a, \; Bel(P(a_2)) = a, \; Bel(P(a_1) \wedge P(a_2)) = b \, \} \in C_2$ *and* $\epsilon_1, \epsilon_2, ..., \epsilon_m \in \{0, 1\}$ *then*

$$\lim_{r \to \infty} ME(K^r)(\bigwedge_{i=1}^{m} P^{\epsilon_i}(a_i))$$

exists. Furthermore these limit values agree with the values given by the complete set of reasons

(i) $Bel(P(a_i) \wedge Q_j) - \frac{1}{2}\{1 + (-1)^j \sqrt{1 - 4(a - b)}\}Bel(Q_j) = 0,$

(ii) $Bel(Q_1 \wedge Q_2) = 0,$

(iii) $Bel(Q_j) = \frac{1}{2}\left(1 + \dfrac{(-1)^j(2a - 1)}{\sqrt{1 - 4(a - b)}}\right),$ *where* $j = 1, 2.$

It is our strong conjecture that this result generalizes to multiple properties and interactions between between multiple a_i, not just two. However, with our current approach the mathematical technicalities appear quite formidable.

What is remarkable (in our view) about Theorem 2 (and its proven generalizations) is that these 'reasons' simply appear, in a sense retrospectively, to explain the observations, or stated constraints in this case.

The fact that the conjecture holds in all cases that we have so far been able to solve, not just for ME but also for the inference processes MD and CM^∞, leads us to further conjecture that it is true too for both of these inference processes, and indeed that with the obvious meaning it extends also to C_s for $s > 2$. [Note however the example given in [29] of a reasonably well behaved inference process for which the limits do exist yet do not correspond to any set of reasons.]

3   An Example

On such a small scale almost all examples are of course totally artificial, but with this indulgence consider the following situation. I am in the changing room of a swimming pool and I notice that pairs of kids keep coming in from the pool.

I form the impression that in about 5/18 of the cases[10] the kids are smiling and sharing a bar of chocolate. The remaining times they are looking glum and there is no chocolate in sight. Curious as to what bizarre game is being played out in the pool I ask the attendant and she tells me that it the climax of a swimming lesson. Pairs of kids are chosen, apparently at random, and they have to dive in, swim to the bottom, touch hands and then resurface. As an incentive the pair get a chocolate bar if they manage it, nothing if they don't.

I start to wonder how likely it is that any one kid can swim down to the bottom, or how likely that three randomly chosen kids will be able to pull off a similar manoeuvre.

In this case, letting $P(a_i)$ stand for "kid a_i is able to swim to the bottom", my knowledge base $K(a_1, a_2)$ consists of just

$$\{ Bel(P(a_1) \wedge P(a_2)) = 5/18 \}.$$

Now it turns out as a special (proven) case of the general Conjecture that in this case

$$\lim_{r \to \infty} ME(K^r)(\bigwedge_{i=1}^{m} P^{\epsilon_i}(a_i)) = \lim_{r \to \infty} ME(\overline{K}^r)(\bigwedge_{i=1}^{m} P^{\epsilon_i}(a_i)) \qquad (4)$$

where

$$\overline{K}^r = \{ Bel(P(a_1)) = Bel(P(a_2)) = 1/2, \; Bel(P(a_1) \wedge P(a_2)) = 5/18 \}.$$

We can now work out the values of the right hand side of (4) using Theorem 2. In this case $b = 5/18$, $a = 1/2$ and these values are precisely the ones given by the complete set of reasons Q_1, Q_2 with

(i) $Bel(P(a_i) \wedge Q_1) - \frac{1}{3} Bel(Q_1) = 0$,
     $Bel(P(a_i) \wedge Q_2) - \frac{2}{3} Bel(Q_2) = 0$,
(ii) $Bel(Q_1 \wedge Q_2) = 0$,
(iii) $Bel(Q_1) = \frac{1}{2}$,
      $Bel(Q_2) = \frac{1}{2}$.

In particular were I to follow 'common sense' (i.e. adopt the method described in this paper) then on the basis of my knowledge I should assign belief $\frac{1}{2} \cdot \frac{1}{3} + \frac{1}{2} \cdot \frac{2}{3} = \frac{1}{2}$ to $P(a_i)$, i.e. to any kid being able to swim to the bottom[11], and belief

$$\frac{1}{2} \cdot \left(\frac{1}{3}\right)^3 + \frac{1}{2} \cdot \left(\frac{2}{3}\right)^3 = \frac{1}{6}$$

to $P(a_i) \wedge P(a_j) \wedge P(a_k)$, i.e. to any three kids being able to pull off the manoeuvre.

---

[10] The figure of 5/18 has been chosen here to ensure that all subsequent figures are simple fractions.
[11] This, of course, quite rightly agrees with the value given by the $\overline{K}^r$.

Effectively then the situation *may* be explained as follows. There are two equally probable 'reasons' or 'conditions' (Q_1, Q_2) either of which might prevail out there in the pool at any one time. Given that it is Q_1 the probability that any randomly chosen kid will be capable of reaching the bottom is $1/3$. On the other hand if it is Q_2 that prevails then with probability two thirds the kid will be so capable. I of course have no idea what Q_1 and Q_2 might be, but given what a satisfying explanation they provide I might nevertheless feel like giving them names, say, 'troll' and 'grand bleu' respectively!

4   Conclusion

In this paper we have generalized the 'common sense approach' of [29] and [28] to inductive reasoning about events $P(a_i)$ based on general knowledge about isolated individuals a_i (the $s = 1$ case) to the case where we also have knowledge of pairwise interactions between individuals (the $s = 2$ case). Analogously to the results in those earlier papers we can show in the $s = 2$ case with a single property P that the limits as $r \to \infty$ exist and agree with those given by a certain complete set of reasons. The fact that reasons emerge in this way, a posteriori, seems to us remarkable and of some philosophical, and perhaps even practical, import.

Furthermore the evidence to date would lead us to conjecture that this same conclusion continues to hold if we also allow knowledge of higher order interactions $(s > 2)$, not only for ME but also such other inference processes as MD and CM^∞.

We conclude by briefly remarking that whilst the proven generalization of Theorem 2 was given in terms induction on the $P(a_i)$ a straightforward corollary of this result is that it holds equally well for any sentence $\theta(a_1)$ from L^1. That is, there are some

$$0 \le \lambda_1, \lambda_2, ..., \lambda_t, \beta_1, \beta_2, ..., \beta_t \le 1$$

such that

$$\lim_{r \to \infty} ME(K^{(r)})(\theta^{\epsilon_1}(a_1) \wedge \theta^{\epsilon_2}(a_2) \wedge ... \wedge \theta^{\epsilon_m}(a_m)) = \sum_{i=1}^{t} \lambda_i \beta_i^{\sum \epsilon_j} (1 - \beta_i)^{t - \sum \epsilon_j}.$$

5   Appendix

We shall prove Theorem 2 via a series of lemmas. Let $K = K(a_1, a_2)$ be as in the statement of the theorem so K^r is the set of constraints

$$\{Bel(P(a_i)) = a, \ Bel(P(a_i) \wedge P(a_j)) = b \mid 1 \le i, j \le r, \ i \ne j\}. \qquad (5)$$

We first notice that the requirement that K^r be consistent for all r forces $a^2 \le b \le a$. That $b \le a$ is of course obvious. To show that $a^2 \le b$ let[12]

---

[12] These W_i are of course functions of r. However to keep the notation simple we shall, throughout, avoid explicit mention of such dependencies whenever possible.

$$W_i = ME(K^r) \left(\bigvee_{\substack{\epsilon \in \{0,1\}^r \\ \sum \epsilon_j = i}} \bigwedge_{1 \le t \le r} P^{\epsilon_t}(a_t) \right).$$

Then since $ME(K^r)$ satisfies the constraints in K^r,

$$\sum_{i=0}^{r} W_i = 1, \quad \sum_{i=0}^{r} \frac{i}{r} W_i = a \text{ and } \sum_{i=0}^{r} \frac{i(i-1)}{r(r-1)} W_i = b. \tag{6}$$

Expanding the inequality

$$0 \le \sum_{i=0}^{r} (i - ra)^2 W_i$$

and using the identities in (6) gives

$$a^2 - \frac{a(1-a)}{r-1} \le b$$

and hence $a^2 \le b$ since this must hold for all r.

The plan now, hardly surprisingly, is to conduct a rather detailed analysis of the $ME(K^r)$ as a function of r. It turns out that this behavior depends markedly on which of the three cases, $b = a, b = a^2, a^2 < b < a$ holds[13]. We first dispose of the edge cases $b = a, a^2$. If $a = b$ then it is easy to check that the (in this case unique) solutions to the K^r give the same values as the complete set of reasons Q_1, Q_2 with $\beta_1 = 1$, $\beta_2 = 0$, $\lambda_1 = a$, $\lambda_2 = 1 - a$.

Turning to the case $b = a^2$, it follows by standard results about ME (see [28], we are now back to the $s = 1$ case notice!) that

$$ME(K^r)(\bigwedge_{i=1}^{m} P^{\epsilon_i}(a_i)) = ME(K_0^r)(\bigwedge_{i=1}^{m} P^{\epsilon_i}(a_i))$$

$$= ME(K_1^r)(\bigwedge_{i=1}^{m} P^{\epsilon_i}(a_i))$$

where

$$K_0^r = \{ Bel(P(a_1)) = a \}$$

and K_1^r is the complete set of reasons Q_1, Q_2 with $\beta_1 = 1 - a$, $\beta_2 = a$, $\lambda_1 = 0$, $\lambda_2 = 1$.

It now only remains to prove the theorem under the standing assumption that $a^2 < b < a$. The first step in the proof is to employ the Lagrange Multiplier method to show that,

$$ME(K^r) \left(\bigwedge_{j=1}^{i} P(a_j) \wedge \bigwedge_{j=i+1}^{r} \neg P(a_j) \right) = xy^i z^{i^2} \tag{7}$$

---

[13] It is, apparently, this trichotomy which makes the proof of the Conjecture so messy.

for some $x, y, z > 0$. To see this notice that by the Renaming Principle (see [24] or [26]), $ME(K^r)\left(\bigwedge_{j=1}^r P^{\epsilon_j}(a_j)\right)$ depends only on $\sum_{j=1}^r \epsilon_j$, so our system of constraints corresponds to

$$\sum_{i=0}^r \binom{r}{i} u_i = 1, \quad \sum_{i=0}^r \binom{r}{i}\frac{i}{r}u_i = a \quad \text{and} \quad \sum_{i=0}^r \binom{r}{i}\frac{i(i-1)}{r(r-1)}u_i = b \quad (8)$$

where u_i denotes $Bel\left(\bigwedge_{j=1}^r P^{\epsilon_j}(a_j)\right)$ for $\sum_{j=1}^r \epsilon_j = i$. Furthermore these equations have a strictly positive solution, namely the canonical solution to the complete set of reasons in the statement of Theorem 2, so by the Open Mindedness Principle (see [24] or [26]), their maximum entropy solution must also be strictly positive. Thus the Lagrange Multiplier method can correctly be applied to show that

$$-1 - \log u_i - \lambda - \frac{i}{r}\mu - \frac{i(i-1)}{r(r-1)}\nu = 0$$

for some Lagrange Multipliers λ, μ, ν. Hence

$$u_i = e^{-1-\lambda}\left(e^{-\frac{\mu}{r}}e^{\frac{\nu}{r(r-1)}}\right)^i \left(e^{-\frac{\nu}{r(r-1)}}\right)^{i^2}$$

which gives the above identity (7).

For x, y, z as specified in (7) and $i = 0, 1, ...r$, define

$$Z_i = \binom{r}{i} y^i z^{i^2}.$$

We shall now spend some time investigating the properties of these Z_i for large r. For that reason we shall assume from now on that r is large.

Lemma 1. Let $0 \le i_0 \le r/2$. If the Z_i have a (local) maximum at i_0 then the Z_i are increasing for $0 \le i \le i_0$.

Proof. Suppose not. Then there is $1 \le i_1 \le i_0 - 1$ such that $Z_{i_1} \le Z_{i_1-1}$ and $Z_{i_1} \le Z_{i_1+1}$, so

$$\frac{Z_{i_1}}{Z_{i_1-1}} \le 1 \le \frac{Z_{i_1+1}}{Z_{i_1}}$$

and consequently

$$z^2 \ge \frac{(i_1+1)(r-i_1+1)}{i_1(r-i_1)}.$$

However, since the Z_i have a maximum at i_0,

$$\frac{Z_{i_0}}{Z_{i_0-1}} \ge 1 \ge \frac{Z_{i_0+1}}{Z_{i_0}}$$

and consequently

$$z^2 \le \frac{(i_0+1)(r-i_0+1)}{i_0(r-i_0)}.$$

This is however impossible because the function

$$f(x) = \frac{(x+1)(r-x+1)}{x(r-x)}$$

is (strictly) decreasing on $[0, \frac{r}{2})$.

Lemma 2. *Let $r/2 \le i_0 \le r$. If the Z_i have a (local) maximum at i_0 then the Z_i are decreasing for $i_0 \le i \le r$.*

Proof. We have

$$Z_{r-i} = \left(z^{r^2}y^r\right)\binom{r}{i}\left(\frac{1}{z^{2r}y}\right)^i z^{i^2} \tag{9}$$

so the result follows from the previous Lemma.

Corollary 1. *The Z_i can have at most two local maxima, at most one in each of $[0, \frac{r}{2}]$ and $[\frac{r}{2}, r]$.*

Next we shall aim to describe how 'peaked' these maxima must be.

Lemma 3. *Let $0 < \delta < \frac{1}{2}$ and let*

$$f_\delta(x) = 1 - \delta x - (1-\delta)xe^{\frac{1-x}{1-\delta}}.$$

Then $f_\delta(x) > 0$ for $x \in (-\infty, 1) \cup (1, 2-2\delta)$.

Proof. We have

$$f_\delta'(x) = -\delta - (1-\delta-x)e^{\frac{1-x}{1-\delta}} \quad \text{and} \quad f_\delta''(x) = \left(2 - \frac{x}{1-\delta}\right)e^{\frac{1-x}{1-\delta}}.$$

It follows that f_δ' increases on $(-\infty, 2-2\delta)$. Also, $f_\delta'(1) = 0$ so it follows that f_δ is negative (f_δ decreasing) on $(-\infty, 1)$ and f_δ' is positive (f_δ increasing) on $(1, 2-2\delta)$. Since $f_\delta(1) = 0$, the result follows.

Lemma 4. *Let $0 < \alpha < \beta < \frac{1}{2}$ be such that $\beta < 2(1-\alpha)\alpha$. Then*

$$e^{\frac{\beta-\alpha}{\beta(1-\beta)}} < \frac{\beta(1-\alpha)}{\alpha(1-\beta)} < e^{\frac{\beta-\alpha}{\alpha(1-\alpha)}}.$$

Proof. Since $1 < \beta/\alpha < 2(1-\alpha)$ and $0 < \alpha/\beta < 1$, by the previous Lemma, $f_\alpha\left(\frac{\beta}{\alpha}\right) > 0$ and $f_\beta\left(\frac{\alpha}{\beta}\right) > 0$. The first inequality gives

$$1 - \beta - (1-\alpha)\frac{\beta}{\alpha}e^{\frac{\alpha-\beta}{\alpha(1-\alpha)}} > 0, \quad \text{i.e.} \quad \frac{\beta(1-\alpha)}{\alpha(1-\beta)} < e^{\frac{\beta-\alpha}{\alpha(1-\alpha)}},$$

and the second inequality gives

$$1 - \alpha - (1-\beta)\frac{\alpha}{\beta}e^{\frac{\beta-\alpha}{\beta(1-\beta)}} > 0, \quad \text{i.e.} \quad e^{\frac{\beta-\alpha}{\beta(1-\beta)}} < \frac{\beta(1-\alpha)}{\alpha(1-\beta)},$$

as required.

Lemma 5. *Let $0 < \alpha < \beta \le \frac{1}{2}$. If the Z_i are decreasing for $[\alpha r] \le i \le \beta r$ then $Z_{[\alpha r]} > r^2 Z_{[\beta r]}$, (assuming r large).*

Proof. In the proof, we will write αr for $[\alpha r]$ etc.. Clearly we may assume that $\beta < \frac{1}{2}, 2(1-\alpha)\alpha$ since otherwise we could work instead with a suitable $\alpha < \beta' < \beta$. Let $t = z^2$, $u = yz^{2\alpha r + 1}$. We have

$$\frac{Z_{i+1}}{Z_i} = ut^{i-\alpha r} \frac{r-i}{i+1} < 1 \quad (i = \alpha r, ..., \beta r - 1). \tag{10}$$

Note that for $i = \alpha r$ this yields $u < (\alpha r + 1)/(r - \alpha r)$. Since the function

$$f(x) = t^x \frac{r-x}{x+1}$$

has at most two local extrema we may assume that it is monotone on $[\alpha r, \beta r]$ (otherwise, we would consider a subinterval).

First consider the case when f is decreasing on $[\alpha r, \beta r]$. We have, in particular then that

$$t^{\beta r} \frac{r - \beta r}{\beta r + 1} < t^{\beta r - 1} \frac{r - \beta r + 1}{\beta r}$$

so

$$t < \left(\frac{\beta r + 1}{\beta r}\right) \left(\frac{(1-\beta)r + 1}{(1-\beta)r}\right).$$

Let $\alpha < \gamma < \beta$ (γ fixed). For $\gamma r \le i \le \beta r$,

$$ut^{i - \alpha r} \frac{r-i}{i+1} \le ut^{\gamma r - \alpha r} \frac{r - \gamma r}{\gamma r + 1}$$

and

$$ut^{\gamma r - \alpha r} \frac{r - \gamma r}{\gamma r + 1} < \left(\frac{\alpha r + 1}{r - \alpha r}\right) \left(\frac{r - \gamma r}{\gamma r + 1}\right) \left(\left(\frac{\beta r + 1}{\beta r}\right) \left(\frac{(1-\beta)r + 1}{(1-\beta)r}\right)\right)^{\gamma r - \alpha r}.$$

As r tends to ∞, the expression on the right hand side above tends to

$$\left(\frac{\alpha}{1 - \alpha}\right) \left(\frac{1 - \gamma}{\gamma}\right) e^{\frac{\gamma - \alpha}{\beta(1-\beta)}}.$$

By continuity and Lemma 4, this is less than some $c < 1$ for γ close to β. Consequently, $Z_{\beta r}/Z_{\gamma r} < c^{\beta r - \gamma r}$. If r is sufficiently large, $c^{\beta r - \gamma r} < r^{-2}$ and the result follows since $Z_{\gamma r} < Z_{\alpha r}$.

If f is increasing on $[\alpha r, \beta r]$ then

$$t^{\alpha r} \frac{r - \alpha r}{\alpha r + 1} < t^{\alpha r + 1} \frac{r - \alpha r - 1}{\alpha r + 2}$$

so

$$\frac{1}{t} < \left(\frac{\alpha r + 1}{\alpha r + 2}\right) \left(\frac{r - \alpha r - 1}{r - \alpha r}\right).$$

Fix $\alpha < \gamma < \beta$. For $\alpha r \leq i \leq \gamma r$ we have

$$ut^{i-\alpha r}\frac{r-i}{i+1} \leq ut^{\gamma r - \alpha r}\frac{r-\gamma r}{\gamma r + 1} = ut^{\beta r - \alpha r}\left(\frac{r-\beta r}{\beta r + 1}\right)\left(\frac{\beta r + 1}{r - \beta r}\right)\left(\frac{r - \gamma r}{\gamma r + 1}\right)t^{\gamma r - \beta r}$$

and since by (10), $ut^{\beta r - \alpha r}\frac{r-\beta r}{\beta r + 1} < 1$,

$$ut^{\gamma r - \alpha r}\frac{r - \gamma r}{\gamma r + 1} \leq \left(\frac{\beta r + 1}{r - \beta r}\right)\left(\frac{r - \gamma r}{\gamma r + 1}\right)\left(\left(\frac{\alpha r + 1}{\alpha r + 2}\right)\left(\frac{r - \alpha r - 1}{r - \alpha r}\right)\right)^{\beta r - \gamma r}.$$

As r tends to ∞, the expression on the right hand side above tends to

$$\left(\frac{\beta}{1-\beta}\right)\left(\frac{1-\gamma}{\gamma}\right)e^{-\frac{\beta-\gamma}{\alpha(1-\alpha)}}.$$

By continuity and Lemma 4 again, this is less than some $c < 1$ for γ close to α. Consequently, $Z_{\gamma r}/Z_{\alpha r} < c^{\gamma r - \alpha r}$. The result follows as in the previous case since $Z_{\beta r} < Z_{\gamma r}$ and if r is sufficiently large, $c^{\gamma r - \alpha r} < r^{-2}$.

Lemma 6. *Let $0 < \beta < \alpha < \frac{1}{2}$. If the Z_i are increasing for $[\beta r] \leq i \leq \alpha r$ then $Z_{[\alpha r]} > r^2 Z_{[\beta r]}$, provided that r is large.*

Proof. The proof is similar to the proof of Lemma 5. As before, we will write αr for $[\alpha r]$ etc. and set $t = z^2$, $u = yz^{2\alpha r + 1}$. We have

$$\frac{Z_i}{Z_{i+1}} = \frac{1}{ut^{i-\alpha r}}\frac{i+1}{r-i} < 1 \quad (i = \beta r, ..., \alpha r - 1). \tag{11}$$

Note that for $i = \alpha r - 1$ this yields $t/u < (r - \alpha r + 1)/\alpha r$. Since the function

$$g(x) = t^{-x}\frac{x+1}{r-x}$$

has at most two local extrema, we may assume that it is monotone on $[\beta r, \alpha r]$ (otherwise, we would again consider a subinterval). Initially we shall consider the case when g is increasing on $[\beta r, \alpha r]$. We have, in particular,

$$t^{-\alpha r + 2}\frac{\alpha r - 1}{r - \alpha r + 2} < t^{-\alpha r + 1}\frac{\alpha r}{r - \alpha r + 1},$$

so

$$t < \left(\frac{\alpha r}{\alpha r - 1}\right)\left(\frac{r - \alpha r + 2}{r - \alpha r + 1}\right).$$

Let $\beta < \gamma < \alpha$ (γ fixed). For $\beta r \leq i \leq \gamma r - 1$,

$$\frac{1}{ut^{i-\alpha r}}\frac{i+1}{r-i} \leq \frac{1}{ut^{\gamma r - 1 - \alpha r}}\frac{\gamma r}{r - \gamma r + 1} = \frac{t}{u}\frac{1}{t^{\gamma r - \alpha r}}\frac{\gamma r}{r - \gamma r + 1}$$

and

$$\frac{t}{u}\frac{1}{t^{\gamma r-\alpha r}}\frac{\gamma r}{r-\gamma r+1} < $$
$$\left(\frac{r-\alpha r+1}{\alpha r}\right)\left(\frac{\gamma r}{r-\gamma r+1}\right)\left(\left(\frac{\alpha r}{\alpha r-1}\right)\left(\frac{(1-\alpha)r+2}{(1-\alpha)r+1}\right)\right)^{\alpha r-\gamma r}.$$

As r tends to ∞, the expression on the right hand side tends to

$$\left(\frac{1-\alpha}{\alpha}\right)\left(\frac{\gamma}{1-\gamma}\right)e^{\frac{\alpha-\gamma}{\alpha(1-\alpha)}}.$$

By Lemma 4, this is less than some $c < 1$. Consequently, $\frac{Z_{\beta r}}{Z_{\gamma r}} < c^{\gamma r-\beta r}$. If r is sufficiently large, $c^{\beta r-\gamma r} < r^{-2}$ and the result follows since $Z_{\gamma r} < Z_{\alpha r}$.
If g is decreasing on $[\beta r, \alpha r]$ then

$$t^{-\beta r}\frac{\beta r+1}{r-\beta r} > t^{-\beta r-1}\frac{\beta r+2}{r-\beta r-1}$$

so

$$\frac{1}{t} < \left(\frac{\beta r+1}{\beta r+2}\right)\left(\frac{r-\beta r-1}{r-\beta r}\right).$$

Fix $\beta < \gamma < \alpha$. For $\gamma r \le i \le \alpha r$ we have

$$\frac{1}{ut^{i-\alpha r}}\frac{i+1}{r-i} \le \frac{1}{ut^{\gamma r-\alpha r}}\frac{\gamma r+1}{r-\gamma r} = \frac{1}{ut^{\beta r-\alpha r}}\left(\frac{\beta r+1}{r-\beta r}\right)\left(\frac{r-\beta r}{\beta r+1}\right)\left(\frac{\gamma r+1}{r-\gamma r}\right)t^{\beta r-\gamma r}$$

and since by (11),

$$\frac{1}{ut^{\beta r-\alpha r}}\frac{\beta r+1}{r-\beta r} < 1,$$

$$\frac{1}{ut^{\gamma r-\alpha r}}\frac{\gamma r+1}{r-\gamma r} \le \left(\frac{r-\beta r}{\beta r+1}\right)\left(\frac{\gamma r+1}{r-\gamma r}\right)\left(\left(\frac{\beta r+1}{\beta r+2}\right)\left(\frac{r-\beta r-1}{r-\beta r}\right)\right)^{\gamma r-\beta r}.$$

As r tends to ∞, the expression on the right hand side tends to

$$\left(\frac{1-\beta}{\beta}\right)\left(\frac{\gamma}{1-\gamma}\right)e^{-\frac{\gamma-\beta}{\beta(1-\beta)}}.$$

By Lemma 4 again, this is less than some $c < 1$. Consequently, $Z_{\beta r}/Z_{\gamma r} < c^{\gamma r-\beta r}$. Again, the result follows since $Z_{\gamma r} < Z_{\alpha r}$ and if r is sufficiently large, $c^{\gamma r-\beta r} < r^{-2}$.

Using (9) we can see that properties described in Lemma 5 and Lemma 6 hold also in $\left[\frac{1}{2}, 1\right]$.

We now return to considering $ME(K^r)$. Notice that, as introduced early at (6), $W_i = xZ_i$. Applying the above lemmas let m_1 and m_2 be the two values of i at which W_i has a local maximum (if there is only one, take that for m_1 and leave m_2 undefined). Let

$e_1 = m_1/r$ and $e_2 = m_2/r$ and let $\delta(r)$ be such that $r\delta(r)$ is the smallest (positive) natural number satisfying

$$Z_i \leq \frac{1}{r^2} \max\{Z_{re_1}, Z_{re_2}\}$$

whenever $|i - re_1|, |i - re_2| > r\delta(r)$. Note that by using Lemmas 5 and 6 and (9) we can now argue that $\delta(r) \to 0$ as $r \to \infty$. In more detail, by using these lemmas and compactness we can show that for any given k there is an r_k such that for $r \geq r_k$, if Z_i is increasing (decreasing) on $[\beta r, \alpha r]$ ($[\alpha r, \beta r]$) and $\frac{1}{k} \leq \alpha, \beta \leq 1 - \frac{1}{k}$, $|\alpha - \beta| \geq \frac{1}{k}$ then $Z_{\beta r} < r^{-2} Z_{\alpha r}$. Also by these lemmas if $e_1 < \frac{1}{k}$ then $Z_{re_1} r^{-2} Z_{3r/k}$ for sufficiently large r and similarly if $1 - \frac{1}{k} < e_2$. Putting together these observations and Lemma 1 gives the required conclusion.

Let

$$Y_1 = \sum_{|i-re_1| \leq r\delta(r)} W_i, \qquad Y_2 = \sum_{\substack{|i-re_2| \leq r\delta(r) \\ i-re_1 > r\delta(r)}} W_i.$$

From (6) we obtain

$$Y_1 + Y_2 = 1 + O(1/r), \tag{12}$$
$$e_1 Y_1 + e_2 Y_2 = a + O(\delta(r)), \tag{13}$$
$$e_1^2 Y_1 + e_2^2 Y_2 = b + O(\delta(r)). \tag{14}$$

In the case that e_2 is not defined we obtain these same equations but without the terms in Y_2.

Lemma 7. *Under the standing assumption that that $a > b > a^2$, e_2 is defined for large r and*

$$e_2 = 1 - e_1 + O(\delta(r)). \tag{15}$$

Proof. First note that if there were arbitrarily large r for which e_2 were not defined then from (12)-(14) we would have that for such r $|e_1 - a|, |e_1^2 - b| = O(\delta(r))$, forcing that $|a^2 - b| = O(\delta(r))$, and hence that $a^2 = b$, in contradiction to our standing assumptions.

A similar argument shows that there must be some $\eta > 0$ such that $e_2 - e_1 > \eta$ for large r. Since if not, then, given $\eta > 0$, by there would be, by (12), arbitrarily large r for which

$$e_1 Y_1 + e_2 Y_2 = e_1(Y_1 + Y_2) + (e_2 - e_1)Y_2 = e_1 + \eta\theta_1(r),$$

$$e_1^2 Y_1 + e_2^2 Y_2 = e_1^2(Y_1 + Y_2) + (e_2^2 - e_1^2)Y_2 = e_1^2 + \eta\theta_2(r),$$

for some $|\theta_1(r)|, |\theta_2(r)| \leq 2$. By (13) and (14) then, there would be arbitrarily large r for which $|e_1 - a|, |e_1^2 - b| < 3\eta$. Again this would force the contradictory $a^2 = b$ since η can be chosen arbitrarily small.

Having established the existence of such an $\eta > 0$, by Corollary 1, $e_1 < 1/2$ and $e_2 > 1/2$.

The proof now turns on looking at the W_i around $r/2$. At this point the W_i may be increasing, decreasing or 'level', by which we mean that the relation $\circ$ below is $<, >$ or $=$ respectively:

$$W_{\frac{r-1}{2}} \circ W_{\left(\frac{r-1}{2}+1\right)} \qquad \text{if } r \text{ is odd}$$

$$W_{\left(\frac{r}{2}-1\right)} \circ W_{\left(\frac{r}{2}+1\right)} \qquad \text{if } r \text{ is even.}$$

Note that the above implies $1 \circ yz^r$ (whether r is odd or even). First suppose that the W_i are increasing (around $r/2$) for arbitrarily large r. That means that (for such r)

$$\frac{W_{((r-i)+1)}}{W_{(r-i)}} = y^2 z^{2r} \frac{W_{(i-1)}}{W_i} > \frac{W_{(i-1)}}{W_i} \tag{16}$$

so $e_2 \geq 1 - e_1$ (the $W_{\left(\frac{r}{2}+i\right)}$, or $W_{\left(\frac{r+1}{2}+i\right)}$ if r is odd, cannot start descending before the $W_{\left(\frac{r}{2}-i\right)}$, or $W_{\left(\frac{r-1}{2}-i\right)}$, start descending). On the other hand, if $e_2 > 1 - e_1 + 2\delta(r)$ then

$$m_2 > r - m_1 + 2r\delta(r). \tag{17}$$

Now by induction on $i \leq r/2$ (16) gives us that $W_{\frac{r}{2}+i} \geq W_{\frac{r}{2}-i}$ for r even (and the corresponding result for r odd) since if this holds for i then

$$W_{\frac{r}{2}+i+1} = y^2 z^{2r} W_{\frac{r}{2}-i-1} \frac{W_{\frac{r}{2}+i}}{W_{\frac{r}{2}-i}} \geq y^2 z^{2r} W_{\frac{r}{2}-i-1} \geq W_{\frac{r}{2}-i-1}.$$

In particular then $W_{(r-m_1)} \geq W_{m_1}$. Together with (17) and the definitions of $\delta(r)$, m_1, m_2 and Y_1, this gives that

$$Y_2 \geq W_{m_2} > r^2 W_{r-m_1} \geq r^2 W_{m_1} \geq r Y_1$$

for arbitrarily large r. Again by a similar argument to that which showed the existence of e_2 this gives with (12)-(14) that $b = a^2$, the required contradiction. The same obtains similarly when the W_i are decreasing and e_2 is not within $2\delta(r)$ of $1 - e_1$. (When the W_i are level, they are entirely symmetric around $r/2$ so e_2 actually equals $1 - e_1$.)

Lemma 8. *Under the standing assumption that $a^2 < b < a$,*

$$e_1 = \frac{1}{2}(1 - \sqrt{1 - 4(a - b)} + O(\delta(r)) \tag{18}$$

$$e_2 = \frac{1}{2}(1 + \sqrt{1 - 4(a - b)} + O(\delta(r)) \tag{19}$$

$$Y_1 = \frac{1}{2}\left(1 - \frac{2a - 1}{\sqrt{1 - 4(a - b)}}\right) + O(\delta(r)) \tag{20}$$

$$Y_2 = \frac{1}{2}\left(1 + \frac{2a - 1}{\sqrt{1 - 4(a - b)}}\right) + O(\delta(r)) \tag{21}$$

Proof. From (12), (13), (14) and (15) by solving for e_1 we obtain

$$(e_1)^2 - e_1 + (a - b) = O(\delta(r)) \tag{22}$$

which along with (15) yields (18) and (19). (20) and (21) now follow from (12) and (13) since our assumptions guarantee that $a - b < \frac{1}{4}$.

We can now complete the proof of Theorem 2. From Lemma 8 it follows that e_1, e_2, Y_1 and Y_2 tend to

$$\beta_1 = \tfrac{1}{2}(1 - \sqrt{1 - 4(a - b)}), \qquad \beta_2 = \tfrac{1}{2}(1 + \sqrt{1 - 4(a - b)}),$$

$$\lambda_1 = \tfrac{1}{2}\left(1 - \frac{2a - 1}{\sqrt{1 - 4(a - b)}}\right), \quad \lambda_2 = \tfrac{1}{2}\left(1 + \frac{2a - 1}{\sqrt{1 - 4(a - b)}}\right),$$

respectively. Hence, since

$$ME(K^r)\left(\bigwedge_{j=1}^{i} P(a_j) \wedge \bigwedge_{j=i+1}^{m} \neg P(a_j)\right)$$

$$= \sum_{j=0}^{r-m} \binom{r - m}{j} xy^{(j+i)} z^{(j+i)^2}$$

$$= \sum_{j=i}^{r-m+i} \binom{r - m}{j - i} xy^j z^{j^2}$$

$$= \sum_{j=0}^{r} \binom{r}{j} \frac{j(j-1)...(j-i+1)}{r(r-1)...(r-i+1)} \frac{(r-j)(r-j-1)...(r-j-m+i+1)}{(r-i)(r-i-1)...(r-m+1)} xy^j z^{j^2}$$

$$= \sum_{j=0}^{r} \frac{j(j-1)...(j-i+1)}{r(r-1)...(r-i+1)} \frac{(r-j)(r-j-1)...(r-j-m+i+1)}{(r-i)(r-i-1)...(r-m+1)} W_j,$$

we have

$$\lim_{r \to \infty} ME(K^r)\left(\bigwedge_{j=1}^{i} P(a_j) \wedge \bigwedge_{j=i+1}^{m} \neg P(a_j)\right) = \lambda_1 \beta_1^i (1-\beta_1)^{m-i} + \lambda_2 \beta_2^i (1-\beta_2)^{m-i}.$$

This completes the proof of Theorem 2.

References

1. Bacchus, F., Grove, A.J., Halpern, J.Y. & Koller, D., From statistical knowledge bases to degrees of belief, *Artificial Intelligence* **87**:1-2, 1996, pp 75-143.
2. Carnap, R.: *Logical foundations of probability,* University of Chicago Press, Chicago, and Routledge & Kegan Paul Ltd., 1950.
3. Carnap, R.: *The continuum of inductive methods,* University of Chicago Press, 1952.

4. Carnap, R.: *Replies and systematic expositions,* in: The Philosophy of Rudolf Carnap, ed. P.A.Schlipp, La Salle, Illinois, Open Court, 1963.
5. Carnap, R. & Jeffrey, R.C. eds.: *Studies in inductive logic and probability,* University of California Press, 1971.
6. Carnap, R.: A basic system for inductive logic, part 2, in *Studies in Inductive Logic and Probability (Volume II),* Ed. R.C.Jeffrey, University of California Press, Berkeley and Los Angeles, 1980.
7. Carnap, R. & Stegmüller, W.: Induktive Logik und Wahrscheinlichkeit, Springer-Verlag, Vienna, 1958, pp 243-249.
8. Eells, E.: On the alleged impossibility of inductive probability, *British Journal for the Philosophy of Science,* **39**, 1988, pp 111-116.
9. Elby, A.: Contentious contents - for inductive probability, *British Journal for the Philosophy of Science,* **45**, 1994, pp 193-200.
10. Fagin, R. & Halpern, J.Y., Reasoning about Knowlege and Probability, *Journal of the ACM,* **41**:2, 1994, pp 340-367.
11. Fagin, R., Halpern, J.Y., Moses, Y. & Vardi, M.Y., *Reasoning about Knowledge,* MIT Press, 1995.
12. Gillies, D.: Discussion: In defense of the Popper-Miller argument, *Philosophy of Science,* **53**, 1986, pp 110-113.
13. Glaister, S., Inductive Logic, in *A Companion to Philosophical Logic,* Ed. D.Jacquette, Blackwell Publishers, 2001.
14. Good, I.J.: The impossibility of inductive probability, *Nature,* **310**, 1984, pp 434.
15. Grove, A.J., Halpern, J.Y. & Koller, D., Asymptotic conditional probabilities: the unary case, *SIAM Journal of Computing* **25**:1, 1996, pp 1-51.
16. Grove, A.J., Halpern, J.Y. & Koller, D., Asymptotic conditional probabilities: the non-unary case, *Journal of Symbolic Logic* **61**:1, 1996, pp 250-275.
17. Jeffrey, R.C.: The impossibility of inductive probability, *Nature,* **310**, 1984, pp 433.
18. Johnson, W.E.: Probability: The deductive and inductive problems, *Mind,* **49**, 1932, pp 409-423.
19. Kemeny, J.G.: A contribution to Inductive Logic, *Philosophy and Phenomenological Research,* **13**, 1953, pp 371-374.
20. Kemeny, J.G.: Carnap's theory of probability and induction, *The Philosophy of Rudolf Carnap,* Open Court Publishing Company, LaSalle, Ill., 1963, pp 711-738.
21. Levi, I.: The impossibility of inductive probability, *Nature,* **310**, 1984, pp 433.
22. Miller, D., Induction: a problem solved, *Karl Poppers kritischer Rationalismus heute,* Eds. J.M.Bohm, H.Holweg & C.Hoock, Tübingen: Mohr Siebeck, 2002, pp 81-106.
23. Musgrave, A., How to do without inductive logic, *Science and Education,* **8**, 1999, pp 395-412.
24. Paris, J.B., *The Uncertain Reasoner's Companion: A Mathematical Perspective,* Cambridge university Press, 1994.
25. Paris, J.B., Common sense and maximum entropy, *Synthese,* **117**, 1999, pp 75-93.
26. Paris, J.B. & Vencovská, A., A note on the inevitability of maximum entropy, *International Journal of Approximate Reasoning,* **4**(3), 1990, pp 183-224.
27. Paris, J.B. & Vencovská, A., Common sense and stochastic independence. *Foundations of Bayesianism,* Eds. D.Corfield & J.Williamson, Kluwer Academic Press, 2002, pp 203-240.
28. Paris, J.B., Vencovská, A & Wafy, M., Some limit theorems for ME, CM^∞ and MD. Technical Report of the Manchester Centre for Pure Mathematics, University of Manchester, UK., no. 2001/9, ISSN 1472-9210, 2001.

29. Paris, J.B. & Wafy, M., On the emergence of reasons, *Journal of the IGPL*, **9(2)**, 2001, pp 207-216. [In electronic form at http://www3.oup.co.uk/igpl/Volume_09/ Issue_02/#Paris.]

30. Popper, K. & Miller D.W.: A proof of the impossibility of inductive probability, *Nature*, **302**, 1983, pp 687-688.

31. Popper, K. & Miller D.W.: The impossibility of inductive probability, *Nature*, **310**, 1984, pp 434.

32. Popper, K. & Miller, D.W.: Has inductive probability been proved impossible? *Nature*, **315**, 1985, pp 461.

33. Popper, K.: Why probabilistic support is not inductive, *Philosophical Transactions of the Royal Society (Series A)*, **321**, 1987, pp 569-591.

34. Redhead, M.: On the impossibility of inductive probability, *British Journal for the Philosophy of Science*, **36**, 1985, pp 185-191.

35. Wafy, M., *A study of an inductive problem using inference processes*, Ph.D. Thesis, Manchester University, 2000.

36. Wise, J. & Landsberg, P.T.: Has inductive probability been proved impossible? *Nature*, **315**, 1985, pp 461.

37. Wise, J. & Landsberg, P.T.: On the possibility of inductive probability, *Nature*, **316**, 1985, pp 22.

38. Zabell, S.L.: Predicting the unpredictable, *Synthese,* **90**, 1992, pp 205-232.

Completing Incomplete Bayesian Networks

Manfred Schramm[1] and Bertram Fronhöfer[2]

[1] Institut für Informatik, Technische Universität München
schramm@pit-systems.de
[2] KI Institut, Fakultät Informatik, Technische Universität Dresden
fronhoefer@pit-systems.de

Abstract. For reasoning with uncertain knowledge the use of probability theory has been broadly investigated. Two main approaches have been developed: Bayesian Networks and MaxEnt Completion.

- **Bayesian Networks** allow to describe large probability distributions, but require the specification of a graph structure on the involved variables together with a complete specification of a set of conditional probabilities according to this graph. Both the graph and the set of conditional probabilities may be cumbersome to work out.
- **MaxEnt Completion** is able to cope, in addition to uncertain knowledge, also with incomplete knowledge. Moreover, it does not depend on an ordering of knowledge (graph), which reduces the effort of specification.

In this paper we investigate ways to combine these two approaches: We consider two kinds of incomplete Bayesian Networks — thus coping with incomplete (uncertain) knowledge — and study the usefulness of some variations of the MaxEnt Completion for processing or completing them.

This analysis detected limits of the use of so-called update method *'ground'*.

Keywords: Probabilistic Reasoning, Maximum Entropy, Bayesian Networks, Conditionals.

1 Introduction

In recent years Bayesian Networks (**BN**) have become very popular in many application domains where dealing with uncertain knowledge is an everyday task. Although quite successful, BN suffer from the drawback that their use requires the respective uncertain knowledge to be known completely with respect to the underlying graph, a requirement which quite often cannot be met in practice.[1]

---

[1] Note that the specification of a BN may also be cumbersome due to the required right arrangement of this knowledge, i.e., the construction of this underlying graph. In case of large sets of knowledge, this graph construction may become so complex that it cannot be performed by hand. This is a further problem which, however, lies outside the scope of this paper.

G. Kern-Isberner, W. Rödder, and F. Kulmann (Eds.): WCII 2002, LNAI 3301, pp. 200–218, 2005.
© Springer-Verlag Berlin Heidelberg 2005

For this reason it is worth while to study what can be done when this uncertain knowledge is not completely known. We will refer to this case quite informally by speaking about *incomplete BN.*

Remark 1. As some special forms of incompleteness are sufficient to present our technical results, we assume, that someone who wants to specify probabilistic knowledge in the spirit of Bayesian Networks has a clear view of the dependencies existing among the involved variables, which allows him to specify these dependencies completely in form of a graph. However, we will allow him to have no complete information about the probabilities of these dependencies and therefore the specification of the conditional knowledge base may remain incomplete. It is this incompleteness — more precise concepts will be defined later — of the conditional knowledge base on which we will focus in this paper.   ∎

When a BN is incomplete, the application of some kind of completion procedure is indispensable for obtaining a (complete) probability distribution. A natural suggestion is to look at completion procedures which heed the Principle of Maximum Entropy (**MaxEnt**), because of the minimality properties guaranteed by this principle (see e.g. [9]). Fortunately, most of the conditional knowledge occurring in BN is in form of linear constraints, a type of knowledge for which efficient algorithms for MaxEnt Completion have been developed ([8,11]).

Unfortunately, the naive application of a MaxEnt Completion to the conditional knowledge base of a BN — complete or incomplete — is problematic. As well known — see e.g. [2,4] — already in the case of a complete BN, the mere application of MaxEnt Completion to a BN's conditional knowledge base of probabilistic constraints (= conditional probabilities) yields a probability distribution which might differ considerably from the one intended by the BN specification, i.e., from the one which would be computed by the 'standard BN algorithm' (BNA) (see Remark 5).

This discrepancy results from certain independence requirements, which are implicitly contained in the graph of a BN (see e.g. [7]), but not represented by the BN's conditional knowledge base, and therefore not taken into account by standard MaxEnt Completion. This is due to the fact, that probabilistic systems based on the MaxEnt Method are tailored in such a way that they lead to unique probability distributions, which is important for obtaining unique probabilistic judgments. To assure unique probability distributions under MaxEnt Completion, MaxEnt-Systems (like [8,11]) are restricted to linear constraints. This precludes in particular the specification of certain independence relations (as typically present in BN) whose specification by constraints would result in a set of non-linear ones. [2]

In this paper we discuss two ways to repair this behavior, i.e., to make MaxEnt-Systems respecting these independencies:

---

[2] Of course, MaxEnt-Systems deal *implicitly* with (conditional) independencies: They derive independency relations automatically from the constraints in the conditional knowledge base and use these relations to increase the performance of the MaxEnt Completion algorithms. Cf. [9,11,12]

– A method which changes the way the constraints are understood (the so-called 'ground' update algorithm [5]) without using the graph and
– a method, which mainly uses the standard interpretation of the constraints (the so-called 'floating' update method), but which is applied iteratively following the graph (joint distribution method [2, 4]).

'*Ground*' *update* and *joint distribution method* are two promising candidate algorithms for a respective generalization of complete BN to incomplete ones. Both methods are also compatible with the 'standard BN algorithm' as they produce the same probability distribution in the complete case, although with different costs.

However, if we generalize incompleteness in direction of conditional constraints whose antecedents are no longer conjunctions of elementary events of all parent variables, 'ground' update is no more applicable, because the calculations may become dependent of the order in which the constraints are processed and the independences inherent in the underlying graph may not always be respected. But this kind of incompleteness of a BN is interesting as such constraints may occur when only single influences of one node/variable on another one are known. We discuss such an incomplete BN — so-called atomic Bayesian Structures — and show that only the joint distribution method works to our satisfaction.

The paper is organized as follows:

In section 2 we introduce important concepts and define in 2.2 the two incompleteness concepts **atomic** and **parental Bayesian Structures**. In 2.3 we define some concepts of consistency and as a consequence, the (standard) 'complete' BN, in 2.4. In section 3 we present different ways of using probabilistic constraints within the MaxEnt Completion algorithm: update via the methods 'floating' and 'ground'. Section 4 explains principally different ways to proceed for computing the MaxEnt probability distribution on an incomplete BN. In section 5 we apply the methods from section 3 to a complete BN. In the sections 6 and 7 we discuss completion of the two types of incomplete BN mentioned above. In section 8, we summarize the results. Section 9 presents an example for updating by the method 'ground' and an example for the difference between the local or global application of MaxEnt.

Let us finally mention that all the examples in this paper are calculated with the system **PIT** and available on our webpage ([8]). An earlier version of this paper was published as [10].

2   Technical Preliminaries

In this section we introduce basic terminology of MaxEnt Completion and Bayesian Networks including different types of incompleteness of the latter.

2.1   Conditional Knowledge Bases (CKB)

– Let $\mathcal{V}$ be a finite set $\{V_1, \ldots, V_n\}$ of (discrete) **variables**. Each $V_i \in \mathcal{V}$ can be identified with its set of possible **values** $\{v_{i1}, \ldots, v_{ik_i}\}$ $(k_i > 1)$, thus referring to the **size** of the variable V_i by k_i.

In our examples we will assume these variables as two-valued. This allows to understand a variable V as the set consisting of a value and its negation, e.g., $V = \{a, \neg a\}$.

- In the terminological tradition of probability theory, we consider $\Omega_V := \times V$ (Cartesian product) as an **event space** with its power set, denoted by $\mathcal{A}_V$, as **set of events** or **event algebra**.

  Furthermore, we consider probability distributions $P : \mathcal{A}_V \to [0, 1]$ which assign probabilities to events in compliance with the laws of probability theory.

  As usual we denote **conditional probabilities** by $P(b \,|\, a) := \frac{P(a \wedge b)}{P(a)}$, where $P(a) > 0$, and we call a the **antecedent** of the conditional $P(b \,|\, a)$. [3]

- A **conditional knowledge base (CKB)** (over V) is a set of statements of the form $[P^c(b|a) = p_{b,a}]$ with $p_{b,a} \in [0, 1]$ and $a, b, \subseteq \Omega_V$.

Remark 2.

- In order to avoid problems with the case $P(a) = 0$, a CKB is in fact composed of the constraints $[P(a \wedge b) = p_{b,a} \cdot P(a)]$ instead of $[P^c(b|a) = p_{b,a}]$, which yield $P(a \wedge b) = 0$ in case of $P(a) = 0$.
- Since the conditional probabilities are also called **constraints** we will mark the statements in a CKB by ^{c}. This allows to distinguish them from probabilities derived from them by some method.
- The type of constraint just defined is **linear**, i.e., a linear combination of probabilities.
- Since P refers to a probability distribution, for all $a, b \subseteq \Omega_V$ must not exist constraints $[P^c(b|a) = p_1]$ and $[P^c(b|a) = p_2]$ in a CKB with $p_1 \neq p_2$.

- The **Method of Maximum Entropy** or **MaxEnt Completion** — see [3] and [9] for its justification — if applied to a CKB $\mathcal{K}$ (over V), selects from the set of all probability distributions which fulfill all the constraints in $\mathcal{K}$, the subset $S_{\mathcal{K}}$ of those probability distributions which have maximal **(Shannon) entropy**, where the (Shannon) entropy H of a probability distribution P is defined as[4]

$$H(P) := - \sum_{\omega \in \Omega_V} P(\omega) \cdot \log P(\omega) \tag{1}$$

As well known, if all the constraints in $\mathcal{K}$ are linear, $S_{\mathcal{K}}$ consists of a unique probability distribution to which we refer by P^*.

2.2   Bayesian Structures (BS)

We will consider pairs consisting of special types of CKB together with an underlying graph on the variables. The weakest one will be called a Bayesian Structure, where — in accordance with Remark 1 — we require the CKB to consist of

---

[3] We will often write $a \wedge b$ resp. $a \vee b$ instead of $a \cap b$ resp. $a \cup b$ and $\neg a$ for the complement of the event a.

[4] The logarithm's base does not matter, usually $\ln$ is taken. (Note that $0 \cdot \log 0 = 0$).

probabilities of 'son-events' conditioned by 'parent-events'. By stepwise adding further conditions we finally arrive at the standard notion of a (complete) BN.

- Let $\mathcal{G} = (\mathcal{V}, K)$ be a directed acyclic graph (**dag**) with a set $\mathcal{V}$ of nodes and a set of edges $K \subseteq \mathcal{V} \times \mathcal{V}$. We denote by $pa(V) = \{W \in \mathcal{V} \mid (W, V) \in K\}$ the **set of parent nodes** of a node $V \in \mathcal{V}$.
- A **Bayesian Structure (BS)** is a pair $(\mathcal{G}, \mathcal{K})$, where $\mathcal{G} = (\mathcal{V}, K)$ is a dag with $\mathcal{V}$ a finite set of discrete variables and $\mathcal{K}$ is a CKB over $\mathcal{V}$ such that for each constraint $[P^c(v \mid y) = p_{v,y}] \in \mathcal{K}$ holds: $v \subseteq V \in \mathcal{V}$ and $y \subseteq \times pa(V))$.

We next define different types of constraints in the CKB of a BS.

- A constraint $[P^c(v|y) = p_{v,y}]$ is called **atomic** iff there is a $V_i \in \mathcal{V}$ with $v \in V_i$ and there is a $W \subseteq \mathcal{V}$ with $y \in \times W$.[5]
- An atomic constraint $[P^c(v \mid y) = p_{v,y}]$ (where $v \in V \in \mathcal{V}$ and $y \in \times W$ with $W \subset \mathcal{V}$) in the CKB of a BS $(\mathcal{G}, \mathcal{K})$ — with dag $\mathcal{G} = (\mathcal{V}, K)$ — is called **parental** iff $W = pa(V)$.
  If $pa(V) = \emptyset$ then constraints of the form $[P^c(v) = p_v]$ are parental.
- A CKB is called **atomic/parental** iff all its constraints are atomic/parental and a BS with an atomic/parental CKB will also be called **atomic/parental**.
- A parental BS $(\mathcal{G}, \mathcal{K})$ — with dag $\mathcal{G} = (\mathcal{V}, K)$ — is **exhaustive parental** for a value $v \in V \in \mathcal{V}$ iff
  - in case of $pa(V) \neq \emptyset$:
    $\forall y \in \times pa(V)$ exists a constraint $[P^c(v \mid y) = p_{v,y}] \in \mathcal{K}$ and
  - in case of $pa(V) = \emptyset$ exists a constraint $[P^c(v) = p_v] \in \mathcal{K}$.
  A parental BS $(\mathcal{G}, \mathcal{K})$ — with dag $\mathcal{G} = (\mathcal{V}, K)$ — is **exhaustive parental** for a $V \in \mathcal{V}$ if it is exhaustive parental for all $v \in V$ and it is **exhaustive parental** iff it is exhaustive parental for all $V \in \mathcal{V}$.

Remark 3. As consequence of these definitions we have
exhaustive parental BS $\subset$ parental BS $\subset$ atomic BS $\subset$ BS   ∎

Example 1. Let us consider the dag $\mathcal{G}$ of Fig. 1 with 4 two-valued variables A, B, C and D and let us denote their elements by $\{a, \neg a\}$, $\{b, \neg b\}$ etc.
We will give examples of different kinds of constraints:

Constraints *not* admissible for a BS over $\mathcal{G}$:

not atomic: $[P^c((a \wedge \neg b) \vee (b \wedge c) \mid \neg d \wedge b) = 0.3]$, $[P^c(c \wedge d) = 0.8]$
atomic: $[P^c(a \mid b) = 0.3]$, $[P^c(\neg b \mid d \wedge c) = 0.8]$
              $[P^c(d \mid \neg a \wedge c) = 0.4]$, $[P^c(\neg d \mid a \wedge b) = 0.2]$

Constraints admissible for a BS over $\mathcal{G}$:

not atomic: $[P^c(\neg d \mid (b \wedge \neg c) \vee (\neg b \wedge c)) = 0.38]$
atomic, not parental: $[P^c(\neg d \mid c) = 0.3]$, $[P^c(\neg d) = 0.58]$
parental: $[P^c(\neg b \mid a) = 0.4]$, $[P^c(d \mid \neg b \wedge c) = 0.28]$, $[P^c(a) = 0.9]$

---

[5] Of course, $v \in V_i$ is no event. More correctly we would have to write $V_1 \times \cdots \times V_{i-1} \times \{v\} \times V_{i+1} \times \cdots \times V_n$. Also $y \in \times W$ is an analogous shorthand notation.

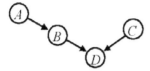

Fig. 1. A dag $\mathcal{G}$ of variables

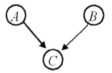

Fig. 2. A dag of variables, representing marginal independence of the variables A and B in a BN

2.3   Consistency Concepts

– Quite generally we call a BS $(\mathcal{G}, \mathcal{K})$ **consistent** if there is a probability distribution which satisfies all constraints in $\mathcal{K}$ and respects the independence relations expressed by ([7]) the dag $\mathcal{G}$.

We will also make use of the following special consistency concepts.

– A BS $(\mathcal{G}, \mathcal{K})$ — with dag $\mathcal{G} = (\mathcal{V}, K)$ — is **weakly node consistent** for a variable $V \in \mathcal{V}$ iff there is a probability distribution on $\Omega^V := V \times (\times pa(V))$ which satisfies the constraints on Ω^V and **strongly node consistent** for a variable $V \in \mathcal{V}$ iff each probability distribution on $pa(V)$ can be extended to a probability distribution on Ω^V which satisfies the constraints on Ω^V. A BS $(\mathcal{G}, \mathcal{K})$ — with dag $\mathcal{G} = (\mathcal{V}, K)$ — is **weakly/strongly node consistent** iff it is weakly/strongly node consistent for all $V \in \mathcal{V}$.

– An atomic BS $(\mathcal{G}, \mathcal{K})$ — with dag $\mathcal{G} = (\mathcal{V}, K)$ — is **sum-consistent** iff for all $V \in \mathcal{V}$ holds:

  - in case of $pa(V) \neq \emptyset$: $\forall y \in \times \mathcal{W}$ with $\mathcal{W} \subseteq pa(V)$ holds:
    * $\exists v \in V \, \exists [P^c(v \mid y) = p_{v,y}] \in \mathcal{K} \implies \sum_{v \in V} p_{v,y} \leq 1$ and
    * $\forall v \in V \, \exists [P^c(v \mid y) = p_{v,y}] \in \mathcal{K} \implies \sum_{v \in V} p_{v,y} = 1$
  - in case of $pa(V) = \emptyset$:
    * $\exists v \in V \, \exists [P^c(v) = p_v] \in \mathcal{K} \implies \sum_{v \in V} p_v \leq 1$ and
    * $\forall v \in V \, \exists [P^c(v) = p_v] \in \mathcal{K} \implies \sum_{v \in V} p_v = 1$

  where undefined $p_{v,y}$ resp. p_v are 0-summands.

Obviously, a strongly node consistent BS is consistent, but not vice versa, which we demonstrate by the following examples.

Example 2. Using the dag in Fig. 2 with the two-valued variables A, B and C and the following CKB (of parental constraints)

$\mathcal{K}_2 := \{[P^c(a) = 0], [P^c(b) = 0], [P^c(c \mid a \wedge b) = 0.8], [P^c(\neg c \mid a \wedge b) = 0.7]\}$.

Then this BS not exhaustive parental and is not sum-consistent for the node C. Moreover this BS is weakly node consistent — there is a probability distribution with $P(a \wedge b) = 0$ which makes it consistent — but not strongly node consistent (cf. Remark 2). ∎

Example 3. Using again the dag in Fig. 2 together with the CKB
$$\mathcal{K}_3 := \{[P^c(c\,|\,a) = 0.5], [P^c(c\,|\,\neg a) = 0.6], [P^c(c\,|\,b) = 0.4], [P^c(c\,|\,\neg b) = 0.5]\}$$
which is atomic, but not parental, and which is sum-consistent.
From $\mathcal{K}_3$ follows that $0.5 \le P(c) \le 0.6$ resp. $0.4 \le P(c) \le 0.5$ which is only possible with $P(a) = 1, P(\neg a) = 0, P(b) = 0, P(\neg b) = 1$. Consequently, we have a weakly node consistent BS, but not a strongly node consistent one.   ∎

2.4   (Standard) Complete BN

 – A **(complete) BN** is a BS which is exhaustive parental and sum-consistent.

Obviously, a BN is strongly node consistent, as the distribution on $V \times (\times pa(V))$ can be calculated by the constraints in the CKB and the joint distribution of the parents.

Remark 4. The requirement of sum-consistency restricts the definition of BN, because sum-consistency would not be necessary on parent events of probability zero (compare example 2). However, in this case the conditional probabilities in the CKB (with zero probability for their antecedents) are unimportant and can be replaced by arbitrary sum-consistent ones without any effect on the resulting probability distribution.   ∎

Remark 5. A BN $(\mathcal{G}, \mathcal{K})$ — with dag $\mathcal{G} = (\mathcal{V}, K)$ — determines a unique probability distribution $P_{(\mathcal{G},\mathcal{K})}$ over $\Omega_{\mathcal{V}}$ by:

$$P_{(\mathcal{G},\mathcal{K})}(v_1, ..., v_n) = \prod_{i=1}^{n} P^c(v_i \,|\, y_i) \qquad \textbf{(BNA)}$$

where $(v_1, ..., v_n) \in \Omega_{\mathcal{V}}$ and $y_i = (v_{l_1}, ..., v_{l_{|pa(V_i)|}}) \in \times pa(V_i)$ with $v_{l_j} \in V_{l_j} \in pa(V_i)$ and $v_{l_j} \in \{v_1, ..., v_n\}$.   ∎

Remark 6. Our definition of a BN is not free from redundancy: Since $\sum_{v \in V} p_{v,y} = 1$, for one $v' \in V$ the value $p_{v',y}$ needs not to be specified, because it can be computed from the others as long as their sum is not greater than 1.   ∎

Remark 7. Note that a probability distribution computed according to (BNA) will not satisfy constraints of the form $[P^c(b|a) = p_{b,a}]$ in case of $P(a) = 0$. However, it will satisfy the slightly different constraint $[P(a \wedge b) = p_{b,a} \cdot P(a)]$ presented in Remark 2, which we prefer for this reason.   ∎

2.5   Parental and Atomic BS

As already discussed in Remark 1, in this paper we will only consider incomplete BN whose incompleteness stems from an incomplete CKB, while we always assume the graph to be complete in the sense of our definition of a BS. Precisely, we will consider **atomic BS** and **parental BS**. Of course, they are not the only cases of incompleteness which can be imagined. A further step might be to allow in a constraint $v \subset V$ instead of $v \in V$ or to allow arbitrary conditional probabilities on the set of variables $\{V\} \cup pa(V)$. Equivalently we might allow probability intervals[6] with the constraints (but not for the information about the distribution of the parents which will later be denoted by $Q(...)$) without changing the qualitative results on the application of certain completion methods. However, since all these generalizations will not falsify our negative results on certain completion methods, we decided to focus on interesting special cases.

Remark 8. The intuition underlying a parental BS is that it is obtained from a complete BN by deleting some of the constraints in the BN's CKB.
Note however, that due to the redundancy of our BN definition — see Remark 6 — just deleting at most one constraint for each tuple of values of parent nodes allows immediately to recompute the missing constraint.   ∎

Remark 9. The intuition underlying an atomic BS is that it is also obtained from a complete BN by deleting some of the constraints in the BN's CKB. but, in addition, more or less events in the antecedents of the remaining constraints have been deleted as well (while adapting the probabilities in order to preserve consistency).

Since this complete BN will of course not be known in practice, we do not have an easy to check criterion for the consistency in case of atomic constraints, which meets the situation for a CKB in general.   ∎

For a parental BS sum-consistency is sufficient for strong node consistency which is of course sufficient for consistency (cf. Remark 4).

3   MaxEnt Update Algorithms

Instead of solving the problem of maximizing entropy — MaxEnt Problem for short — directly, e.g., by means of the well known Lagrange multiplier method, some algorithms to calculate the distribution with maximal entropy are based on an fixpoint calculation, which is described, for instance, in [5] and which relies heavily on fundamental investigations by [1].

In this section we will first present this fixpoint algorithm and then, in order to calculate the probability distribution of a BN and to receive the same probability values as using BNA, we describe a suitable modification (the method 'ground'). Both algorithms make no use of the dag $\mathcal{G}$.

3.1   'Floating' Application of Conditional Probabilities

Given a set of m linear constraints $C_0, \ldots, C_{m-1}$, with this efficient MaxEnt Completion algorithm the problem of **maximizing entropy** is solved by calculating a sequence of distributions $(P^0, P^1, \ldots)$ which converges — as has been

---

[6] as discussed in [4] and also implemented in the system [8].

shown in [1] — to the MaxEnt distribution P^* which satisfies the constraints $C_0, \ldots, C_{m-1}$. The uniform distribution is taken as starting point P^0. The distribution P^{k+1} is calculated from P^k by taking the set of all distributions which fulfill the constraint $C_{[k \bmod m]}$ and by calculating from this set the distribution P^{k+1} for which the information theoretic distance $CR(P^{k+1}, P^k)$ to the distribution P^k is minimal, where this distance is measured by the cross-entropy function (also called Kullback-Leibler number or I-divergence)

$$CR(P^{k+1}, P^k) := \sum_{\omega \in \Omega} P^{k+1}(\omega) \cdot \log \frac{P^{k+1}(\omega)}{P^k(\omega)}$$

With the algorithm just described (**standard MaxEnt Completion**) we solve a sequence of problems where we have to minimize the cross-entropy $CR(P^{k+1}, P^k)$ — cross-entropy Problems for short. This is an advantage, because the computation of $CR(P^{k+1}, P^k)$ is very simple and efficient for certain types of linear constraints as, e.g., conditional probability statements.

For instance, for a conditional constraint

$$P^c(b|a) = x$$

(with a and b arbitrary events) the distribution P^{k+1} with minimal $CR(P^{k+1}, P^k)$ can be computed from P^k by means of the following formula (thus considering $P^c(b|a) = x$ as the constraint $C_{[k \bmod m]}$):

$$P^{k+1}(\omega) = \begin{cases} P^k(\omega) \cdot \alpha \cdot \beta^{1-x} & \text{for } \omega \in (a \wedge b) \\ P^k(\omega) \cdot \alpha \cdot \beta^{-x} & \text{for } \omega \in (a \wedge \neg b) \\ P^k(\omega) \cdot \alpha & \text{for } \omega \in (\neg a) \end{cases} \qquad (2)$$

where

$$\beta = \frac{x \cdot P^k(a \wedge \neg b)}{(1 - x) \cdot P^k(a \wedge b)}$$

and α is the factor needed for renormalization. (This factor may also be omitted and normalization can be done easily in a subsequent independent step.) We also call this algorithm the **standard Update Algorithm**.

Example 4. The following CKB $\mathcal{K}_4$, consisting of just one constraint, yields (approximately) the following MaxEnt distribution P_4^*:

$\mathcal{K}_4$	$P_4^*(a \wedge b)$	$P_4^*(a \wedge \neg b)$	$P_4^*(\neg a \wedge b)$	$P_4^*(\neg a \wedge \neg b)$
$P^c(b\,|\,a) = 0.9$	0.36	0.04	0.30	0.30

∎

Remark 10. In view of the objectives of this paper, it is important to notice that the probability of a has changed from its initial value $P_4^0(a) = 0.5$ (in the uniform distribution) to the value $P_4^*(a) = P_4^1(a) \approx 0.4$. The constraint $P^c(b|a)$ therefore has changed the probability of its antecedent a, i.e., the probability is **'floating'**. ∎

3.2   'Ground' Application of Conditional Probabilities

This floating of the probability of a in example 4 is necessary for really obtaining the distribution with maximum entropy. However, this has been criticized by people who construe conditional probabilities as causalities and — as we will show in section 5.1 — it is also in conflict with the intentions underlying the specification of a BN by a set of linear constraints.[7] To cope with the demands of these critics a different *update method* for computing P^{k+1} from P^k has been proposed where the probability of the antecedent remains stable or *'ground'*. [8]

The method 'ground' implements the idea, that the probability of the antecedent should *not* be changed by the update (in other words: the update should respect the probabilities of the parents). [2] uses this idea to define probabilistic counterfactuals, [5] mentions that this method could be used to implement BN.[9] With the method 'ground' we obtain for a conditional constraint $P^c(b|a) = x$:

$$P^{k+1}(\omega) = \begin{cases} P^k(\omega) \quad \cdot \quad \dfrac{x}{P^k(b\,|\,a)} & \text{for } \omega \in (a \wedge b) \\[3mm] P^k(\omega) \quad \cdot \quad \dfrac{(1-x)}{P^k(\neg b\,|\,a)} & \text{for } \omega \in (a \wedge \neg b) \quad\quad (3) \\[3mm] P^k(\omega) & \text{for } \omega \in (\neg a) \end{cases}$$

Remark 11. This **ground Update Algorithm** is also optimal with respect to cross-entropy, i.e., if we start with a distribution P^k and a conditional probability $P^c(b|a) = x$ together with the condition $P^{k+1}(a) = P^k(a)$ the solution found by minimization of cross-entropy will be the same as applying the update rule given in Eq. (3).[10]   ∎

Example 5. Processing the constraints with the method 'ground', we obtain the following probability distribution:

---

[7] See for instance, the discussion in [2] as counterargument to a critique of Pearl concerning MaxEnt or see [4] for a counterintuitive dependency on a metalevel by MaxEnt.

[8] This terminology goes back to [5] where the algorithm for *'floating'* update is distinguished from *'ground'* update.

[9] This method is also offered as an option by the system SPIRIT (described in [5]). For PIT users only: In the syntax of PIT [8], the default processing of constraints is *'floating'*, while a constraint has to be marked by 'm' (for respecting marginal independence) for being processed by the method *'ground'*. Example: $Pm(c|a) = 0.7$ instead of the standard case $P(c|a) = 0.7$ for the method *'floating'*.

[10] As possible elements in the subspaces $(a \wedge b)$, $(a \wedge \neg b)$ and $(\neg a)$ are not distinguished by a constraint, the Lagrange solution will use the same factors for every element inside these sets. But given that we have to use the same factors inside the subsets, the problem can be reduced to three variables (one for each case). Moreover, we can conclude that the factor for the case $\neg a$ is 1 (as the mode *'ground'* demands the probability of the antecedent (resp. its complement) to remain the same), the (unique) solution (presented in Eq. (3)) can be derived from the conditions.

$\mathcal{K}_5$	$P_5^*(a \wedge b)$	$P_5^*(a \wedge \neg b)$	$P_5^*(\neg a \wedge b)$	$P_5^*(\neg a \wedge \neg b)$
$P^c(b \mid a) = 0.9$	0.45	0.05	0.25	0.25

We get $P_5^* = P_5^1$. Consequently, P_5^* is the distribution P^* such that the distance $CR(P^*, P^0)$ is minimal provided the *'ground'* application of the constraint $P^c(b \mid a) = 0.9$. ∎

Remark 12. Since the two update methods — 'floating' resp. 'ground' — may be selected individually for each constraint, we will write $P_f^c(x \mid y)$ for a constraint which shall be processed by the (standard) update method *'floating'* and write $P_g^c(x \mid y)$ for a constraint which shall be processed by the update method *'ground'*. If all constraints are marked as 'floating' then MaxEnt Completion is performed (see Eq. (2)). ∎

Remark 13. Let us finally point out that it makes no difference for unconditional constraints whether they are processed as 'floating' or as 'ground', since $P(a) \equiv P(a \mid \Omega)$ and $P(\Omega) = 1$ cannot be changed. ∎

4 Local Versus Global MaxEnt

Independently of 'ground' or 'floating' application of constraints we can have different principal views of the working of MaxEnt Completion on a BS.

- Given the conditional probabilities and the independencies expressed by the graph, we may look directly, for instance, by an iteration over the entire knowledge, for the **global** probability distribution(s) which have maximal entropy.
- An other possibility (which we pursue in this paper) is to follow the 'causal' resp. conditional structure of the dag of the BS and to calculate the sequence of **local, node specific** probability distributions of maximal entropy, given the (already calculated) distribution of the node's parents and using the part of the CKB corresponding to the node.

In our examples we will call the latter approach the **joint distribution method**. We will mark the (new, local) constraints, which encode the joint distribution of the parents, by Q (as in $Q(a \wedge b)$ with $a \in A$ and $b \in B$), as this distribution can be thought as obtained by **Q**uerying that part of the 'network' which has been computed already (see also section 9.2 for the difference between the results with a global and a local approach).

5 MaxEnt-Like Algorithms Applied to (Complete) BN

Since we want to apply algorithms inspired by MaxEnt Completion, it is reasonable to analyze first which of these algorithms are applicable to compute the probability of an event in case of a complete BN, i.e., whether they can replace (BNA). Technically, we can restrict our focus for this investigation to the most simple (and standard) example of a BN which includes marginal independencies: A two-valued node/variable C with two two-valued parent nodes/variables A and B (see Fig. 2).

5.1   Using 'Floating' Conditional Probabilities

If we just take the CKB of a BN specification and compute the MaxEnt distribution satisfying these constraints, then in nearly all cases the computed MaxEnt distribution will contain additional dependencies which would not be contained in the distribution computed by the ordinary (BNA).

Example 6. To recall this effect, we use a specification of a (simple) BN with the graph from Fig. 2 and with A, B and C two-valued variables. The result on $\mathcal{K}_6$ is the following probability distribution P_6^*:

$\mathcal{K}_6$	
$P^c(a)$	$= Q(a) = 0.5;$
$P^c(b)$	$= Q(b) = 0.5;$
$P_f^c(c \mid a \wedge b)$	$= 0.1;$
$P_f^c(c \mid a \wedge \neg b)$	$= 0.2;$
$P_f^c(c \mid \neg a \wedge b)$	$= 0.3;$
$P_f^c(c \mid \neg a \wedge \neg b)$	$= 0.4;$

$P_6^*(a \wedge b \wedge c)$	$= 0.024$
$P_6^*(a \wedge b \wedge \neg c)$	$= 0.219$
$P_6^*(a \wedge \neg b \wedge c)$	$= 0.051$
$P_6^*(a \wedge \neg b \wedge \neg c)$	$= 0.206$
$P_6^*(\neg a \wedge b \wedge c)$	$= 0.077$
$P_6^*(\neg a \wedge b \wedge \neg c)$	$= 0.180$
$P_6^*(\neg a \wedge \neg b \wedge c)$	$= 0.097$
$P_6^*(\neg a \wedge \neg b \wedge \neg c)$	$= 0.146$

For checking the independence of the parents A and B we calculate:

$$P_6^*(a) \cdot P_6^*(b) = 0.25, \ \ P_6^*(a \wedge b) = 0.243$$

Although the dependence effect is small in this example, we see that the (marginal) independence of the parents A and B (originally present in the (a priori) uniform distribution) is lost.   ∎

5.2   Using 'Ground' Conditional Probabilities

As mentioned in [5], we get the same result as with the (BNA) if we consider the constraints as 'ground'.

By definition — see Eq. (3) — 'ground' conditional probabilities do not change the probability of their antecedents. Because an antecedent is an elementary event of the joint distribution of the parents, the update does not change this distribution. Therefore using the conditionals of a BN in the mode 'ground' will keep the parents in their state of independence.

Example 7. Continuing the example from before and assuming the constraints as 'ground' — as given in $\mathcal{K}_7$ — we obtain the probability distribution P_7^*.

$\mathcal{K}_7$	
$P^c(a)$	$= Q(a) = 0.5;$
$P^c(b)$	$= Q(b) = 0.5;$
$P^c_g(c \mid a \wedge b)$	$= 0.1;$
$P^c_g(c \mid a \wedge \neg b)$	$= 0.2;$
$P^c_g(c \mid \neg a \wedge b)$	$= 0.3;$
$P^c_g(c \mid \neg a \wedge \neg b)$	$= 0.4;$

$P^*_7(a \wedge b \wedge c)$	$= 0.025$
$P^*_7(a \wedge b \wedge \neg c)$	$= 0.225$
$P^*_7(a \wedge \neg b \wedge c)$	$= 0.050$
$P^*_7(a \wedge \neg b \wedge \neg c)$	$= 0.200$
$P^*_7(\neg a \wedge b \wedge c)$	$= 0.075$
$P^*_7(\neg a \wedge b \wedge \neg c)$	$= 0.175$
$P^*_7(\neg a \wedge \neg b \wedge c)$	$= 0.100$
$P^*_7(\neg a \wedge \neg b \wedge \neg c)$	$= 0.150$

We calculate $P^*_7(a) \cdot P^*_7(b) = 0.25 = P^*_7(a \wedge b)$ and see that the independency of the parents is preserved.   ∎

5.3   Using the Joint Distribution of the Parents

As mentioned in section *Discussion* in [2], an equivalent result can be assured by including into the CKB the already calculated joint distribution of the parents of the node (here denoted as $Q(...)$). This taking into account of past calculations ensures that the result of the local update problem does not introduce any dependency between the parents.

Remark 14. Note that it makes no difference whether the constraints are processed as 'ground' or as 'floating', since the parents are fully constrained and cannot float. For this reason we don't mark our constraints as 'ground' or 'floating'.   ∎

Example 8. We replace in the CKB of the preceding example 7 the constraints on $P(a)$ and $P(b)$ by the uniform distribution on $A \times B$.

$\mathcal{K}_8$				
$P^c(a \wedge b)$	$= Q(a \wedge b)$	$= 0.25;$	$P^c(c \mid a \wedge b)$	$= 0.1;$
$P^c(a \wedge \neg b)$	$= Q(a \wedge \neg b)$	$= 0.25;$	$P^c(c \mid a \wedge \neg b)$	$= 0.2;$
$P^c(\neg a \wedge b)$	$= Q(\neg a \wedge b)$	$= 0.25;$	$P^c(c \mid \neg a \wedge b)$	$= 0.3;$
$P^c(\neg a \wedge \neg b)$	$= Q(\neg a \wedge \neg b)$	$= 0.25;$	$P^c(c \mid \neg a \wedge \neg b)$	$= 0.4;$

As already mentioned, the resulting distribution is equivalent to P^*_7.   ∎

Remark 15. It is not sufficient to insert the marginal probabilities $Q(a)$ and $Q(b)$ of the parents instead of the complete distribution of the parents in $\mathcal{K}_8$. With the marginal probabilities, A and B would become dependent (cf. $\mathcal{K}_6$ above).   ∎

Discussion: For a BN with a standard set of conditional probabilities the method of using *'ground'* conditionals (as in $\mathcal{K}_7$) fulfills the demand of preserving the independencies (among the parents). The method is easy to specify and is applied locally (i.e. in every node). It does not introduce any additional constraints.

The alternative of using additional constraints (as in $\mathcal{K}_8$) does not seem to be of interest, as the necessary number of additional constraints is exponential in the number n of parents and their respective sizes k_i $((\prod_{i=1}^{n} k_i) - 1)$.[11] Moreover, the calculation of the queries may also become expensive (even if done only once as proposed in [4]), if it is necessary to condition on other nodes for their calculation.

6   Parental BS

Having shown how algorithms derived from the MaxEnt Completion algorithm can be used to deal with complete BN specifications, we will now study their application to incomplete BN.

We consider first the case of parental BS and the method 'ground'.

Example 9. Using Fig. 2 as graph, we get for $\mathcal{K}_9$ the distribution P_9^*:

$\mathcal{K}_9$	
$P_g^c(c \mid a \wedge b)$	$= 0.1$
$P_g^c(c \mid \neg a \wedge \neg b)$	$= 0.4$

$P_9^*(a \wedge b \wedge c)$	$= 0.025$	$P_9^*(\neg a \wedge b \wedge c)$	$= 0.125$
$P_9^*(a \wedge b \wedge \neg c)$	$= 0.225$	$P_9^*(\neg a \wedge b \wedge \neg c)$	$= 0.125$
$P_9^*(a \wedge \neg b \wedge c)$	$= 0.125$	$P_9^*(\neg a \wedge \neg b \wedge c)$	$= 0.100$
$P_9^*(a \wedge \neg b \wedge \neg c)$	$= 0.125$	$P_9^*(\neg a \wedge \neg b \wedge \neg c)$	$= 0.150$

As with the complete BN with $\mathcal{K}_7$, we get $P_9^*(a) \cdot P_9^*(b) = P_9^*(a \wedge b) = 0.25$.   ■

This situation is indeed covered by the discussion of complete BN, which shows that the method 'ground' (as well as the joint distribution method) will perfectly do the job, as updating via the method 'ground' does not change the antecedents nor influence their complements. If all the antecedents (of the constraints of a node) are disjoint — but see section 9 for a different case — the application of the method 'ground' is safe and we inherit the properties of the complete case of section 5.2.

7   Atomic BS

We now discuss the completion of atomic BS, which intuitively spoken, are more incomplete than parental ones.

For our running example we take a close look at the case where the events we are conditioning on are not elementary in the parents' event space, for instance:
$$\{P^c(c \mid a) = x_1, P^c(c \mid b) = x_2\}$$
How should we handle this situation, where we have knowledge about single influence ('causal' or not), but no information about the joint influence ?

7.1   Using (*'Floating'*) Conditional Probabilities ?

If preservation of the possible marginal independence of the parents is not demanded, we can use the standard method of *'floating'* constraints; compare the complete case 5.1 above.

---

[11] We get '-1' because of redundancies among the constraints (cf. Remark 6).

Example 10. For $\mathcal{K}_{10}$ we obtain the probability distribution P_{10}^*:

$\mathcal{K}_{10}$
$P_f^c(c\,|\,a) = 0.9$
$P_f^c(c\,|\,b) = 0.9$

$P_{10}^*(a \wedge b \wedge c)$	$= 0.221$	$P_{10}^*(\neg a \wedge b \wedge c)$	$= 0.188$
$P_{10}^*(a \wedge b \wedge \neg c)$	$= 0.009$	$P_{10}^*(\neg a \wedge b \wedge \neg c)$	$= 0.037$
$P_{10}^*(a \wedge \neg b \wedge c)$	$= 0.188$	$P_{10}^*(\neg a \wedge \neg b \wedge c)$	$= 0.160$
$P_{10}^*(a \wedge \neg b \wedge \neg c)$	$= 0.037$	$P_{10}^*(\neg a \wedge \neg b \wedge \neg c)$	$= 0.160$

In the distribution P_{10}^* we have (as expected) a dependency between A and B:
$$P_{10}^*(a \wedge b) = 0.23, \quad \text{but } P_{10}^*(a) \cdot P_{10}^*(b) = 0.455^2 = 0.207 \qquad \blacksquare$$

Since there are arguments for accepting a dependence as in example 10 — for a example see [2] — we mention this possibility. For all other cases, we have to look for different approaches.

7.2   Using *'Ground'* Conditional Probabilities ?

Example 11. Using the method 'ground' we obtain for $\mathcal{K}_{11}$ the probability distribution P_{11}^*:

$\mathcal{K}_{11}$
$P_g^c(c\,|\,a) = 0.9$
$P_g^c(c\,|\,b) = 0.9$

$P_{11}^*(a \wedge b \wedge c)$	$= 0.280$	$P_{11}^*(\neg a \wedge b \wedge c)$	$= 0.163$
$P_{11}^*(a \wedge b \wedge \neg c)$	$= 0.0125$	$P_{11}^*(\neg a \wedge b \wedge \neg c)$	$= 0.037$
$P_{11}^*(a \wedge \neg b \wedge c)$	$= 0.214$	$P_{11}^*(\neg a \wedge \neg b \wedge c)$	$= 0.125$
$P_{11}^*(a \wedge \neg b \wedge \neg c)$	$= 0.0425$	$P_{11}^*(\neg a \wedge \neg b \wedge \neg c)$	$= 0.125$

Within this distribution we obtain $P_{11}^*(a) = 0.550$ and $P_{11}^*(b) = 0.493$, which is a shock if compared to the symmetric specification. $\qquad \blacksquare$

Indeed, the result **depends on the ordering** of the constraints, which is the worst case to happen. (If we switch the two constraints in $\mathcal{K}_{11}$, the probabilities will switch as well!). (For some steps of the iteration process see section 9). This means that the method 'ground' only works in certain cases. More precisely, the concept of 'ground' constraints may not work with sets of conditionals where the antecedents are not disjoint. In such cases, applying the update rule to one constraint may change the antecedent of some other constraints (as, e.g., the update of $P_g^c(c\,|\,a)$ may change $P(b)$), which is of course not intended by the method *'ground'*.

7.3   Using *'Ground'* Conditional Probabilities and Marginal Probabilities of the Parents ?

This concept is presented in the next example. Here it is sufficient to consider the $\mathcal{K}_{12}$ with $Q(a) = Q(b) = 0.5$.

Example 12. For $\mathcal{K}_{12}$ we obtain the probability distribution P_{12}^*:

$\mathcal{K}_{12}$	
$P^c(a)$	$= Q(a) = 0.5$
$P^c(b)$	$= Q(b) = 0.5$
$P_g^c(c\,|\,a)$	$= 0.9$
$P_g^c(c\,|\,b)$	$= 0.9$

$P_{12}^*(a \wedge b \wedge c)$	$= 0.261$	$P_{12}^*(\neg a \wedge b \wedge c)$	$= 0.189$
$P_{12}^*(a \wedge b \wedge \neg c)$	$= 0.011$	$P_{12}^*(\neg a \wedge b \wedge \neg c)$	$= 0.039$
$P_{12}^*(a \wedge \neg b \wedge c)$	$= 0.189$	$P_{12}^*(\neg a \wedge \neg b \wedge c)$	$= 0.136$
$P_{12}^*(a \wedge \neg b \wedge \neg c)$	$= 0.039$	$P_{12}^*(\neg a \wedge \neg b \wedge \neg c)$	$= 0.136$

Though $P^*(a)$ and $P^*(b)$ are no longer dependent on the ordering of the constraints, A and B are probabilistically dependent, as $P^*_{12}(a \wedge b) = 0.272$, but $P^*_{12}(a) \cdot P^*_{12}(b) = 0.25$. Again this may be acceptable (see [2]) or not.   ■

7.4   Using Conditional Probabilities ('Ground' or 'Floating') and the Joint Distribution of the Parents

Example 13. For working with the method of complete parent distributions — mentioned by [2, 4] — it is sufficient to consider the *local* problem $\mathcal{K}_{13}$ with $Q(a \wedge b) = Q(a \wedge \neg b) = Q(\neg a \wedge b) = Q(\neg a \wedge \neg b) = 0.25$. The difference between 'ground' and 'floating' conditional probabilities is not relevant, as the complete distribution of the parents is fixed. For $\mathcal{K}_{13}$ we obtain the probability distribution P^*_{13}:

$\mathcal{K}_{13}$		
$P^c(a \wedge b)$	$= Q(a \wedge b)$	$= 0.25$
$P^c(a \wedge \neg b)$	$= Q(a \wedge \neg b)$	$= 0.25$
$P^c(\neg a \wedge b)$	$= Q(\neg a \wedge b)$	$= 0.25$
$P^c(\neg a \wedge \neg b)$	$= Q(\neg a \wedge \neg b)$	$= 0.25$
$P^c(c \mid a)$		$= 0.9$
$P^c(c \mid b)$		$= 0.9$

$P^*_{13}(a \wedge b \wedge c)$	$= 0.241$
$P^*_{13}(a \wedge b \wedge \neg c)$	$= 0.009$
$P^*_{13}(a \wedge \neg b \wedge c)$	$= 0.209$
$P^*_{13}(a \wedge \neg b \wedge \neg c)$	$= 0.041$
$P^*_{13}(\neg a \wedge b \wedge c)$	$= 0.209$
$P^*_{13}(\neg a \wedge b \wedge \neg c)$	$= 0.041$
$P^*_{13}(\neg a \wedge \neg b \wedge c)$	$= 0.125$
$P^*_{13}(\neg a \wedge \neg b \wedge \neg c)$	$= 0.125$

By specification, A and B remain independent.   ■

8   Conclusion

In contrast to parental BS, where we can use the *'ground'* method, with atomic BS we seem to be bound to the expensive method of querying after each iteration step the (complete) joint distribution of the parents, as described in [2, 4], in order to preserve their possible independence.

A second result is that the update algorithm *'ground'* gets suspect, as we conclude that a constraint of type *'ground'* is not a linear constraint on the possible distributions. If the constraint were linear, the MaxEnt solution would be unique and equivalent to the fixpoint of solving the cross-entropy problems. As the cross-entropy solution is ordering dependent (and therefore especially not unique), the constraint $P^c_g(v \mid y) = x$, i.e. $P^c(v \mid y) = x$ **and** 'stable' antecedent y, cannot be linear. This argument is not falsified by the observation, that in every node a *'ground'* constraint is mapped into linear constraints: some of the corresponding probabilities are calculated *on the fly* (expressed by $Q(...)$) which seems to generate a kind of non-linear statements in view of the global MaxEnt Problem. This view is also supported by the observation that a constraint of type *'ground'* cannot be expressed in the matrix of linear constraints given for the MaxEnt Problem. This matrix only works on linear vectors of P^*; the equation $P^{k+1}(x) = P^k(x)$ with $x \subset \Omega$ cannot be expressed on this level.

9   Appendix

9.1   Updating 'Ground' Constraints, Stepwise

To help understanding the situation described in section 7.2, we show some changes in the probability distribution while updating a constraint, using the method *'ground'*. This analysis proves that updating a certain constraint may change the probability of the antecedent of another constraint, if the antecedents are not disjoint.

Example 14. Given the CKB $\mathcal{K}_{14} = \{P_g^c(c\,|\,a) = 0.9, P_g^c(c\,|\,b) = 0.9;\}$.
Using the update method 'ground', described in Eq. (3), we will show two steps in the update process. We start with the uniform distribution $P_{14}^0(\omega) = 0.125$ for all $\omega \in A \times B \times C$.
    We calculate $P_{14}^0(c\,|\,a) = 0.5$ and update all $P(\omega)$ ($\omega \in A \times B \times C$) with

$$\omega \in (a \wedge c) \quad \text{by the factor } \tfrac{0.9}{0.5} = 1.8 \text{ and}$$
$$\omega \in (a \wedge \neg c) \text{ by the factor } \tfrac{0.1}{0.5} = 0.2$$

We receive the distribution P_{14}^1:

$P_{14}^1(a \wedge b \wedge c)$	$P_{14}^1(a \wedge b \wedge \neg c)$	$P_{14}^1(a \wedge \neg b \wedge c)$	$P_{14}^1(a \wedge \neg b \wedge \neg c)$
0.225	0.025	0.225	0.025
$P_{14}^1(\neg a \wedge b \wedge c)$	$P_{14}^1(\neg a \wedge b \wedge \neg c)$	$P_{14}^1(\neg a \wedge \neg b \wedge c)$	$P_{14}^1(\neg a \wedge \neg b \wedge \neg c)$
0.125	0.125	0.125	0.125

We calculate $P_{14}^1(b) = 0.5$, which means that updating with the first constraint has not changed the probability of the antecedent of the second constraint, and calculate $P_{14}^1(c\,|\,b) = \tfrac{0.35}{0.5} = 0.7$.
We update all $P(\omega)$ with

$$\omega \in (b \wedge c) \quad \text{by the factor } \tfrac{0.9}{0.7} \approx 1.286 \text{ resp.}$$
$$\omega \in (b \wedge \neg c) \text{ by the factor } \tfrac{0.1}{0.3} \approx 0.333.$$

Applying these factors yields P_{14}^2:

$P_{14}^2(a \wedge b \wedge c)$	$P_{14}^2(a \wedge b \wedge \neg c)$	$P_{14}^2(a \wedge \neg b \wedge c)$	$P_{14}^2(a \wedge \neg b \wedge \neg c)$
0.2893	0.0083	0.225	0.025
$P_{14}^2(\neg a \wedge b \wedge c)$	$P_{14}^2(\neg a \wedge b \wedge \neg c)$	$P_{14}^2(\neg a \wedge \neg b \wedge c)$	$P_{14}^2(\neg a \wedge \neg b \wedge \neg c)$
0.1607	0.0417	0.125	0.125

By this step, $P_{14}^2(b) = 0.5$ has been preserved, but $P_{14}^2(a) = 0.5476 \neq 0.5 = P_{14}^1(a)$ has changed, which was not intended.   ∎

9.2   Global Versus Local MaxEnt

To show the difference between the local (stepwise) and a global approach, including non-linked parents, we use a dag consisting of (two-valued) nodes A, B and C as in Fig. 2 together with a (single) constraint $P(c|a \wedge b) = 0.9$.

Example 15. The set of CKB for the local approach looks as follows:

| CKB for node A : (empty) |
| CKB for node B : (empty) |

CKB for node C : $P^c(c \mid a \wedge b)$	$= 0.9$
$P^c(a \wedge b)$	$= Q(a \wedge b)$
$P^c(a \wedge \neg b)$	$= Q(a \wedge \neg b)$
$P^c(\neg a \wedge b)$	$= Q(\neg a \wedge b)$
$P^c(\neg a \wedge \neg b)$	$= Q(\neg a \wedge \neg b)$

The expressions $Q(\ldots)$ again indicate, that the probability is inherited from earlier computation steps (via Queries to the already calculated part of the BS) and is consequently fixed in the CKB for node C.

The joint distribution method will first derive $P^*_{15}(a) = P^*_{15}(b) = 0.5$ and then calculate P^*_{15} within node C:

$P^*_{15}(a \wedge b \wedge c)$	$= 0.225$	$P^*_{15}(\neg a \wedge b \wedge c)$	$= 0.125$
$P^*_{15}(a \wedge b \wedge \neg c)$	$= 0.025$	$P^*_{15}(\neg a \wedge b \wedge \neg c)$	$= 0.125$
$P^*_{15}(a \wedge \neg b \wedge c)$	$= 0.125$	$P^*_{15}(\neg a \wedge \neg b \wedge c)$	$= 0.125$
$P^*_{15}(a \wedge \neg b \wedge \neg c)$	$= 0.125$	$P^*_{15}(\neg a \wedge \neg b \wedge \neg c)$	$= 0.125$

Example 16.

On the other hand, the knowledge base $\mathcal{K}_{16}$ for the global specification (where the symbol $\perp\!\!\!\perp$ represents independence) is:

$\mathcal{K}_{16}$
$P^c(c \mid a \wedge b) = 0.9;$
$A \perp\!\!\!\perp B$

The solution of the global problem seems to approach distribution P^*_{16} with entropy 1.995 which (as to be expected) is higher than with the sequential solution (1.987). Since non linear constraints are present, this model is not necessarily unique, but here it suffices to show that there exists a distribution of higher entropy which also satisfies the constraints.

$P^*_{16}(a \wedge b \wedge c)$	$= 0.18876$
$P^*_{16}(a \wedge b \wedge \neg c)$	$= 0.02097$
$P^*_{16}(a \wedge \neg b \wedge c)$	$= 0.12412$
$P^*_{16}(a \wedge \neg b \wedge \neg c)$	$= 0.12412$
$P^*_{16}(\neg a \wedge b \wedge c)$	$= 0.12412$
$P^*_{16}(\neg a \wedge b \wedge \neg c)$	$= 0.12412$
$P^*_{16}(\neg a \wedge \neg b \wedge c)$	$= 0.14690$
$P^*_{16}(\neg a \wedge \neg b \wedge \neg c)$	$= 0.14690$

References

1. I. Csiszár. I-divergence geometry of probability distributions and minimization problems. *The Annals of Probability*, 3(1):146–158, 1975.
2. Daniel Hunter. Causality and maximum entropy updating. *Int. Journal of approximate reasoning*, 3:87–114, 1989.
3. J.B.Paris and A. Vencovská. A note on the inevitability of maximum entropy. *Int. Journal of approximate reasoning*, 4:183–223, 1990.
4. Thomas Lukasiewicz. Credal networks under maximum entropy. In *Uncertainty in Artificial Intelligence: Proceedings of the Sixteenth Conference (UAI-2000)*, pages 363–370, San Francisco, CA, 2000. Morgan Kaufmann Publishers.
5. Carl-Heinz Meyer and Wilhelm Rödder. Probabilistic knowledge representation and reasoning at maximum entropy by SPIRIT. In G. Görz and Steffen Hölldobler, editors, *KI 96: Advances in Artificial Intelligence*, Dresden, 1996. LNAI 1137, pages 273–286.

6. J.B. Paris. On filling-in missing information in causal networks. Technical Report, Dept. of Mathematics, University of Manchester, April 2003.
7. Judea Pearl. Probabilistic Reasoning in Intelligent Systems: Networks of Plausible Inference. Kaufmann,1988.
8. Homepage of PIT. http://www.pit-systems.de.
9. Manfred Schramm and Bertram Fronhöfer. PIT: A System for Reasoning with Probabilities. In Emil Weydert Gabriele Kern-Isberner, Thomas Lukasiewicz, (ed): *KI-2001 Workshop: Uncertainty in Artificial Intelligence*, Informatik Berichte 287 - 8/2001, p. 109 – 123, September 2001. Fernuniversität Hagen.
10. Manfred Schramm and Bertram Fronhöfer. Completing incomplete Bayesian Networks. In Gabriele Kern-Isberner and Wilhelm Roedder (ed), *Conditionals, Information, and Inference*, p. 231–243, May 2002. Fernuniversität Hagen.
11. Homepage of SPIRIT. http://www.xspirit.de.
12. Jon Williamson. Maximising Entropy Efficiently. *Linköping Electronic Articles in Computer and Information Science*, 7, 2002.

Author Index

Lecture Notes in Artificial Intelligence (LNAI)

Vol. 3238: S. Biundo, T. Frühwirth, G. Palm (Eds.), KI 2004: Advances in Artificial Intelligence. XI, 467 pages. 2004.

Vol. 3230: J.L. Vicedo, P. Martínez-Barco, R. Muñoz, M. Saiz Noeda (Eds.), Advances in Natural Language Processing. XII, 488 pages. 2004.

Vol. 3229: J.J. Alferes, J. Leite (Eds.), Logics in Artificial Intelligence. XIV, 744 pages. 2004.

Vol. 3228: M.G. Hinchey, J.L. Rash, W.F. Truszkowski, C.A. Rouff (Eds.), Formal Approaches to Agent-Based Systems. VIII, 290 pages. 2004.

Vol. 3215: M.G.. Negoita, R.J. Howlett, L.C. Jain (Eds.), Knowledge-Based Intelligent Information and Engineering Systems, Part III. LVII, 906 pages. 2004.

Vol. 3214: M.G.. Negoita, R.J. Howlett, L.C. Jain (Eds.), Knowledge-Based Intelligent Information and Engineering Systems, Part II. LVIII, 1302 pages. 2004.

Vol. 3213: M.G.. Negoita, R.J. Howlett, L.C. Jain (Eds.), Knowledge-Based Intelligent Information and Engineering Systems, Part I. LVIII, 1280 pages. 2004.

Vol. 3209: B. Berendt, A. Hotho, D. Mladenic, M. van Someren, M. Spiliopoulou, G. Stumme (Eds.), Web Mining: From Web to Semantic Web. IX, 201 pages. 2004.

Vol. 3206: P. Sojka, I. Kopecek, K. Pala (Eds.), Text, Speech and Dialogue. XIII, 667 pages. 2004.

Vol. 3202: J.-F. Boulicaut, F. Esposito, F. Giannotti, D. Pedreschi (Eds.), Knowledge Discovery in Databases: PKDD 2004. XIX, 560 pages. 2004.

Vol. 3201: J.-F. Boulicaut, F. Esposito, F. Giannotti, D. Pedreschi (Eds.), Machine Learning: ECML 2004. XVIII, 580 pages. 2004.

Vol. 3194: R. Camacho, R. King, A. Srinivasan (Eds.), Inductive Logic Programming. XI, 361 pages. 2004.

Vol. 3192: C. Bussler, D. Fensel (Eds.), Artificial Intelligence: Methodology, Systems, and Applications. XIII, 522 pages. 2004.

Vol. 3191: M. Klusch, S. Ossowski, V. Kashyap, R. Unland (Eds.), Cooperative Information Agents VIII. XI, 303 pages. 2004.

Vol. 3187: G. Lindemann, J. Denzinger, I.J. Timm, R. Unland (Eds.), Multiagent System Technologies. XIII, 341 pages. 2004.

Vol. 3176: O. Bousquet, U. von Luxburg, G. Rätsch (Eds.), Advanced Lectures on Machine Learning. IX, 241 pages. 2004.

Vol. 3171: A.L.C. Bazzan, S. Labidi (Eds.), Advances in Artificial Intelligence – SBIA 2004. XVII, 548 pages. 2004.

Vol. 3159: U. Visser, Intelligent Information Integration for the Semantic Web. XIV, 150 pages. 2004.

Vol. 3157: C. Zhang, H. W. Guesgen, W.K. Yeap (Eds.), PRICAI 2004: Trends in Artificial Intelligence. XX, 1023 pages. 2004.

Vol. 3155: P. Funk, P.A. González Calero (Eds.), Advances in Case-Based Reasoning. XIII, 822 pages. 2004.

Vol. 3139: F. Iida, R. Pfeifer, L. Steels, Y. Kuniyoshi (Eds.), Embodied Artificial Intelligence. IX, 331 pages. 2004.

Vol. 3131: V. Torra, Y. Narukawa (Eds.), Modeling Decisions for Artificial Intelligence. XI, 327 pages. 2004.

Vol. 3127: K.E. Wolff, H.D. Pfeiffer, H.S. Delugach (Eds.), Conceptual Structures at Work. XI, 403 pages. 2004.

Vol. 3123: A. Belz, R. Evans, P. Piwek (Eds.), Natural Language Generation. X, 219 pages. 2004.

Vol. 3120: J. Shawe-Taylor, Y. Singer (Eds.), Learning Theory. X, 648 pages. 2004.

Vol. 3097: D. Basin, M. Rusinowitch (Eds.), Automated Reasoning. XII, 493 pages. 2004.

Vol. 3071: A. Omicini, P. Petta, J. Pitt (Eds.), Engineering Societies in the Agents World. XIII, 409 pages. 2004.

Vol. 3070: L. Rutkowski, J. Siekmann, R. Tadeusiewicz, L.A. Zadeh (Eds.), Artificial Intelligence and Soft Computing - ICAISC 2004. XXV, 1208 pages. 2004.

Vol. 3068: E. André, L. Dybkjær, W. Minker, P. Heisterkamp (Eds.), Affective Dialogue Systems. XII, 324 pages. 2004.

Vol. 3067: M. Dastani, J. Dix, A. El Fallah-Seghrouchni (Eds.), Programming Multi-Agent Systems. X, 221 pages. 2004.

Vol. 3066: S. Tsumoto, R. Słowiński, J. Komorowski, J.W. Grzymała-Busse (Eds.), Rough Sets and Current Trends in Computing. XX, 853 pages. 2004.

Vol. 3065: A. Lomuscio, D. Nute (Eds.), Deontic Logic in Computer Science. X, 275 pages. 2004.

Vol. 3060: A.Y. Tawfik, S.D. Goodwin (Eds.), Advances in Artificial Intelligence. XIII, 582 pages. 2004.

Vol. 3056: H. Dai, R. Srikant, C. Zhang (Eds.), Advances in Knowledge Discovery and Data Mining. XIX, 713 pages. 2004.

Vol. 3055: H. Christiansen, M.-S. Hacid, T. Andreasen, H.L. Larsen (Eds.), Flexible Query Answering Systems. X, 500 pages. 2004.

Vol. 3048: P. Faratin, D.C. Parkes, J.A. Rodríguez-Aguilar, W.E. Walsh (Eds.), Agent-Mediated Electronic Commerce V. XI, 155 pages. 2004.

Vol. 3040: R. Conejo, M. Urretavizcaya, J.-L. Pérez-de-la-Cruz (Eds.), Current Topics in Artificial Intelligence. XIV, 689 pages. 2004.

Vol. 3035: M.A. Wimmer (Ed.), Knowledge Management in Electronic Government. XII, 326 pages. 2004.

Vol. 3034: J. Favela, E. Menasalvas, E. Chávez (Eds.), Advances in Web Intelligence. XIII, 227 pages. 2004.

Vol. 3030: P. Giorgini, B. Henderson-Sellers, M. Winikoff (Eds.), Agent-Oriented Information Systems. XIV, 207 pages. 2004.

Vol. 3029: B. Orchard, C. Yang, M. Ali (Eds.), Innovations in Applied Artificial Intelligence. XXI, 1272 pages. 2004.

Vol. 3025: G.A. Vouros, T. Panayiotopoulos (Eds.), Methods and Applications of Artificial Intelligence. XV, 546 pages. 2004.

Vol. 3020: D. Polani, B. Browning, A. Bonarini, K. Yoshida (Eds.), RoboCup 2003: Robot Soccer World Cup VII. XVI, 767 pages. 2004.

Vol. 3012: K. Kurumatani, S.-H. Chen, A. Ohuchi (Eds.), Multi-Agents for Mass User Support. X, 217 pages. 2004.

Vol. 3010: K.R. Apt, F. Fages, F. Rossi, P. Szeredi, J. Váncza (Eds.), Recent Advances in Constraints. VIII, 285 pages. 2004.